PRINCIPLES OF CHEMICAL KINETICS

PRINCIPLES OF CHEMICAL KINETICS

GORDON G. HAMMES

Cornell University

ACADEMIC PRESS New York San Francisco London
A Subsidiary of Harcourt Brace Jovanovich, Publishers

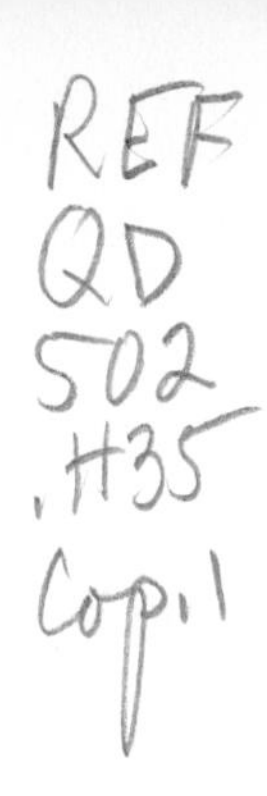

ACADEMIC PRESS, INC.
111 Fifth Avenue, New York, New York 10003

United Kingdom Edition published by
ACADEMIC PRESS, INC. (LONDON) LTD.
24/28 Oval Road, London NW1 7DX

Library of Congress Cataloging in Publication Data

Hammes, Gordon G Date
Principles of chemical kinetics.

Based on Chemical kinetics, by I. Amdur and G. G. Hammes, published in 1966.
Includes bibliographical references and index.
1. Chemical reaction, Rate of. I. Amdur, Isadore, Date Chemical kinetics. II. Title.
QD502.H35 541'.39 77-92240
ISBN 0-12-321950-7

PRINTED IN THE UNITED STATES OF AMERICA

CONTENTS

Chapter 4 Unimolecular Decompositions in the Gas Phase

Chapter 5 Chemical Reactions in Molecular Beams

Chapter 6 Energy Transfer and Energy Partitioning in Chemical Reactions

Chapter 7 Kinetics in Liquid Solutions

PREFACE

This work is based on an earlier book, ''Chemical Kinetics: Principles and Selected Topics,'' by I. Amdur and myself. The untimely death of I. Amdur has prevented a joint revision, but the present book retains the same philosophy of presentation and organization. However, the many recent innovations in chemical kinetics, especially in gas-phase dynamics, have required rather extensive revision of the earlier book. I am greatly indebted to Mrs. Alice Amdur and to the McGraw-Hill Book Company for permission to use portions of the older text.

The purpose of this book is to present the principles of chemical kinetics along with modern applications. Thus the student will learn not only the basic formulations but also will be stimulated (hopefully) by the exciting current research in chemical kinetics. Obviously a complete description of modern chemical kinetics would require a volume (or volumes) considerably larger than this one. Therefore, only a selection of topics is possible. Many extremely interesting aspects are not covered, but I believe that a student who understands the material presented will have no trouble in going directly to the literature for further information. The phenomenology and commonly used theories of chemical kinetics are presented in a critical manner, with particular emphasis on collision dynamics. How and what mechanistic information can be obtained from various experimental approaches is stressed throughout the book.

The concise presentation is designed to stimulate both students and teachers, and it is expected that the references cited at the end of each chapter will have to be consulted. Most of the problems are quite challenging and are designed both to test the students' comprehension of the subject matter and to complement the textual material.

The material in this book has been used as the basis for a one-semester course offered to seniors and graduate students at Cornell University. I firmly believe that at this level a general course in chemical kinetics, covering reactions in both gases and liquids, is needed. Students all too soon reach the stage where their interests are confined to more specialized topics. Although some introductory material is reviewed, this text is intended for students with a general college background in chemistry, physics, and mathematics, and with a typical undergraduate course in physical chemistry.

I am indebted to my colleagues at Cornell for many stimulating discussions of chemical kinetics, especially S. Bauer and B. Widom. I am particularly grateful to P. Houston for his critical comments and many references related to the application of lasers to kinetic problems. I also would like to thank Miss Connie Wright, not only for typing several drafts of the manuscript, but for patiently correcting many errors and for organizing much of the material associated with this project.

PRINCIPLES OF CHEMICAL KINETICS

CHAPTER 1

EMPIRICAL ANALYSIS OF REACTION RATES

1-1 INTRODUCTION

Chemical kinetics is concerned with the dynamics of chemical reactions. In particular, it deals with the rates of chemical reactions and how these rates can be explained in terms of a reaction mechanism. Ideally, a complete reaction mechanism would involve a knowledge of all the molecular details of the reaction, including the energetics and stereochemistry, e.g., interatomic distances and angles throughout the course of the reaction, of the individual molecular steps involved in the mechanism. In practice, experiments involve the determination of an average rate of reaction of a large number of molecules; therefore proposed mechanisms usually present a sequence of steps in which the molecules are presumed to be in some sort of average energy and stereochemical state. Throughout its historical development, however, chemical kinetics has become increasingly concerned with presenting a more detailed molecular picture, although the ultimate goal has not yet been achieved, even for relatively simple reactions.

Thermodynamics gives little information about the mechanism of chemical reactions, but chemical kinetics provides an approach for obtaining a reaction mechanism. In principle, kinetic properties should be predictable from quantum and statistical mechanics and kinetic theory; however, the difficulty of treating time-dependent problems, in general, and of knowing the nonequilibrium energy distributions has made it extremely difficult to develop a rigorous and a generally usable theory of chemical kinetics. Several useful approximate theories have been developed, but the elucidation of reaction mechanisms proceeds almost exclusively through experimental work at the present time.

In this book we shall be concerned with two problems:

(1) a critical (but concise) presentation of currently accepted methods of proceeding from rate measurements to mechanisms and of interpreting the results in terms of existing theories, and

(2) a presentation of some of the currently active research topics in chemical kinetics.

These selected topics should provide a firm foundation for further study of modern chemical kinetics.

1-2 DEFINING KINETIC SYSTEMS

Before the rate of a reaction can be meaningfully discussed, the system under consideration must be precisely defined. In particular, the system can be either open or closed. In a closed system, matter is neither gained nor lost, while in an open system, matter may be exchanged with the surroundings. For the most part, we shall be concerned with closed constant-volume systems both because they are much easier to handle experimentally and theoretically and because they are most frequently encountered in the laboratory.

Another factor to be considered is the thermal state of the system. In practice, an effort is usually made to work with isothermal systems; however, if the reaction of interest is sufficiently exothermic, it may be impossible to maintain an isothermal state. In the extreme case of rapid exothermic reactions, e.g., flames, explosions, etc., the system is best described as adiabatic rather than isothermal.

A final factor to be considered is the homogeneity of the system. Although heterogeneous reactions are of considerable interest, e.g., surface catalysis, they will not be discussed here. The experimental problem of establishing the homogeneity of a system is not trivial, since often the walls of the vessel containing the reaction mixture can cause heterogeneous effects to be of importance.

1-3 REACTION RATES AND RATE LAWS

In a closed constant-volume system the rate of a chemical reaction can be defined simply as the rate of change with time of the concentration of any of the reactants or products. The concentration can be expressed in any convenient units of quantity per unit volume, e.g., moles per liter, moles per cubic centimeter, or grams per cubic centimeter. The rate will be defined as a positive quantity, regardless of the component whose concentration change is measured. As an example, consider the generalized chemical reaction

$$aA + bB \rightarrow cC + dD \tag{1-1}$$

The rate can be expressed as $-dA/dt$, $-dB/dt$, dC/dt, or dD/dt, where A, B, C, and D designate the concentrations in arbitrary units. (The parentheses usually used to designate concentrations will be omitted if no confusion is caused by this omission; for the sake of convenience, we shall usually use the units of moles per liter.) For the rate of a reaction to be independent of the component used to describe the rate, the stoichiometry of the reaction must be considered. Thus a unique rate of reaction consistent with the specified stoichiometry can be readily defined for reaction (1-1) as

$$R = -\frac{1}{a}\frac{dA}{dt} = -\frac{1}{b}\frac{dB}{dt} = \frac{1}{c}\frac{dC}{dt} = \frac{1}{d}\frac{dD}{dt} \tag{1-2}$$

Although this definition of reaction rate has not been universally adopted, it will be used throughout this book.

If the restriction of constant volume is removed, two additional effects (neglecting convection) cause the concentration c to change: diffusion through the system and a change in volume V of the system. This can be expressed mathematically by the equation

$$dc = \left(\frac{\partial c}{\partial t}\right)_{V,D} dt + \left(\frac{\partial c}{\partial t}\right)_{V,R} dt + \left(\frac{\partial c}{\partial V}\right)_{R,D} dV \tag{1-3}$$

where the subscripts V, D, and R designate no volume change, no diffusion, and no chemical reaction, respectively. If c is the number of moles per unit volume n/V,

$$\frac{dc}{dt} = \frac{1}{V}\frac{dn}{dt} - \frac{n}{V^2}\frac{dV}{dt} \quad \text{and} \quad \left(\frac{\partial c}{\partial V}\right)_{R,D} = -\frac{n}{V^2}$$

From the definition of reaction rate in a closed system

$$\left(\frac{\partial c}{\partial t}\right)_{V,D} = \pm aR$$

where a is the stoichiometric coefficient from the equation for the overall chemical reaction [cf. Eqs. (1-1) and (1-2)], and where the plus sign applies to products, the minus sign to reactants. Also, Fick's second law states that

$$\left(\frac{\partial c}{\partial t}\right)_{R,V} = \nabla^2(Dc)$$

where D is the diffusion coefficient. Combining all these equations, we obtain

$$\frac{1}{V}\frac{dn}{dt} = \pm aR + \nabla^2(Dc) \tag{1-4}$$

In the case of a constant-volume system without diffusion,

$$\frac{1}{V}\frac{dn}{dt} = \frac{dc}{dt} = \pm aR$$

exactly as previously given. For a flow system of constant cross-sectional area A, the specific volume is certainly not constant in gaseous systems, and even in liquids it is not necessarily constant. Therefore for a steady state in which the mass flow does not change with time,

$$\frac{dn}{dt} = \frac{dn}{dx}u$$

and

$$\frac{1}{V}\frac{dn}{dx}u = \frac{1}{Ax}\frac{dn}{dx}u = \pm aR + \nabla^2(Dc) \tag{1-5}$$

where u is the linear flow velocity. In such a system with no diffusion, n can be measured at various values of x, and R can be calculated directly from the value of dn/dx at a fixed value of x. Alternatively, if the dependence of R on the concentrations of reactants and products is known, Eq. (1-5) can often be integrated. Flow experiments are frequently used for kinetic studies in both liquids and gases.

In general, the reaction rate can be a function of all the species present in a reaction mixture. Thus

$$R = f(c_1, c_2, \ldots, c_j) \tag{1-6}$$

This equation is called the rate law for a given reaction. In general, the form of the rate law is determined by experiment and cannot be predicted from the stoichiometric equation for the overall reaction. For example, the rate law for the reaction [1]

$$H_2 + Br_2 \rightarrow 2HBr$$

is

$$R = \frac{1}{2}\frac{d}{dt}(HBr) = \frac{k(H_2)(Br_2)^{1/2}}{1 + k'(HBr)/(Br_2)} \tag{1-7}$$

One of the objectives of experimental kinetics is to propose a reasonable mechanism which will conform to the experimentally observed rate law.

In many cases, the rate equation takes the particularly simple form of a constant times the product of powers of concentrations so that

$$R = f(c_1, c_2, \ldots, c_j) = kc_1^{n_1}c_2^{n_2}\cdots c_j^{n_j} \tag{1-8}$$

In such cases the reaction order is defined as the sum of the exponents $(\sum_{i=1}^{j} n_i)$, and the reaction order with respect to any one component is simply the exponent associated with the particular component in question. Of course, the order of the reaction gives no information per se concerning the molecularity or number of molecules involved in the various steps in the reaction mechanism. The constant k appearing in the rate equation (1-8) is called the rate constant, and simple dimensional analysis shows that it must have the units of concentration$^{1-\Sigma n_i}$ time^{-1}. For example, if $\sum n_i = 1$, the units of the rate constant are time^{-1} (sec^{-1}, min^{-1}, etc.); if $\sum n_i = 2$, the units are concentration^{-1} time^{-1} (M^{-1} sec^{-1}, etc.).

The reaction order is not necessarily an integer. For example, under certain conditions, the reaction

$$CH_3CHO \rightarrow CH_4 + CO$$

has a rate law with reaction order $\frac{3}{2}$ [2]:

$$-\frac{d}{dt}(CH_3CHO) = k(CH_3CHO)^{3/2}$$

In the case of the H_2–Br_2 reaction, the concept of reaction order is clearly without meaning, except under limiting conditions where one or the other of the terms in the denominator of the rate law [Eq. (1-7)] is dominant.

The rate of a chemical reaction is usually not measured directly; instead, the concentration of one of the reactants or products is determined as a function of time. A common procedure for determining the reaction order is to compare the experimental results with integrated rate equations for reactions of different orders. This can be best illustrated by integrating several particular rate equations as examples.

For a first-order rate equation,

$$-\frac{dc}{dt} = kc \tag{1-9}$$

This equation can be integrated easily by separating the variables and using integration limits such that at $t = 0$, $c = c_0$ and at $t = t$, $c = c$. The result is

$$\ln(c_0/c) = kt \tag{1-10}$$

If the reaction being studied is first order, a plot of $\ln c$ versus time should be a straight line with a slope of $-k$. Often the dependent variable chosen is the decrease in concentration of reactant. If this variable is designated as x and c_0 is the initial concentration,

$$\frac{dx}{dt} = k(c_0 - x) \qquad \text{and} \qquad \ln \frac{c_0}{c_0 - x} = kt$$

For a reaction second order with respect to one component,

$$-\frac{dc}{dt} = kc^2 = \frac{dx}{dt} = k(c_0 - x)^2 \tag{1-11}$$

and the integrated rate equation is

$$\frac{1}{c} - \frac{1}{c_0} = kt = \frac{1}{c_0 - x} - \frac{1}{c_0} \tag{1-12}$$

so that a plot of $1/c$ versus t should be linear for a second-order reaction. If the reaction is second order overall but first order with respect to each of two reactants consumed in equimolar amounts, that is, $A + B \rightarrow$ products,

$$\frac{dx}{dt} = k(A_0 - x)(B_0 - x) \tag{1-13}$$

For $A_0 = B_0 = c_0$, the integrated equation is exactly the same as given above, and for $A_0 \neq B_0$

$$\frac{1}{B_0 - A_0} \ln \frac{A_0(B_0 - x)}{B_0(A_0 - x)} = kt \tag{1-14}$$

In this particular case, the advantage of using the dependent variable x is clearly indicated. In a similar manner, integrated rate equations can be obtained for more complex reaction orders. For example, a reaction of nth order ($n > 0$, $\neq 1$) with respect to a single reactant has the rate law

$$\frac{dx}{dt} = k(c_0 - x)^n \tag{1-15}$$

and integration yields

$$\frac{1}{n-1}\left[\frac{1}{(c_0 - x)^{n-1}} - \frac{1}{c_0^{n-1}}\right] = kt \tag{1-16}$$

An alternative method of determining the reaction order is to measure the half-life $t_{1/2}$ as a function of initial concentrations. The half-life is simply the time necessary for x to become equal to one-half the initial concentration. (Any other fraction of the initial concentration would serve equally well.) From the integrated rate equations, it can be seen that for a first-order reaction [cf. Eq. (1-10)]

$$t_{1/2} = (\ln 2)/k \tag{1-17}$$

while for reactions of order n with $n > 1$ [cf. Eq. (1-16)]

$$t_{1/2} = \frac{1}{k(n-1)} \frac{2^{n-1} - 1}{c_0^{n-1}} \tag{1-18}$$

Thus the dependence of $t_{1/2}$ on the initial concentration determines the reaction order directly.

Still another method of determining reaction order is the method of initial rates. This method involves the direct measurement of the derivative dx/dt as the corresponding ratio of finite increments $\Delta x/\Delta t$. The fraction reacted is kept as small as possible, usually 0.1 or less. If a measurement is made at two different initial concentrations of any one component, the concentrations of the other reactants being held constant, the reaction order with respect to that component can be determined. For example, if the two rates and corresponding initial concentrations are designated as $(\Delta x/\Delta t)_1$, $(\Delta x/\Delta t)_2$, and (A_1), (A_2), then for a rate law of the form given in Eq. (1-8),

$$\left(\frac{\Delta x}{\Delta t}\right)_1 = k A_1^{n_1} B^{n_2} C^{n_3} \cdots, \qquad \left(\frac{\Delta x}{\Delta t}\right)_2 = k A_2^{n_1} B^{n_2} C^{n_3} \cdots$$

Solving for n_1, we obtain

$$n_1 = \frac{\log(\Delta x/\Delta t)_1 - \log(\Delta x/\Delta t)_2}{\log A_1 - \log A_2} \tag{1-19}$$

The procedure outlined above can be used to determine all the exponents $n_1, n_2, n_3, \ldots$, and the rate constant can be evaluated. The advantage of this method is that complex rate equations, which may be difficult to integrate, can be handled in a convenient manner. Also, the reverse reaction can be completely neglected, provided that initial velocities are actually measured or are obtained by an appropriate extrapolation. For reactions having a simple rate law, i.e., first order, second order, etc., the methods discussed previously are more precise.

Another useful approach for determining the experimental rate law is the isolation method. With this method all of the reactants except one are present in excess. The apparent order of the reaction will be with respect to the one "isolated" reactant since the concentrations of those present in excess will be essentially constant during the course of the reaction. The reaction order with respect to each reactant can be determined by this technique. The primary shortcoming of the isolation method is that for a complex mechanism, the mechanism itself may be altered by the large changes in concentration required for "isolation" of each reactant. Other methods exist for determining the rate law but are not discussed here.

1-4 EXPERIMENTAL METHODS

Although general discussions of experimental methods are seldom profitable, some mention of this subject should be made. Throughout the history of kinetics, the development of ingenious new experimental techniques has produced many of the great advances. Some recent examples will be discussed in considerable detail later in this book. The most desirable analytical methods are those which give a continuous and rapid measurement of the concentration of a specific component. However, it is really necessary only to be able to measure some property of the system which indicates the extent of the chemical reaction occurring. Thus, for example, changes in total pressure and volume can be frequently utilized in kinetic studies. In fact, any property linearly related to the concentration can be conveniently used. This can easily be seen by introducing the concept of the fraction reacted y which is equal to $(c_0 - c_t)/(c_0 - c_\infty)$, where the subscripts refer to the time. For example, $y = x/c_0$ and $\ln(1 - y) = -kt$ for an irreversible first-order reaction. Now if the property measured P_t is equal to $mc_t + q$, where m and q are constants, then it can be readily seen that

$$y = (P_0 - P_t)/(P_0 - P_\infty) \tag{1-20}$$

Examples of suitable properties P are conductance, absorbance, and optical rotation.

Often all the concentrations of reactants and products are linear functions of the same physical property, e.g., pressure, conductance. In this case, detailed analysis can become somewhat complex. As a particular example consider the following prototype reaction, which goes to completion:

$$a\text{A} + b\text{B} \rightarrow g\text{G}$$

For the reaction mixture,

$$P_t = m_\text{A} c_{\text{A}t} + m_\text{B} c_{\text{B}t} + m_\text{G} c_{\text{G}t} + q_\text{A} + q_\text{B} + q_\text{G}$$

If we now introduce the reaction variable x and the initial concentrations of A and B, setting that of G equal to zero,

$$P_t = m_\text{A}(\text{A}_0 - ax) + m_\text{B}(\text{B}_0 - bx) + m_\text{G} gx + q_\text{A} + q_\text{B} + q_\text{G}$$

and

$$P_\infty = m_\text{B}\left(\text{B}_0 - \frac{b\text{A}_0}{a}\right) + \frac{m_\text{G} g\text{A}_0}{a} + q_\text{A} + q_\text{B} + q_\text{G}$$

assuming A is the limiting reagent. Then

$$y = \frac{P_0 - P_t}{P_0 - P_\infty} = \frac{m_\text{A} ax + m_\text{B} bx - m_\text{G} gx}{m_\text{B} b\text{A}_0/a - m_\text{G} g\text{A}_0/a + m_\text{A}\text{A}_0/a} = \frac{x}{\text{A}_0/a}$$

Applications to other reaction mechanisms can be carried out in a similar manner.

1-5 SPECIAL INTEGRATED RATE EQUATIONS

Special integrated rate equations have been derived for the situation where neither the initial concentrations nor the equilibrium values of the concentrations need be known to determine the rate constant. For first-order reactions, Guggenheim [3] has described a method in which measurements of some physical property of the system are made at constant time intervals Δ. Then

$$P_t - P_\infty = (P_0 - P_\infty)e^{-kt} \tag{1-21}$$

$$P_{t+\Delta} - P_\infty = (P_0 - P_\infty)e^{-k(t+\Delta)} \tag{1-22}$$

$$P_t - P_{t+\Delta} = (P_0 - P_\infty)e^{-kt}(1 - e^{-k\Delta}) \tag{1-23}$$

A plot of $\ln(P_t - P_{t+\Delta})$ versus t is linear and has a slope equal to $-k$. The interval Δ should be two or three times greater than the half-life of the reaction for good accuracy. Alternatively, division of Eq. (1-21) by Eq. (1-22) gives

$$(P_t - P_\infty)/(P_{t+\Delta} - P_\infty) = e^{k\Delta} \qquad \text{or} \qquad P_t = P_\infty(1 - e^{k\Delta}) + P_{t+\Delta}e^{k\Delta}$$

A plot of P_t versus $P_{t+\Delta}$ is linear with a slope equal to $e^{k\Delta}$ [4]. This latter method of plotting the data has the advantage that small differences between large numbers, which often have large experimental errors, are not used. Similar equations have been derived for more complex mechanisms including higher-order reactions [5, 6].

1-6 COMPLEX REACTIONS

When the differential equations describing a reaction mechanism cannot be described by a simple one-term rate equation, the mathematical problem of integrating the rate equations can become quite complex. No general method of solving such problems can be given, since usually each reaction mechanism is a special case; however, three relatively simple examples will be presented as illustrations. Combinations of these prototypes are capable of describing many reaction mechanisms.

Reversible First-Order Reaction

$$\mathrm{A} \underset{k_r}{\overset{k_f}{\rightleftharpoons}} \mathrm{B} \tag{1-24}$$

In this case $dx/dt = k_f(\mathrm{A}_0 - x) - k_r(\mathrm{B}_0 - x)$, and straightforward integration with the boundary conditions at $t = 0$, $x = 0$ and at $t = t$, $x = x$ yields

$$(k_f + k_r)t = \ln \frac{k_f\mathrm{A}_0 - k_r\mathrm{B}_0}{k_f\mathrm{A}_0 - k_r\mathrm{B}_0 - x(k_f + k_r)} \tag{1-25}$$

Since k_r/k_f is equal to the equilibrium ratio of $(A_e)/(B_e)$, both rate constants can be evaluated from a single kinetic experiment. For the special case $B_0 = 0$, Eq. (1-25) can be arranged to give

$$(k_f + k_r)t = \ln \frac{A_0 - A_e}{A - A_e}$$

Consecutive Reactions

$$A \xrightarrow{k_1} B \xrightarrow{k_2} C \tag{1-26}$$

The differential equations describing this system are

$$\frac{dA}{dt} = -k_1 A, \qquad \frac{dB}{dt} = k_1 A - k_2 B, \qquad \frac{dC}{dt} = k_2 B$$

Only two of these are independent, however, since $dA + dB + dC = 0$. Integration of the first equation yields

$$A = A_0 e^{-k_1 t} \tag{1-27}$$

and substitution in the second equation gives

$$\frac{dB}{dt} = k_1 A_0 e^{-k_1 t} - k_2 B$$

This equation can be integrated directly and the result (with $B_0 = 0$) is

$$B = \frac{A_0 k_1}{k_2 - k_1}(e^{-k_1 t} - e^{-k_2 t}) \tag{1-28}$$

These two equations give a complete kinetic description of the system, but if a similar expression for C is wanted, the equation for mass conservation in the system can be used. Assuming $C_0 = 0$, we obtain

$$C = A_0 - A - B = A_0\left[1 + \frac{1}{k_1 - k_2}(k_2 e^{-k_1 t} - k_1 e^{-k_2 t})\right] \tag{1-29}$$

Very often intermediates in reaction mechanisms are present in very small concentrations. In such a situation, the rate of change of the concentration of the intermediate with time is much smaller than the corresponding quantities for the reactants and products. The intermediate is then said to be in a steady state, and its time derivative can be set equal to zero (to a good approximation). The result is usually a vast simplification of the rate equation. For example, if (B) is in a steady state,

$$\frac{dB}{dt} \approx 0 = k_1 A - k_2 B, \qquad \text{and} \qquad \frac{dA}{dt} = -\frac{dC}{dt} = -k_1 A$$

The reaction then becomes first order, and integration with $C_0 = 0$ yields

$$\begin{aligned} A &= A_0 e^{-k_1 t} \\ B &= A_0(k_1/k_2)e^{-k_1 t} \\ C &= A_0 - A - B = A_0[1 - (1 + k_1/k_2)e^{-k_1 t}] \end{aligned} \tag{1-30}$$

This last equation is identical with the exact solution [Eq. (1-29)] if $t \gg 1/k_2$ and $k_2 \gg k_1$. The first condition ensures that the steady state is established rapidly compared to the rate of the overall reaction, while the second condition fixes the concentration of B at a very low value, which in turn implies $dB/dt \approx 0$.

The concept of the steady state is of such great importance in chemical kinetics that consideration of a somewhat more complex mechanism is profitable. Consider the reaction mechanism

$$A \underset{k_{-1}}{\overset{k_1}{\rightleftharpoons}} B \underset{k_{-2}}{\overset{k_2}{\rightleftharpoons}} C \tag{1-31}$$

In this case,

$$-\frac{dA}{dt} = k_1 A - k_{-1} B, \qquad -\frac{dB}{dt} = (k_{-1} + k_2)B - k_1 A - k_{-2} C$$

Direct integration of this equation is extremely difficult [7]; however, if B is assumed to be in a steady state ($dB/dt = 0$), we find that

$$B = \frac{k_1 A + k_{-2} C}{k_{-1} + k_2} \tag{1-32}$$

and

$$-\frac{dA}{dt} = \frac{k_1 k_2}{k_{-1} + k_2} A - \frac{k_{-1} k_{-2}}{k_{-1} + k_2} C = k_f A - k_r C \tag{1-33}$$

This latter equation is formally the same as that encountered previously for a first-order reversible reaction and can be easily integrated [cf. Eq. (1-25)]. Simplification of rate equations via the steady-state approximation is often adopted, but evidence for the validity of this approximation should be critically examined in each case where it is used.

This example also can be used to illustrate a point often overlooked, namely the distinction between a phenomenological (experimental) rate constant and a flux coefficient [8]. It is very tempting to say that $k_f A$ is the number of molecules making the transition $A \rightarrow C$ and that $k_r C$ is the number of C molecules making the transition $C \rightarrow A$ per unit time so that the rate equation is simply the difference between two opposing fluxes. However, this is not the case: clearly k_f and k_r individually are combinations of rate constants characterizing steps in both the forward and reverse directions of the reaction. To calculate the flux, an arbitrary boundary must be assumed either between A and B or between B and C, and the amount of material passing across this boundary calculated. If the boundary is put between B and C, then the flux into the C state is

$$k_2\mathrm{B} = (\mathrm{B}/\mathrm{A})k_2\mathrm{A} = r\mathrm{A}$$

where the flux coefficient r is defined by the above equation. The flux out of the C state is

$$k_{-2}\mathrm{C} = r'\mathrm{C}$$

The flux coefficients are not uniquely defined. If the boundary is put between A and B, then $r = k_1$ and $r' = k_{-1}(\mathrm{B}/\mathrm{C})$. The flux coefficients must satisfy two conditions: firstly, $r\mathrm{A} - r'\mathrm{C}$ must be equal to $k_\mathrm{f}\mathrm{A} - k_\mathrm{r}\mathrm{C}$ which is the overall rate of the reaction, $-d\mathrm{A}/dt$; secondly, at equilibrium r/r' must be equal to the equilibrium constant. For this simple example, it can be readily shown that the two conditions are satisfied. Use of the steady-state approximation for B gives

$$r = \frac{k_1 + k_{-2}(\mathrm{C}/\mathrm{A})}{k_{-1} + k_2} k_2$$

With this expression for r and with $r' = k_{-2}$, it directly follows that $r\mathrm{A} - r'\mathrm{C} = k_\mathrm{f}\mathrm{A} - k_\mathrm{r}\mathrm{C}$. At equilibrium, $\mathrm{C}/\mathrm{A} = k_1 k_2/k_{-1}k_{-2}$ so that

$$r/r' = k_1 k_2/k_{-1}k_{-2} = k_\mathrm{f}/k_\mathrm{r}$$

and the second condition is satisfied. However, $r \neq k_\mathrm{f}$ and $r' \neq k_\mathrm{r}$. Furthermore the flux coefficient r is time dependent. Finally the values of r and r' are always greater than k_f and k_r, respectively. In fact, at equilibrium

$$r/k_\mathrm{f} = 1 + k_2/k_{-1} = r'/k_\mathrm{r}$$

The deviation of r from k_f must be identical to the deviation of r' from k_r since $r/r' = k_\mathrm{f}/k_\mathrm{r}$ at equilibrium.

This analysis of a three-state system is a simple example of a general phenomenon. Molecules exist in many different energy states, and in general a distinction must be made between flux coefficients and phenomenological (experimental) rate constants. An experimentally determined rate constant is not a simple measure of the flux of material going from one set of states to another. In general it contains microscopic rate constants for processes occurring in both the forward *and* reverse reactions. The rate constant is time independent, whereas the flux coefficient generally is not. An important point is that theories of chemical kinetics calculate fluxes so that comparison with experimental rate constants must be made with care. A general estimate of the difference between flux coefficients and rate constants is not available.

They probably are of the same order of magnitude, but calculations with simple models suggest the differences are not negligibly small. Although we will, because of common practice, use the term "rate constant" to describe both the theoretical and experimental coefficients, the conceptual difference between a rate constant and a flux coefficient should be kept in mind.

Concurrent Reactions

First let us consider a mechanism whereby a single reactant produces several products by first-order processes:

$$\mathrm{A} \xrightarrow{k_1} \mathrm{B}, \qquad \mathrm{A} \xrightarrow{k_2} \mathrm{C}, \qquad \mathrm{A} \xrightarrow{k_3} \mathrm{D} \tag{1-34}$$

These equations can be easily integrated as

$$-\frac{d\mathrm{A}}{dt} = (k_1 + k_2 + k_3)\mathrm{A} = k\mathrm{A}$$

Therefore

$$\mathrm{A} = \mathrm{A}_0 e^{-kt} \tag{1-35}$$

Also

$$\frac{d\mathrm{B}}{dt} = k_1\mathrm{A} = k_1\mathrm{A}_0 e^{-kt}$$

Therefore

$$\mathrm{B} = \mathrm{B}_0 + (k_1\mathrm{A}_0/k)(1 - e^{-kt})$$

Similarly

$$\mathrm{C} = \mathrm{C}_0 + (k_2\mathrm{A}_0/k)(1 - e^{-kt}), \qquad \mathrm{D} = \mathrm{D}_0 + (k_3\mathrm{A}_0/k)(1 - e^{-kt})$$

Note that if $\mathrm{B}_0 = \mathrm{C}_0 = \mathrm{D}_0 = 0$, $\mathrm{B}:\mathrm{C}:\mathrm{D} = k_1:k_2:k_3$; that is, the products are in constant ratio independent of time and the initial concentration. Thus even if the rates were too fast to be measured, relative rate constants could be determined simply by measuring the relative concentrations of products.

We shall now consider a mechanism whereby different reactants produce a common product:

$$\mathrm{A} \xrightarrow{k_1} \mathrm{C}, \qquad \mathrm{B} \xrightarrow{k_2} \mathrm{C} \tag{1-36}$$

The rate equations describing this system are

$$-\frac{d\mathrm{A}}{dt} = k_1\mathrm{A}, \qquad -\frac{d\mathrm{B}}{dt} = k_2\mathrm{B}$$

which can be integrated, yielding

$$\mathrm{A} = \mathrm{A}_0 e^{-k_1 t}, \qquad \mathrm{B} = \mathrm{B}_0 e^{-k_2 t}$$

However,

$$C = A_0 - A + B_0 - B = C_\infty - A_0 e^{-k_1 t} - B_0 e^{-k_2 t}$$

where $C_\infty = A_0 + B_0$. This can be rearranged to give

$$\log(C_\infty - C) = \log(A_0 e^{-k_1 t} + B_0 e^{-k_2 t}) \tag{1-37}$$

If $k_1 = k_2$, a plot of $\log(C_\infty - C)$ versus t is linear. In general, however, a curvature is seen during the initial part of the reaction. After a sufficient length of time, one of the exponential terms becomes negligible, for example, $A_0 e^{-k_1 t} \ll B_0 e^{-k_2 t}$, and the plot becomes linear. From the linear portion of the curve the smaller of the rate constants can be evaluated. The initial portion of the curve then can be easily analyzed to obtain the other rate constant.

Parallel reactions are frequently encountered in practice, although the mechanisms are often of different reaction order. In any case, the treatment of such reaction mechanisms parallels that of the simpler case treated here.

1-7 KINETICS AND THE EQUILIBRIUM STATE

The equilibrium state is a special kinetic state, namely that state in which the net change of all components with respect to time is equal to zero. For the sake of simplicity, in the ensuing discussion we shall assume all activity coefficients equal to unity. This assumption will in no way invalidate the conclusions reached. In addition, the principle of detailed balance must be obeyed. It states that at equilibrium the rate of conversion from a state A to another state B must exactly equal the *direct* conversion rate from state B to A. Although this sounds very much like the kinetic definition of equilibrium given above, the two are not exactly the same [9]. Consider the reaction scheme (1-38).

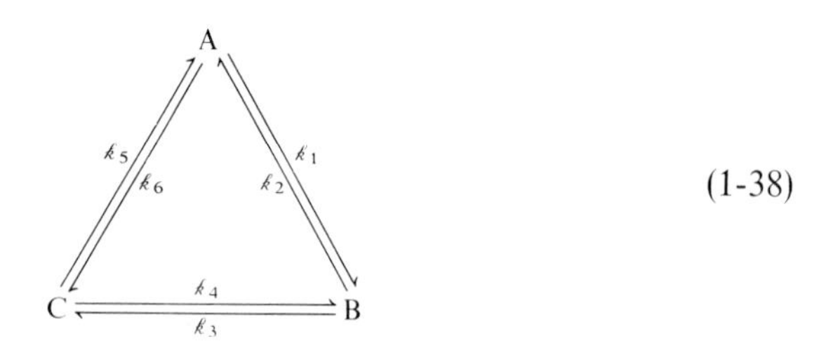

If, as the equilibrium condition, the time derivatives of A, B, and C are now set equal to zero in the usual fashion, for example,

$$\frac{dA_e}{dt} = 0 = -(k_1 + k_6)A_e + k_2 B_e + k_5 C_e$$

where the subscript e designates an equilibrium concentration, the following expressions can be obtained:

$$\frac{B_e}{A_e} = \frac{k_1k_4 + k_1k_5 + k_4k_6}{k_2k_4 + k_2k_5 + k_3k_5}, \qquad \frac{C_e}{A_e} = \frac{k_1k_3 + k_2k_6 + k_3k_6}{k_2k_4 + k_2k_5 + k_3k_5} \tag{1-39}$$

These look quite different from equilibrium constants usually encountered, which are ratios of single rate constants. However, the principle of detailed balance states that *each* individual step must also be in equilibrium

$$\frac{B_e}{A_e} = \frac{k_1}{k_2}, \qquad \frac{C_e}{A_e} = \frac{k_6}{k_5}, \qquad \frac{B_e}{C_e} = \frac{k_4}{k_3}$$

Combination of these equations gives

$$k_1k_3k_5 = k_2k_4k_6$$

This relationship, then, is a direct consequence of the principle of detailed balance, and when it is used to simplify Eqs. (1-39), we find, for example, that

$$\frac{B_e}{A_e} = \frac{k_1}{k_2}\frac{k_2k_4 + k_2k_5 + k_3k_5}{k_2k_4 + k_2k_5 + k_3k_5} = \frac{k_1}{k_2}, \qquad \text{etc.}$$

As a second illustration, consider a reaction which occurs through two reaction paths:

$$A \underset{k_2}{\overset{k_1}{\rightleftharpoons}} B, \qquad A + C \underset{k_4}{\overset{k_3}{\rightleftharpoons}} C + B \tag{1-40}$$

At equilibrium,

$$-\frac{dA_e}{dt} = \frac{dB_e}{dt} = 0 = k_1A_e - k_2B_e + k_3A_eC_e - k_4B_eC_e \quad \text{or} \quad \frac{B_e}{A_e} = \frac{k_1 + k_3C_e}{k_2 + k_4C_e}$$

This unusual result of having a concentration-dependent equilibrium constant can be resolved by remembering that equilibrium also must be attained for each individual path. Therefore

$$k_1A_e = k_2B_e, \qquad k_3A_eC_e = k_4B_eC_e$$

and substitution of these relationships into the equation above results in the elimination of terms containing C_e, as expected.

A condition closely related to detailed balance which also can be used to help understand the second example is that the equilibrium state is independent of the reaction path. This, of course, follows from the definition of thermodynamic equilibrium. A further important consequence of these equilibrium principles is that if the rate law for the reaction in the forward direction is known, the rate law for the reverse reaction can be derived from it and the equilibrium expression for the overall reaction. As an example, consider the reaction

$$2NO_2 \underset{k_r}{\overset{k_f}{\rightleftharpoons}} 2NO + O_2 \tag{1-41}$$

which has the following rate law in the forward direction [10]:

$$-\frac{1}{2}\frac{d}{dt}(NO_2) = k_f(NO_2)^2$$

Since, according to Eq. (1-41),

$$K = \frac{k_f}{k_r} = \frac{(NO)^2(O_2)}{(NO_2)^2}$$

a possible rate law in the reverse direction is

$$+\frac{1}{2}\frac{d}{dt}(NO_2) = k_r(NO)^2(O_2)$$

In order for the above-mentioned relationships between forward and reverse rate laws to be valid, we must be certain that the rate law determined experimentally is applicable at equilibrium. Even so, the above derivation of the rate law for the reverse reaction is not unambiguous, since the equilibrium expression does not take into account the stoichiometry in the actual mechanism. The reaction above, to be completely general, should be written as

$$2nNO_2 \underset{k_r}{\overset{k_f}{\rightleftharpoons}} 2nNO + nO_2 \tag{1-42}$$

where n can be any positive number.

The equilibrium expression is then

$$K = \frac{k_f}{k_r} = \frac{(NO)^{2n}(O_2)^n}{(NO_2)^{2n}} \tag{1-43}$$

and the rate law for the reverse reaction is

$$\frac{1}{2n}\frac{d(NO_2)}{dt} = k_r\frac{(NO)^{2n}(O_2)^n}{(NO_2)^{2n-2}}$$

Ordinarily we set n equal to 1, but such a procedure may not always be valid. Therefore the only unambiguous procedure for determining the rate law for the reverse reaction is by direct kinetic measurements.

1-8 TEMPERATURE DEPENDENCE OF RATE CONSTANTS

The rates of most chemical reactions are very sensitive to temperature changes. The first quantitative formulation of the dependence of reaction rates on temperature was given by Arrhenius [11] and is still extensively used. He proposed the following relationship between the specific rate constant k and the absolute temperature T:

$$k = Ae^{-E_a/RT} \tag{1-44}$$

where A and E_a are constants; A is commonly called the preexponential factor, and E_a the activation energy; R is the gas constant. This equation fits many of the available experimental kinetic data, although sometimes the more general equation

$$k = A'T^n e^{-E_a'/RT} \tag{1-45}$$

is used. Here n can have any value dictated by experiment. Note that A and E_a are *different* constants for the forward and reverse directions of a given reaction. The relationship between these parameters and the equilibrium constant K' is

$$K' = \frac{k_f}{k_r} = \frac{A_f}{A_r} \exp\left[-\frac{(E_{af} - E_{ar})}{RT}\right]$$

At constant pressure and temperature the thermodynamic equilibrium constant K can be written as

$$K = \exp(-\Delta G^0/RT) = \exp(\Delta S^0/R)\exp(-\Delta H^0/RT)$$

where ΔG^0, ΔH^0, and ΔS^0 are the standard-free-energy, standard-enthalpy, and standard-entropy changes for the reaction. For reactions in the gas phase, rate constants are usually expressed in concentration units whereas equilibrium constants are expressed in pressure units with a standard state of 1 atm for each reactant. Therefore, for ideal gases $K' = K(RT)^{-\Delta n}$, where Δn is the change in the number of moles for the reaction, i.e., the number of moles of products minus the number of moles of reactants in the stoichiometric equation. Differentiation of the thermodynamic and kinetic expressions for K' with respect to temperature gives

$$\frac{d \ln K'}{dT} = \frac{\Delta H^0 - \Delta nRT}{RT^2} = \frac{\Delta E^0}{RT^2} = \frac{(E_{af} - E_{ar})}{RT^2}$$

In this equation, ΔE^0 is the standard internal energy change for the reaction. Thus the difference in activation energies corresponds to an energy of reaction, while the ratio of A's is related to the entropy change. For reactions in liquids, both rate constants and equilibrium constants are usually expressed in terms of concentrations with a standard state of 1 M so that $K = K'$; in this case the difference in activation energies corresponds to the enthalpy of reaction and the ratio of A's again is related to the entropy of reaction. [Since $\Delta(PV) \approx 0$, the energy and enthalpy of reaction are essentially identical.] A more detailed analysis taking into account the more accurate form of Arrhenius's equation and the heat-capacity differences between products and reactants can be carried out, but the results are quite similar to the simpler case presented here. In some cases, such an equation as (1-45) will not fit experimental observations; each exception, however, usually requires a special explanation.

1-9 KINETICS AND REACTION MECHANISMS

The primary goal of all kinetic studies, of course, should be to find a reaction mechanism consistent with the kinetic data. Often several mechanisms show this consistency, and the investigator must turn to nonkinetic methods and a general knowledge of chemistry and physics to decide which reaction pathway is most probable.

For example, if a given reactant can form several products, a quantitative analysis of the products often provides information on the mechanism. This is not true if an equilibrium distribution of products is obtained; in this case nothing can be said about the reaction intermediates. However, the product distribution is often determined by the relative rates of the reactions occurring rather than by thermodynamics. Consider the solvolysis of tertiary butyl halides in aqueous ethanol, in which the following reactions occur [12]:

$$(CH_3)_3CX + H_2O \rightarrow (CH_3)_3COH + H^+ + X^-$$
$$(CH_3)_3CX + C_2H_5OH \rightarrow (CH_3)_3COC_2H_5 + H^+ + X^-$$
$$(CH_3)_3CX \rightarrow (CH_3)_2C{=}CH_2 + H^+ + X^-$$

The chloride, bromide, and iodide compounds all react in first-order reactions but at markedly different rates. Nevertheless, the fraction of reaction leading to the unsaturated isobutylene product is the same for each halide and is not the equilibrium fraction. This suggests that all the halide compounds react to form the same reaction intermediate and that the further reaction of this intermediate determines the product distribution. The mechanism proposed involves the intermediate formation of the tertiary butyl carbonium ion $(CH_3)_3C^+$. The rate of forming this intermediate will be different for each halide, but the relative rates of the three product-determining reactions of the carbonium ion will not depend on the particular halide.

Isotopes may be conveniently used to obtain additional information about reaction mechanisms. A particular example is the formation of an ester from benzoic acid and methyl alcohol. By labeling the methyl alcohol with ^{18}O, it can be shown that the carbon–oxygen bond in the alcohol remains intact during the course of the reaction, implying that an OH group is lost from the acid [13].

Many other methods exist for elucidating reaction mechanisms but are not considered here. Later in this book some specific reaction mechanisms will be discussed in considerable detail. In any event, a proposed mechanism should always be a minimal mechanism consistent with the data; i.e., unnecessarily complex mechanisms are to be avoided. The ultimate goal is a unique mechanism in terms of elementary steps, a goal which unfortunately is rarely, if ever, achieved.

Problems

1-1 The rate of the reaction

$$C_2H_5NH_2(g) \rightarrow C_2H_4(g) + NH_3(g)$$

is measured by noting P_{total} as a function of time. At 500°C, the following results were obtained:

Time (sec):	0	60	360	600	1200	1500
P_{total} (mm Hg):	55	60	79	89	102	105

(a) Find the order of the reaction.
(b) Find the specific rate constant.

1-2 The overall stoichiometry of the reaction of nitric oxide with hydrogen is

$$2NO + 2H_2 \rightarrow N_2 + 2H_2O$$

The kinetic data presented in Table P1-2 have been obtained for this reaction at 1099°K [14].

TABLE P1-2

$P^0_{H_2} = P^0_{NO}$ (mm)	$t_{1/2}$ (sec)	$P^0_{H_2} = P^0_{NO}$ (mm)	$t_{1/2}$ (sec)
354	81	288	140
340.5	102	202	224

$P^0_{H_2}$ (mm)	P^0_{NO} (mm)	Initial rate of reaction (mm/sec)	$P^0_{H_2}$ (mm)	P^0_{NO} (mm)	Initial rate of reaction (mm/sec)
289	400	9.6	400	359	9.0
205	400	6.6	400	300	6.2
147	400	4.7	400	152	1.5

(a) Determine the rate law for this reaction.
(b) Calculate the specific rate constant at 1099°K. Specify the units used.
(c) The relative values of the specific rate constant have the following temperature dependence:

T (°K):	956	984	1024	1061	1099
k (relative):	1.00	2.34	5.15	10.9	18.8

Calculate the Arrhenius activation energy for this reaction.

1-3 The kinetics of the formation of ethyl acetate from acetic acid and ethyl alcohol as homogeneously catalyzed by a constant amount of HCl has been studied by titrating 1-cc aliquots of the reaction mixture with 0.0612 *N* base at various times. The data presented in Table P1-3 have been obtained at 25°C [15].

TABLE P1-3

Initial concentrations			
$(CH_3COOH) = 1.000\ M$,		$(C_2H_5OH) = 12.756\ M$,	
$(H_2O) = 12.756\ M$,		$(CH_3COC_2H_5) = 0$	
t (min)	Base (cc)	t (min)	Base (cc)
0	24.37	148	18.29
44	22.20	313	15.15
62	21.35	384	14.50
108	19.50	442	14.09
117	19.26	∞	12.68

The overall reaction can be written as

$$CH_3COOH + C_2H_5OH \underset{k_{-1}}{\overset{k_1}{\rightleftharpoons}} CH_3COC_2H_5 + H_2O$$

and the reaction has been found to be first order with respect to each of the four reactants. Calculate the specific rate constants k_1 and k_{-1}.

1-4 The hydrolysis of diethyl acetal in aqueous solution

$$CH_3CH(OC_2H_5)_2 + H_2O \rightarrow CH_3CHO + 2C_2H_5OH$$

proceeds with a volume change, so that the reaction rate can be studied with a dilatometer. The data given in Table P1-4 have been obtained at pH 3.48 and 25°C.

TABLE P1-4

t (min)	r (mm)[a]	t (min)	r (mm)[a]	t (min)	r (mm)[a]
2	4.47	22	6.58	42	7.92
4	4.69	24	6.72	44	8.02
6	4.92	26	6.89	46	8.11
8	5.17	28	7.04	48	8.21
10	5.40	30	7.19	50	8.31
12	5.61	32	7.31	52	8.40
14	5.81	34	7.48	54	8.49
16	6.01	36	7.59	56	8.58
18	6.21	38	7.70	58	8.65
20	6.40	40	7.81	60	8.71

[a] The volume change is directly proportional to the change in r, which is the height of the liquid in a capillary.

Show that the above reaction is first order and calculate the specific rate constant.

1-5 The hydrolysis of a mixture of *two* isomers of diethyl-*t*-butylcarbinyl chloride has been studied in alcohol–water mixtures at 25°C and the data in Table P1-5 have been obtained [16].

TABLE P1-5

t (hr)	Amount reacted[a]	t (hr)	Amount reacted[a]	t (hr)	Amount reacted[a]
0	0	2.00	4.96	4.00	6.22
0.25	1.82	2.50	5.39	5.00	6.56
0.50	2.24	3.00	5.75	6.00	6.78
1.00	3.54	3.50	6.02	∞	7.25
1.50	4.37				

[a] Milliliters NaOH necessary to titrate 5-ml aliquot of reaction mixture.

(a) What is the rate law for the hydrolysis of the two isomers?
(b) What are the values of the specific rate constants associated with the mechanism?
Hint: The hydrolysis of similar single compounds existing in only one isomeric state has been found to be first order.

1-6 The decomposition of acetaldehyde into methane and carbon monoxide is a second-order reaction (mechanism $2A \rightarrow 2B + 2C$) characterized by a specific rate constant of 0.19 M^{-1} sec^{-1} at 791°K.

(a) Find the time required to reach half-decomposition at a *constant pressure* of 1 atm at 791°K in the sense that (1) $n/n_0 = \frac{1}{2}$ (n = number of moles) and (2) $c/c_0 = \frac{1}{2}$ (c = concentration).

(b) Find the time required to reach half-decomposition at *constant volume* for an initial pressure of 1 atm at 791°K.

1-7 Consider the hypothetical, second-order, kinetically irreversible reaction at constant temperature and volume

$$2A \xrightarrow{k} \text{products}$$

Let the concentration c_t of A be given in terms of a measurable property P_t by

$$P_t = B\exp(bc_t)$$

where B and b are constants depending only on temperature and t denotes the time.
Derive an expression for the specific rate constant k in terms of t, P_t, B, and b.

1-8 In an aqueous medium the reaction

$$H_2O_2 + 2H^+ + 3I^- \rightarrow I_3^- + 2H_2O$$

has the rate law

$$\frac{d}{dt}(I_3^-) = k(H_2O_2)(I^-) + k'(H_2O_2)(I^-)(H^+)$$

(a) What does the fact that the rate law consists of two independent terms imply about the mechanism?

(b) Derive a possible rate law for the reverse reaction.

1-9 Consider the reaction mechanism

$$A \xrightarrow{k_1} B \xrightarrow{k_2} C$$

(a) Plot the concentrations of A, B, and C as a function of time for $k_1 = 1\ \text{sec}^{-1}$ and $k_2 = 0.5\ \text{sec}^{-1}$. Assume $B_0 = C_0 = 0$.

(b) Make the same plot for $k_1 = 1\ \text{sec}^{-1}$, $k_2 = 10^2\ \text{sec}^{-1}$. (Does this problem tell you anything about when the steady-state approximation is justified?)

(c) At what time is the concentration of B maximized in each of the preceding cases?

1-10 The decomposition of N_2O_5 ($2N_2O_5 \rightarrow 4NO_2 + O_2$) proceeds by the mechanism

$$N_2O_5 \underset{k_2}{\overset{k_1}{\rightleftharpoons}} NO_2 + NO_3$$

$$NO_2 + NO_3 \xrightarrow{k_3} NO_2 + O_2 + NO$$

$$NO + NO_3 \xrightarrow{\text{fast}} 2NO_2$$

(a) Assuming NO_3 to be in a steady state, derive an expression for the rate of decomposition of N_2O_5 $[-d(N_2O_5)/dt]$.

(b) Predict the form of the rate law for the reverse reaction, i.e., the formation of N_2O_5 from NO_2 and O_2.

1-11 The exchange of radioactive atoms between chemical species can be represented by the reaction

$$AX + BX^* \rightleftharpoons AX^* + BX$$

where * denotes the radioactive atom. Derive an equation for the rate of exchange and show that it is first order regardless of the mechanism. Assume that the amount of radioactive component is small compared to the total concentration of the same component and that the tracer molecules react at the same rate as the corresponding nonradioactive molecules.

References

1. M. Bodenstein and S. C. Lind, *Z. Phys. Chem.* (*Leipzig*) **57**, 168.
2. M. Letort, *J. Chim. Phys. Physicochim. Biol.* **34**, 267, 355, 428 (1937).
3. E. A. Guggenheim, *Philos. Mag.* **2**, 538 (1926).
4. F. J. Kezdy, J. Jaz, and A. Bruylants, *Bull. Soc. Chim. Belg.* **67**, 687 (1958); E. S. Swinbourne, *J. Chem. Soc.* p. 237 (1960).
5. W. E. Roseveare, *J. Am. Chem. Soc.* **53**, 1651 (1931).
6. J. M. Sturtevant, *J. Am. Chem. Soc.* **59**, 699 (1937).
7. T. M. Lowry and W. T. John, *J. Chem. Soc.* **97**, 2634 (1910).
8. B. Widom, *Science* **148**, 1555 (1965).
9. L. Onsager, *Phys. Rev.* **37**, 405 (1931).
10. M. Bodenstein and I. Ramstetter, *Z. Phys. Chem.* (*Leipzig*) **100**, 68, 106 (1928).
11. S. Arrhenius, *Z. Phys. Chem.* (*Leipzig*) **4**, 226 (1889).
12. R. A. Cooper, E. D. Hughes, and C. K. Ingold, *J. Chem. Soc.* p. 1280 (1937).
13. I. Roberts and H. C. Urey, *J. Am. Chem. Soc.* **60**, 2391 (1938).
14. C. N. Hinshelwood and T. E. Green, *J. Chem. Soc.* p. 730 (1926).
15. O. Knoblauch, *Z. Phys. Chem.* (*Leipzig*) **22**, 268 (1897).
16. H. C. Brown and R. S. Fletcher, *J. Am. Chem. Soc.* **71**, 1845 (1949).

General Bibliography

S. W. Benson, "Foundations of Chemical Kinetics." McGraw-Hill, New York, 1960.
A. A. Frost and R. G. Pearson, "Kinetics and Mechanism," 2d ed. Wiley, New York, 1961.
S. Glasstone, K. J. Laidler, and H. Eyring, "The Theory of Rate Processes." McGraw-Hill, New York, 1941.
H. S. Johnston, "Gas Phase Reaction Rate Theory." Ronald Press, New York, 1966.
V. N. Kondrat'ev, "Chemical Kinetics of Gas Reactions" (translated by J. M. Crabtree and S. N. Carruthers). Addison-Wesley, Reading, Massachusetts, 1964.
K. J. Laidler, "Theories of Chemical Reaction Rates." McGraw-Hill, New York, 1969.
R. D. Levine and R. B. Bernstein, "Molecular Reaction Dynamics." Oxford Univ. Press, London and New York, 1974.
R. E. Weston, Jr., and H. A. Schwarz, "Chemical Kinetics." Prentice-Hall, Englewood Cliffs, New Jersey, 1972.

CHAPTER **2**

THEORIES OF CHEMICAL KINETICS

2-1 INTERACTIONS BETWEEN MOLECULES

Several different approaches have been utilized to develop molecular theories of chemical kinetics which can be used to interpret the phenomenological description of a reaction rate. A common element in all approaches is an explicit formulation of the potential energy of interaction between reacting molecules. Since exact quantum-mechanical calculations are not yet available for any system, this inevitably involves the postulation of specific models of molecules which only approximate the real situation. The ultimate test of the usefulness of such models is found in the number of independent macroscopic properties which can be correctly explained or predicted. Even so, it must be remembered that it is possible for incorrect models to predict reasonably correct macroscopic properties because of fortuitous cancellation of errors, insensitivity of the properties to the nature of the model, relatively large uncertainties in the magnitudes of the properties, or combinations of such effects.

The simplest types of models assume structureless molecules (i.e., no internal degrees of freedom and a potential energy of interaction $U(r)$ that depends only on the internuclear separation r (i.e., central force fields). Furthermore, only interactions between a pair of molecules are considered, limiting this approach to fairly dilute gases. The following models, each identified in Fig. 2-1 in terms of $U(r)$, seem worthy of mention.

1. *Forceless mass points* A gas of such molecules obeys $PV = RT$ at all densities, but intermolecular collisions do not occur so that this model is not useful for chemical kinetics.
2. *Elastic hard spheres* This model obeys $P(V - b) = RT$ at moderate densities and $PV = RT$ at low densities.
3. *Elastic hard spheres with superposed central attractive forces* This is the so-called van der Waals model for which the equation of state is

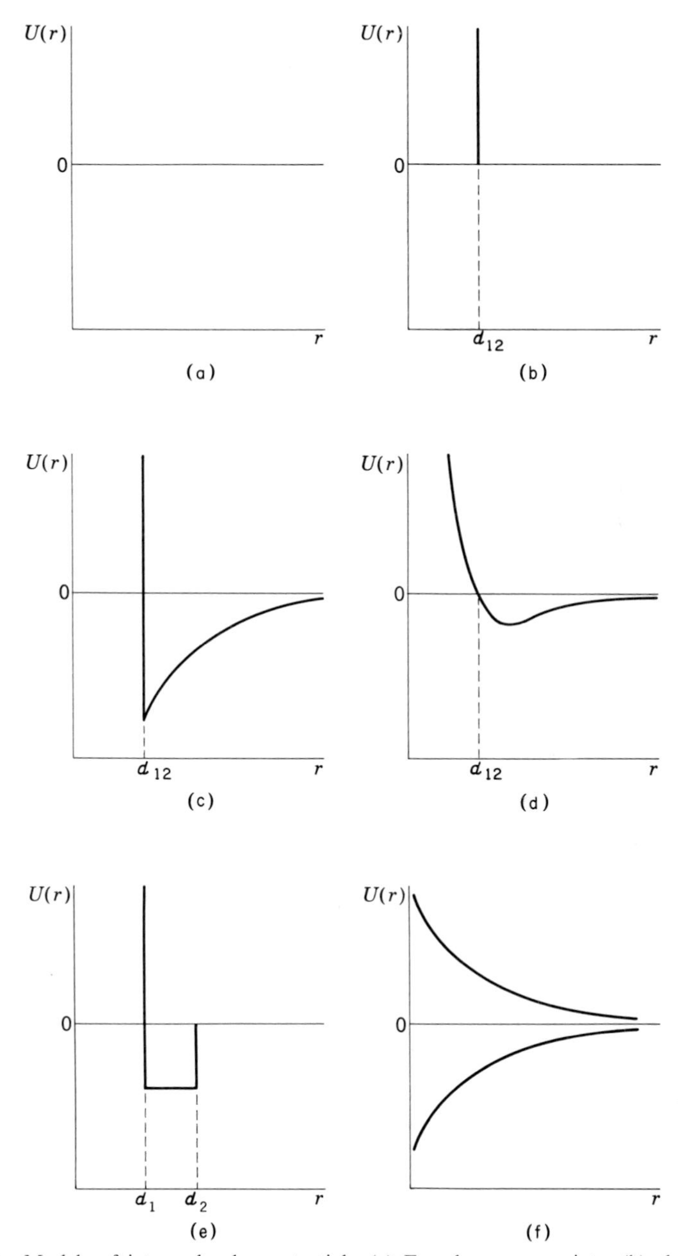

Fig. 2-1 Models of intermolecular potentials. (a) Forceless mass points; (b) elastic hard spheres; (c) elastic hard spheres with superposed central attractive forces; (d) molecules with central finite repulsive and attractive forces; (e) square-well model; (f) point centers of inverse-power repulsion or attraction.

$(P + a/V^2)(V - b) = RT$ at moderate densities (or $PV = RT$ at sufficiently low densities).

4. *Molecules with central finite repulsive and attractice forces* The equation of state for this model is best written in the virial form

$$\frac{PV}{RT} = 1 + \frac{\mathrm{B}(T)}{V} + \frac{\mathrm{C}(T)}{V^2} + \frac{\mathrm{D}(T)}{V^3} + \cdots$$

where $B(T)$, $C(T)$, etc., are known as the second, third, etc., virial coefficients. If the specific form of the molecular interaction is known, these coefficients can be calculated from statistical mechanics.

5. *Square well model* This is a mathematical simplification of the more realistic model possessing finite repulsions and attractions. The region of attraction is restricted to a range bounded by discontinuities at two internuclear separations; at the smaller of these, infinite repulsion occurs, and beyond the larger no interaction exists.

6. Two models, which are of more interest for their mathematical tractability than for their physical plausibility, are *point centers of inverse power repulsion* and *point centers of inverse power attraction.* Models with both central finite repulsion *and* attraction are more general cases of these interactions.

These models are useful in applying simple kinetic theory to chemical kinetics, but a more fundamental approach utilizing quantum-mechanical principles is desirable. In particular, what is desired is a function describing the potential energy of the reactants as they approach each other from infinite separation and then separate as different and distinguishable products molecules as they return to a state of zero potential energy. The construction of such a multidimensional potential-energy surface was first carried out by Eyring and Polanyi in 1931 [1], and attempts have been made continuously since then. However, at this time, calculation of an exact energy surface is still not possible so that semiempirical methods must be utilized.

To illustrate, we consider a displacement reaction of three atoms of the type

$$\mathrm{X + YZ \rightarrow XY + Z} \tag{2-1}$$

As X approaches YZ, the attraction between Y and Z decreases while that between X and Y increases until the activated complex X··Y··Z is formed, after which, if reaction occurs, the internuclear attraction between X and Y continues to increase until a stable XY product is formed, and the atom Z is separated from XY according to

$$\mathrm{X + YZ \rightleftharpoons X\cdot\cdot Y\cdot\cdot Z \rightarrow XY + Z} \tag{2-2}$$

A method for determining the complete potential-energy surface for the three-atom system involves the following procedure.

We first find the potential energy as a function of internuclear distance for all the possible diatomic molecules that can be made from the atoms X, Y, and Z. This can be conveniently done by using spectroscopic data to obtain the constants in the Morse equation for the potential of a diatomic molecule:

$$U(r) = D[e^{-2a(r-r_e)} - 2e^{-a(r-r_e)}] \tag{2-3}$$

where D is the spectroscopic energy of dissociation and is equal to the thermodynamic energy of dissociation plus the vibrational zero-point energy, a is a constant related to the fundamental vibrational frequency, and r_e is the equilibrium bond distance. Eyring and Polanyi [1], extending earlier work by London [2], assumed that $U(r)$ could be represented by

$$U(r)_{XY} = A + \alpha, \qquad U(r)_{YZ} = B + \beta, \qquad U(r)_{XZ} = C + \gamma \tag{2-4}$$

where A, B, and C are the coulombic contributions to the energy, and α, β, and γ are the exchange portions which give rise to the valence forces. Both the coulombic and exchange energies are, of course, functions of r. For the three-atom system, London has shown that the total potential energy is given by

$$U_{XYZ} = A + B + C - (\alpha^2 + \beta^2 + \gamma^2 - \alpha\beta - \beta\gamma - \alpha\gamma)^{1/2} \tag{2-5}$$

corresponding to 60° vector addition of α, β, and γ and direct addition of A, B, and C. In order to calculate U_{XYZ}, we must therefore know A and α, B and β, and C and γ separately. Since A, B, and C are rather small, they are rather arbitrarily taken to be about 14% of the magnitude of $U(r)_{\text{diatomic}}$ for all values of r of interest. This choice is made because the calculations were first carried out for the reaction

$$H + H_2(\text{ortho}) \rightarrow H_2(\text{para}) + H \tag{2-6}$$

and the best fit with experiment was obtained with this apportionment between the coulombic and exchange energies. Approximate calculations for the hydrogen molecule show that the coulombic portion of the energy lies between about 5 and 20%, depending upon the value of r.

The actual labor of constructing the three-atom potential-energy surface was much reduced by London, who showed that for three identical nuclei bound by three s electrons, the linear configuration involves the lowest energy of activation. Because of the exponential dependence of the rate on the energy of activation, the linear configuration has the greatest probability of transforming reactants into products. We can therefore specify the potential energy of each of the diatomic molecules in terms of a single interaction distance such as r_1 or r_2 or $r_1 + r_2$, as in $X\overset{r_1}{—}Y\overset{r_2}{—}Z$. For the central potential of the linear three-atom system U_{XYZ}, therefore, a potential-energy surface results from plotting U_{XYZ} against any two independent internuclear separations in a three-dimensional space. In the absence of a three-dimensional space, the same information may be represented by a potential-energy contour diagram, as illustrated in Fig. 2-2, where $r_1 = r_{XY}$ and $r_2 = r_{YZ}$. A multidimensional energy surface constructed in this manner is often called a LEP (London–Eyring–Polanyi) surface. It may be noted that when parallel cuts are taken through Fig. 2-2 at a large value of r_{XY} (line *ab*) or at a correspondingly large value of r_{YZ} (line *bd*), we have essentially the potential-energy curves for diatomic molecules whose constituent atoms exert both repulsive and attractive forces, as shown in Fig. 2-1d. It can be shown from classical mechanics that the problem of determining the energy of activation becomes one of finding how a frictionless ball of mass $m_X + m_Y + m_Z$ will roll on such a surface. To illustrate, consider the conversion of H_2(ortho) to H_2(para) catalyzed by hydrogen atoms [Eq. (2-6)]. The easiest way for the system to get from the initial state $H + H_2$(ortho) to the final state $H + H_2$(para) is for the ball to roll along the valley, corresponding to the dotted line *acd*. At one extreme of the surface, near *a*, $r_{XY} \cong \infty$ and $r_{YZ} =$ 0.74 Å, but as the ball rolls along the valley, a slight saddle is met in a crater

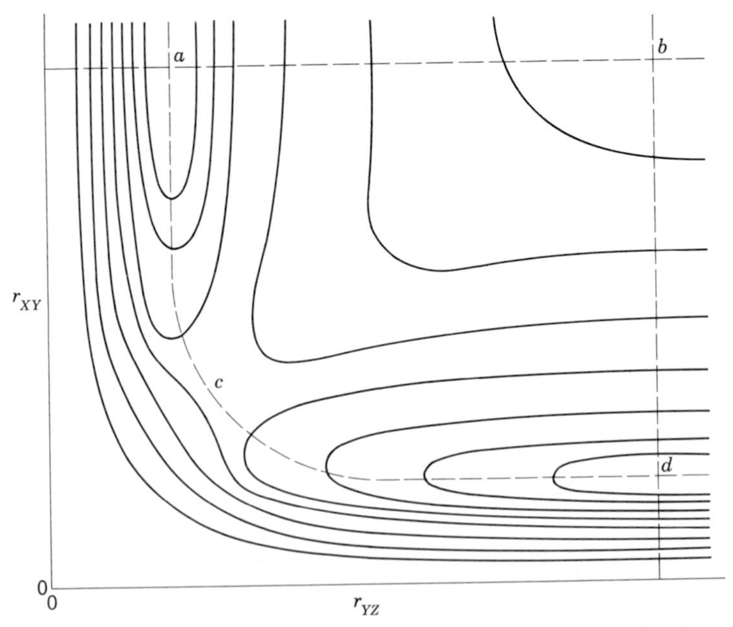

Fig. 2-2 Potential-energy contour diagram for the linear *XYZ* system as a function of internuclear distance. *a* and *d* are potential-energy minima; *b*, a maximum; and *c*, a saddle point.

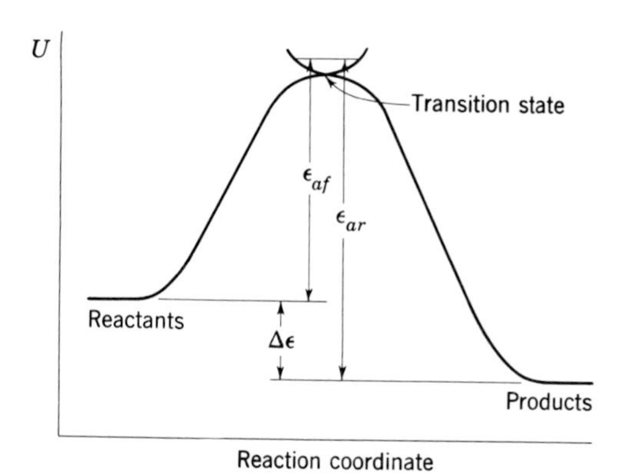

Fig. 2-3 The energy barrier for a generalized system of reactants and products.

at about $r_{XY} = r_{YZ} \cong 0.9$ Å. (The saddle is probably an artifact of the semiempirical assumptions.)

The height of the crater, including the zero-point energy, is about 14 kcal/mole, and since the zero-point energy of the initial system is about 6 kcal/mole, the activation energy is of the order of 8 kcal/mole, in reasonable agreement with the experimental energy of activation obtained by Farkas [3]. As the ball continues its travels beyond the crater, it rolls along the pass in the valley until, when it has reached point *d*, where $r_{XY} = 0.74$ Å and $r_{YZ} \cong \infty$, reaction has occurred, with formation of the products, H + H_2(para). For this illustrative reaction the dotted path *acd* is referred to as the reaction coordinate. Figure 2-3 illustrates the transition of a generalized system from the reactant state to the product state along the reaction coordinate. It shows the relation between the energies of activation for the forward and reverse reactions and the change in energy content for the overall reaction. Returning to the system of three H atoms, we observe that if the ball (system) is placed on the plateau (whose height is over 100 kcal/mole) near *b*, corresponding to three H atoms at essentially infinite separations, it will roll into the valley *acd*, where the kinetic energy accumulated during its descent will cause it to leave the surface at either *a* or *d* (cf. Fig. 2-2). No activation energy is needed for travel along this path, and we have obviously described the termolecular recombination of atomic hydrogen according to

$$H + H + H \rightarrow H_2 + H \tag{2-7}$$

Several important modifications have been made of the semiempirical procedure of Eyring and Polanyi for calculating the potential-energy surface. Sato [4] has used the Heitler–London formulation to obtain the following expressions for the bonding and antibonding states of H_2:

$$U_{\text{bond}} = (A + \alpha)/(1 + k), \qquad U_{\text{anti}} = (A - \alpha)/(1 - k)$$

where k is the square of the electron-overlap integral. He then uses a Morse function [Eq. (2-3)] to represent the energy of the bonding state and a modified Morse function (where the second term is positive) to represent the energy of the antibonding state. Since these energies can be determined spectroscopically, A and α can be calculated as a function of r_{XY} and k. The total potential energy of the system is

$$U = \frac{1}{1 + k}\left[A + B + C - (\alpha^2 + \beta^2 + \gamma^2 - \alpha\beta - \beta\gamma - \alpha\gamma)^{1/2}\right]$$

and the potential-energy surface can be easily determined. A constant value of k (~ 0.15) is chosen which gives the best agreement between the calculated and experimental activation energies. As in the previously outlined treatment of this problem, a linear configuration of the three hydrogen atoms is of primary importance in the activation process. An important difference is the absence of a shallow basin at the maximum of the potential-energy surface, which was previously found; also, more reasonable results are obtained at small values of the internuclear separations. This method has been used to calculate the potential-energy surfaces of several other systems [4–6]; surfaces constructed in this manner are commonly called LEPS surfaces.

A further refinement of the calculation of the potential-energy surface for three hydrogen atoms has been made by Porter and Karplus [7]. They use a valence bond treatment which includes overlap and multiple-exchange integrals. Although an exact formulation is used, explicit calculation of the potential-energy surface again requires a semiempirical approach. As with Sato's method, a Morse function is used to represent the bonding state and a modified Morse function to represent the antibonding state; in addition, several theoretical parameters must still be specified in a semiempirical manner. The actual computation is considerably more cumbersome than with the other procedures. The results are in reasonably good accord with the LEPS surface and experiment. The work of Porter and Karplus represents a significant step forward in putting the practical calculation of potential energy surfaces on a sound theoretical basis.

None of the models discussed thus far for the potential energy of interaction of reacting particles explicitly considers internal degrees of freedom (rotation and vibration) and multiple electronic states. In fact, useful general formulations of potential-energy surfaces incorporating these features are not yet available although, as we shall see later, internal degrees of freedom play an important role in chemical reactivity.

2-2 SIMPLE KINETIC THEORY OF COLLISION DYNAMICS

Binary Collisions

For dilute gas systems, the kinetic theory of gases has been of great value in providing a better understanding of the molecular details of homogeneous reactions. For the present, the following assumptions are made: densities are sufficiently low so that effects of ternary and higher-order collisions are negligible; the kinetic theory of heat, namely, the identification of heat with mechanical molecular motions, is valid; and classical mechanics can be used to describe the mechanical molecular motions.

Collisions between structureless molecules, i.e., molecules having no active internal degrees of freedom, may be visualized conveniently in terms of the two-dimensional trajectories shown in Fig. 2-4, which have several common features:

(1) the potential $U(r)$ is derived from a spherically symmetric force field;
(2) relative-velocity vectors are not used; instead, one molecule is held fixed at the origin and the other is given an initial-velocity vector equal to the actual initial relative velocity, with the result that the trajectory lies in a plane and corresponds to the description of the collision in terms of center-of-mass rather than laboratory coordinates;
(3) the impact parameter is shown as b, the instantaneous separation between the two molecules as r, the minimum separation as r_0, and the angle of deflection as χ.

The impact parameter is the perpendicular distance between the center of one of the molecules which is held fixed and the straight line which the center of the moving molecule would follow if the fixed molecule was not there to deflect it.

The collision cross section for scattering, namely, for a change in the velocity vectors of the interacting molecules, is given by $S = \pi b_{max}^2$, where b_{max} is the largest value of the impact parameter for which a collision can occur. In other words, the collision cross section is the effective target area presented by the two molecules for an encounter producing the smallest physically significant angle of deflection χ. In the case of elastic hard spheres, the largest value of b which will permit a collision is d_{12}, $(d_1 + d_2)/2$, where d_1 and d_2 are the diameters of the colliding spheres. This value of $b = b_{max}$ corresponds to a grazing collision with a limiting value of $\chi = 0$. Values of b smaller than b_{max} lead to more violent collisions up to $\chi = \pi$ for a head-on collision, where $b = 0$, while values of b greater than b_{max} lead to no collisions at all (cf. Fig. 2-4a).

For elastic hard spheres with weak central attractive forces, the collision cross section is enhanced by the attractive force, since two particles can

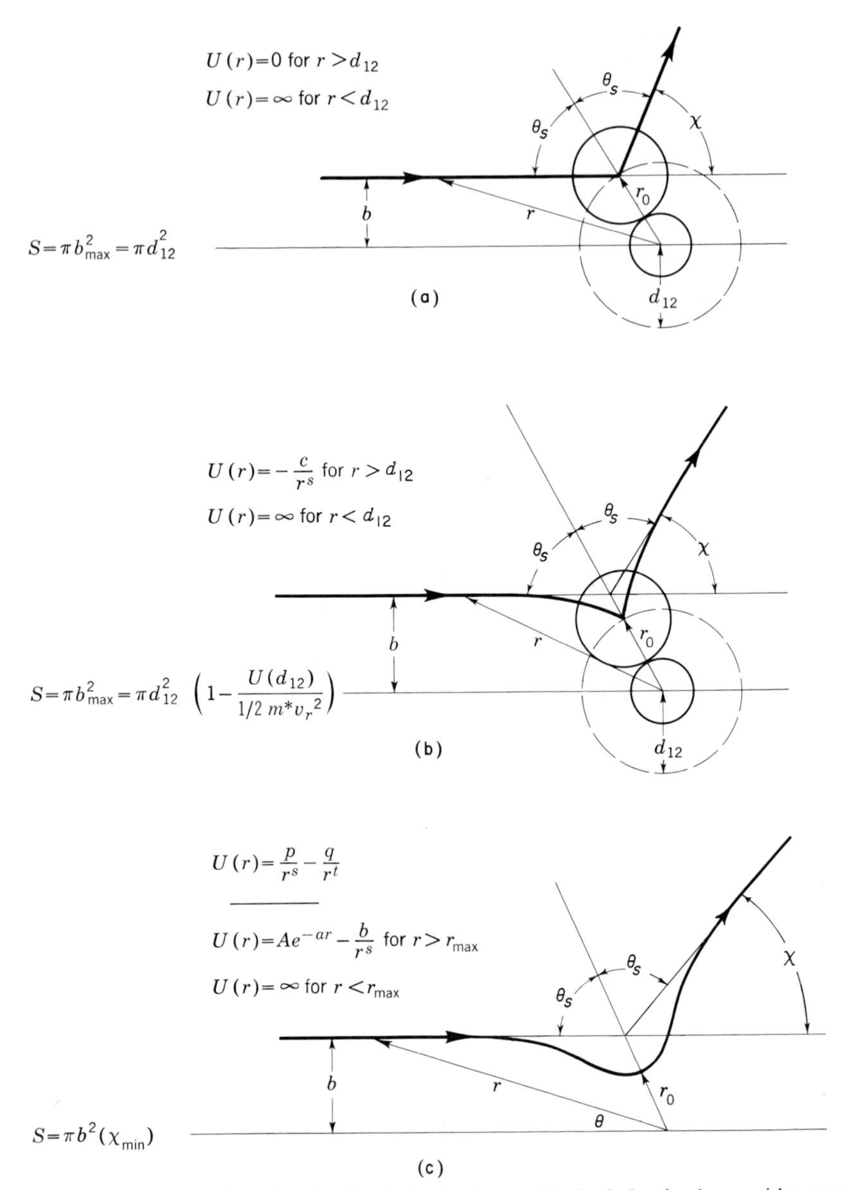

Fig. 2-4 Collision trajectories. (a) Elastic hard spheres; (b) elastic hard spheres with superposed weak central attractions; (c) molecules with central finite repulsive and attractive forces.

make actual contact at r_0 even though b_{max} is greater than d_{12} (cf. Fig. 2-4b). A second-order effect, which, however, may be neglected, is the presence of very small deflections resulting from attractions so weak that the two molecules at r_0 have an internuclear separation greater than d_{12}. Finally, for molecules with central finite repulsive and attractive forces, the concept of a collision becomes simultaneously vaguer and more realistic within the framework of classical mechanics (cf. Fig. 2-4c).

The collision dynamics for two particles whose interaction depends only on the distance between the center of the particles can be readily quantified in terms of classical mechanics. This problem is equivalent to the two-dimensional motion of a particle with a velocity equal to the relative velocity of the two particles, v_r, and a mass equal to the reduced mass of the two particles, $m^* = m_1 m_2/(m_1 + m_2)$. The total energy and angular momentum of this system are constant, and these two quantities can be used to completely define the system. The total energy E is

$$E = U(r) + \tfrac{1}{2}m^*(\dot{r}^2 + r^2\dot{\theta}^2) \qquad (2\text{-}8)$$

where r and θ are the radial and angular coordinates, respectively (see Fig. 2-4c), and $\dot{r}$ and $\dot{\theta}$ are their derivatives with respect to time. The magnitude of the angular momentum L is

$$L = m^* r^2 \dot{\theta} \qquad (2\text{-}9)$$

The initial energy and angular momentum when the particles are infinitely far apart are $\frac{1}{2}m^* v_r^2$ and $m^* v_r b$ and can be equated to Eqs. (2-8) and (2-9), respectively. Rearrangement of the resulting equations leads to the equations of motion for the system:

$$\dot{\theta} = \frac{b v_r}{r^2} \qquad (2\text{-}10)$$

$$\dot{r} = \pm v_r \left[1 - \frac{b^2}{r^2} - \frac{U(r)}{\frac{1}{2}m^* v_r^2}\right]^{1/2} \qquad (2\text{-}11)$$

These equations of motion can now be solved to describe completely the trajectories of the two particles. As the two particles approach, the kinetic energy increases due to the attractive potential energy of interaction and then decreases as the potential energy becomes repulsive until at the point in the trajectory when the two particles are as close as possible, r_0, the radial component of the velocity, $\dot{r}$, is zero. At this turning point, the relationship between the impact parameter, the potential energy of interaction, and the initial energy obtained from Eq. (2-11) is

$$b^2 = r_0^2 \left[1 - \frac{U(r_0)}{\frac{1}{2}m^* v_r^2}\right] \qquad (2\text{-}12)$$

After the turning point is reached, the collision process proceeds by conversion of the potential energy back into kinetic energy and the deflection of the incoming particle through the angle χ. Note that the collision dynamics are symmetrical about the r_0 vector; this is a direct consequence of the reversible nature of the classical equations of motion. The scattering angle θ_S ($\chi = \pi - 2\theta_S$) is a function of the interaction potential $U(r)$ and the initial conditions b and v_r. It also can be determined from the equations of motion:

$$\theta_S = \int_\infty^{r_0} \left(\frac{d\theta}{dr}\right) dr = \int_\infty^{r_0} \frac{b\,dr}{r^2[1 - U(r)/(\frac{1}{2}m^* v_r^2) - b^2/r^2]^{1/2}}$$

Of course, integration of this equation may be difficult in practice, but in principle the scattering is completely described.

If we have two distinguishable groups of molecules, the number of encounters between them in unit volume and unit time is directly proportional to the product of their number densities and to the average velocity with which they approach each other. In addition, each binary collision (which can be treated as though one of the molecules were at rest while the other is approaching it with a velocity equal to the actual relative velocity) will be proportional to a collision cross section which is a measure of the probability that after the collision the moving molecule will be found in an element of solid angle $d\omega$. This cross section is the differential scattering cross section $d\sigma$ and is a function of the velocity of approach, the polar angle of deflection χ, and the azimuthal angle η. The scattering pattern is symmetrical with respect to η for particles whose interaction forces are central, and it can be shown [8] that $d\sigma$ is equal to $\int_0^{2\pi} b\,db\,d\eta$, or $2\pi b\,db$, where b is the impact parameter, previously defined.

The most general derivation of the total collision rate per unit volume starts with two groups of molecules small enough to be treated mathematically as differentials yet large enough for their molecular densities dc_1 and dc_2 to represent a large number of particles per unit volume. Their impact parameters lie between b and $b + db$, their vector velocities lie between $\mathbf{v}_1$ and $\mathbf{v}_1 + d\mathbf{v}_1$ and $\mathbf{v}_2$ and $\mathbf{v}_2 + d\mathbf{v}_2$, and the absolute magnitude of their relative velocity v_r is $|\mathbf{v}_1 - \mathbf{v}_2|$. The collision rate per unit volume between these particular groups is

$$dZ = \kappa 2\pi b\,db\,v_r\,dc_1\,dc_2 \tag{2-13}$$

where the dc_i are given in terms of the molecular densities c_i by the Maxwellian distribution law

$$dc_i = c_i\left(\frac{m_i}{2\pi kT}\right)^{3/2} \exp\left(-\frac{m_i v_i^2}{2kT}\right) dv_{ix}\,dv_{iy}\,dv_{iz} \qquad (i = 1 \text{ or } 2) \tag{2-14}$$

The constant κ is a symmetry number which is 1 for collisions between unlike molecules and $\frac{1}{2}$ for collisions between identical molecules. This avoids double counting of collisions between like molecules. Since b is independent of the components of the $\mathbf{v}_i$, Eq. (2-13) may be integrated over all values of b to give the collision rate between two groups of molecules whose velocity components lie in the range $dv_{ix}\,dv_{iy}\,dv_{iz}$ at $\mathbf{v}_i$:

$$dZ = \kappa\left(\frac{m_1}{2\pi kT}\right)^{3/2}\left(\frac{m_2}{2\pi kT}\right)^{3/2} c_1 c_2 S(v_r) v_r \exp\left(-\frac{m_1 v_1{}^2}{2kT}\right)\exp\left(-\frac{m_2 v_2{}^2}{2kT}\right) \times dv_{1x}\,dv_{1y}\,dv_{1z}\,dv_{2x}\,dv_{2y}\,dv_{2z} \tag{2-15}$$

where $S(v_r) = \int_0^{b_{\max}} 2\pi b\,db$ is the total elastic scattering cross section. In general, it varies with v_r.

In preparation for integration of Eq. (2-15), we transform variables from the cartesian components of the velocities $\mathbf{v}_1$ and $\mathbf{v}_2$ to cartesian coordinates of the velocity of the center of mass of the two molecules $\mathbf{v}_c$ and the cartesian components of the relative velocity $\mathbf{v}_r$ in accordance with the relations

$$\mathbf{v}_c = \frac{m_1\mathbf{v}_1 + m_2\mathbf{v}_2}{m_1 + m_2}, \qquad \mathbf{v}_r = \mathbf{v}_1 - \mathbf{v}_2 \tag{2-16}$$

$$v_{cx} = \frac{m_1 v_{1x} + m_2 v_{2x}}{m_1 + m_2}, \qquad v_{rx} = v_{1x} - v_{2x} \tag{2-17}$$

with corresponding expressions for the other components. Equations (2-17) can be solved to yield

$$v_{1x} = v_{cx} + \frac{m_2}{m_1 + m_2} v_{rx}, \qquad v_{2x} = v_{cx} - \frac{m_1}{m_1 + m_2} v_{rx} \tag{2-18}$$

with analogous expressions for the other components. Equations (2-18) lead directly to

$$\tfrac{1}{2}m_1 v_{1x}^2 + \tfrac{1}{2}m_2 v_{2x}^2 = \tfrac{1}{2}(m_1 + m_2)v_{cx}^2 + \tfrac{1}{2}m^* v_{rx}^2 \tag{2-19}$$

Extension of Eq. (2-19) to the z and y components leads to

$$\tfrac{1}{2}m_1 v_1{}^2 + \tfrac{1}{2}m_2 v_2{}^2 = \tfrac{1}{2}(m_1 + m_2)v_c{}^2 + \tfrac{1}{2}m^* v_r{}^2 \tag{2-20}$$

From Eq. (2-17) and its associated equations for y and z components, we find the following relation between the two sixfold volume elements in cartesian coordinates:

$$dv_{1x}\,dv_{1y}\,dv_{1z}\,dv_{2x}\,dv_{2y}\,dv_{2z} = dv_{cx}\,dv_{cy}\,dv_{cz}\,dv_{rx}\,dv_{ry}\,dv_{rz} \tag{2-21}$$

and Eq. (2-15), after insertion of Eqs. (2-20) and (2-21), becomes

$$dZ = \kappa\left(\frac{m_1}{2\pi kT}\right)^{3/2}\left(\frac{m_2}{2\pi kT}\right)^{3/2} c_1 c_2 \exp\left[-\frac{(m_1 + m_2)v_c{}^2}{2kT}\right]\exp\left(-\frac{m^* v_r{}^2}{2kT}\right) \times S(v_r) v_r\,dv_{cx}\,dv_{cy}\,dv_{cz}\,dv_{rx}\,dv_{ry}\,dv_{rz} \tag{2-22}$$

Equation (2-22) can be integrated over all values of the components of $\mathbf{v}_c$ to give the following relation for the collision rate per unit volume between molecules of type 1 and type 2 with relative velocity components in the range $dv_{rx}\,dv_{ry}\,dv_{rz}$ at v_r:

$$dZ = \kappa\left(\frac{m^*}{2\pi kT}\right)^{3/2} c_1 c_2 \exp\left(-\frac{m^* v_r^{\,2}}{2kT}\right) S(v_r) v_r\, dv_{rx}\, dv_{ry}\, dv_{rz} \tag{2-23}$$

If v_{rx}, v_{ry}, and v_{rz} are transformed to polar coordinates v_r, θ_r, and ϕ_r, Eq. (2-23) becomes

$$dZ = \kappa\left(\frac{m^*}{2\pi kT}\right)^{3/2} c_1 c_2 \exp\left(-\frac{m^* v_r^{\,2}}{2kT}\right) S(v_r) v_r^{\,3} \sin\theta_r\, d\theta_r\, d\phi_r\, dv_r \tag{2-24}$$

This can be integrated over θ_r and ϕ_r to obtain the collision rate per unit volume between molecules of type 1 and type 2 with relative speeds between v_r and $v_r + dv_r$:

$$dZ = \kappa\left(\frac{m^*}{2\pi kT}\right)^{3/2} 4\pi c_1 c_2 \exp\left(-\frac{m^* v_r^{\,2}}{2kT}\right) S(v_r) v_r^{\,3}\, dv_r \tag{2-25}$$

The total number of collisions per unit volume per unit time between molecules of type 1 and type 2 is obtained by integrating Eq. (2-25) over all values of v_r from 0 to infinity:

$$Z = \kappa\left(\frac{m^*}{kT}\right)^{3/2}\left(\frac{2}{\pi}\right)^{1/2} c_1 c_2 \int_0^\infty \exp\left(-\frac{m^* v_r^{\,2}}{2kT}\right) S(v_r) v_r^{\,3}\, dv_r \tag{2-26}$$

For the special case of dissimilar elastic hard spheres, $S(v_r)$ is a constant, πd_{12}^2, and Eq. (2-26) integrates to give

$$\begin{aligned} Z_{12} = c_1 c_2 \pi d_{12}^2 \left[\frac{8kT}{\pi}\left(\frac{1}{m_1} + \frac{1}{m_2}\right)\right]^{1/2} &= c_1 c_2 \pi d_{12}^2 (\bar{v}_1^{\,2} + \bar{v}_2^{\,2})^{1/2} \\ &= c_1 c_2 \pi d_{12}^2 \bar{v}_r \end{aligned} \tag{2-27}$$

while for identical elastic hard spheres, for which $S(v_r)$ is πd^2, the corresponding relation (after inserting $\kappa = \frac{1}{2}$) is

$$Z_{11} = \frac{1}{\sqrt{2}} c^2 \pi d^2 \bar{v} = \frac{1}{2} c^2 \pi d^2 \bar{v}_r \tag{2-28}$$

If an elastic-hard-sphere model is not used, $S(v_r)$ in Eq. (2-26) must be properly expressed as a function of v_r for the specified model. For example, for elastic hard spheres with superposed central attractive forces, the maximum value of r_0 for contact collisions is d_{12}, and if the attractive forces are weak, the slight deflections associated with larger values of r_0 will not contribute significantly to the total reaction cross section. Thus, if r_0 is replaced by d_{12} and b by b_{max} in Eq. (2-12), and it is recalled that $S(v_r) = \pi b_{max}^2$,

$$S(v_r) = \pi d_{12}^2 \left[1 - \frac{U(d_{12})}{\frac{1}{2}m^* v_r^{\,2}} \right]$$

For molecules with central finite attractive and repulsive forces (Fig. 2-4c), we may take $S(v_r) = \pi b^2(\chi_{min})$, where $b(\chi_{min})$ is the impact parameter corresponding to a minimum angle of deflection selected as an arbitrary cutoff to prevent $S(v_r)$ from going to infinity as χ goes to zero when classical collision theory is used. The specific dependence of $b(\chi_{min})$ on v_r will vary with the magnitude of the parameters s and t or a and b in the empirical potential-energy functions. A realistic calculation for this model, i.e., one which avoids an arbitrary cutoff χ_{min}, must be carried out quantum mechanically.

It is possible to combine these results on the frequency of binary collisions with a simple model of the dynamic mechanism of bimolecular reactions to obtain a formal result which agrees well with experiment in a number of cases. The model has been developed by Present [8], whose presentation we follow here.

When two spherically symmetric molecules (whose internal structure and shape are ignored) collide, so that their internuclear separation r is d_{12}, it is assumed that they either leave the collision unchanged or react to form product molecules. In the absence of forces at distances greater than d_{12}, the cross section for reaction, i.e., for the formation of the critical complex, would be simply πd_{12}^2. Actually, repulsions due to interpenetration of the electron clouds (overlap forces) may prevent the molecules from approaching to within the reaction distance d_{12}, and a potential-energy barrier must be overcome before reaction can occur. The height of this barrier is called the activation energy ϵ_a. If $U(r)$ is the potential energy of mutual repulsion of the colliding molecules, then $U(d_{12}) = \epsilon_a$, since $U(\infty) = 0$. For the *repelling* hard spheres of this model, the maximum impact parameter b_{max} is related to the reaction cross section $S_R(\epsilon)$ by

$$\begin{aligned} \pi b_{max}^2 = S_R(\epsilon) &= \pi d_{12}^2[1 - (\epsilon_a/\epsilon)] \qquad \text{for} \quad \epsilon > \epsilon_a \\ S_R(\epsilon) &= 0 \qquad \text{for} \quad \epsilon < \epsilon_a \end{aligned} \tag{2-29}$$

where ϵ is the kinetic energy of relative motion $\frac{1}{2}m^* v_r^{\,2}$. Equation (2-29) states that colliding pairs with kinetic energies of relative motion less than the activation energy will not react. A schematic drawing of the potential, implicit in Eq. (2-29), is given in Fig. 2-5.

If $\epsilon = \frac{1}{2}m^* v_r^{\,2}$ is substituted into Eq. (2-25), the collision frequency per unit volume in which the kinetic energy of relative motion lies between ϵ and

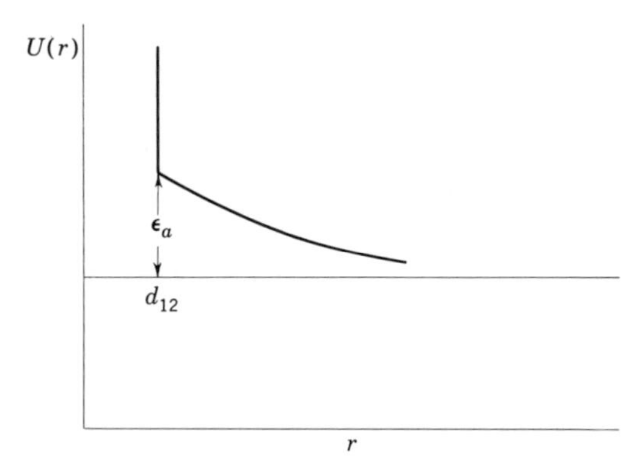

Fig. 2-5 Intermolecular potential for reactive elastic hard spheres with superposed central repulsive forces.

$\epsilon + d\epsilon$ is

$$dZ = \kappa c_1 c_2 (2/kT)^{3/2} (1/\pi m^*)^{1/2} e^{-\epsilon/kT} S(\epsilon)\epsilon \, d\epsilon \tag{2-30}$$

If $S_R(\epsilon)$ from Eq. (2-29) is introduced into Eq. (2-30) and the result is integrated from ϵ_a to ∞, we obtain the collision frequency per unit volume leading to reaction:

$$Z_{\text{reac}} = \kappa c_1 c_2 \pi d_{12}^2 (8kT/\pi m^*)^{1/2} p e^{-\epsilon_a/kT} = Z_{12} p e^{-\epsilon_a/kT} \tag{2-31}$$

An arbitrary steric factor p is generally introduced at this point. This parameter simply corrects for the fact that a collision in which the potential-energy barrier has been overcome does not necessarily lead to reaction unless the reacting molecules are properly oriented with respect to each other. The corresponding expression for identical molecules is readily inferred from Eq. (2-31).

For a simple second-order bimolecular reaction,

$$-\kappa \frac{dc_1}{dt} = Z_{\text{reac}} = k c_1 c_2 \tag{2-32}$$

so that

$$k = \kappa \pi d_{12}^2 (8kT/\pi m^*)^{1/2} p e^{-\epsilon_a/kT} = \kappa \pi d_{12}^2 (8kT/\pi m^*)^{1/2} p e^{-E_a/RT} \tag{2-33}$$

On taking logarithms and differentiating with respect to T, we obtain

$$\frac{d \ln k}{dT} = \frac{\frac{1}{2}RT + E_a}{RT^2} \tag{2-34}$$

If, as is frequently the case, $E_a \gg RT$, Eq. (2-34) reduces to the familiar Arrhenius expression.

The preceding treatment has assumed that only the relative velocity of the reacting molecules determines the energetics of a chemical reaction. Actually an exact treatment for real molecules should also take into account the internal degrees of freedom of the reactants. The exact treatment of this problem is somewhat complex, but if $E_a \gg (n/2)RT$, then [9]

$$Z_{\text{reac}} = Z_{12} \frac{1}{\Gamma(n/2 + 1)} \left(\frac{E_a}{RT} \right)^{n/2} e^{-E_a/RT} \tag{2-35}$$

where n is the number of square terms in the classical expression for the internal energy involved in the activation of the reactants ($n = 1$ for each rotation and 2 for each vibration) and $\Gamma(n/2 + 1)$ is the gamma function. Since Eq. (2-33) provides an adequate representation of most experimental results, Eq. (2-35) is seldom used; it simply provides an additional adjustable parameter in n.

A useful comparison between the predictions of simple collision theory and experiment can be made, since if the activation energy is determined, the experimental frequency factor can be directly compared with that predicted by Eq. (2-33). The hard-sphere diameter can be estimated from transport properties, although the choice of this parameter is somewhat arbitrary. In Table 2-1 a comparison between theory and experiment is presented for several well-studied bimolecular reactions (cf. Benson [10] for a more complete compilation). The tabulated steric factor is that value which makes the experimental and theoretical values coincide. In view of the assumptions involved, many of the steric factors are surprisingly close to unity. However, marked deviations in the form of unreasonably small steric factors do occur, especially if polyatomic molecules are involved. This often indicates that quantum-mechanical effects may be important or that a different classical theory may be required.

If we recall that real molecules do not behave dynamically like elastic hard spheres, it follows that the time of collision is not the zero contact time of such spheres but a longer time which is a function of the distance over which intermolecular forces of significant magnitude are exerted between two colliding molecules. The realistic potentials in Fig. 2-1 are all characterized by short-range character, i.e., by associated forces which diminish rapidly with increasing nuclear separation. For example, for nonpolar spherically symmetric molecules the attractive force decreases as r^{-7}, and the repulsive force more rapidly except at very small separations. A reasonable estimate of the distance Δr over which interaction forces are appreciable is 1 Å (10^{-8} cm). Since $\bar{v}_r$ is of the order of 10^5 cm sec^{-1}, the effective collision time, which is $\Delta r/\bar{v}_r$ at ordinary temperatures, is about 10^{-13} sec. The effective collision time, it is interesting to note, is of the same order of magnitude as the period of a molecular vibration.

TABLE 2-1

Comparison between Theory and Experiment for Biomolecular Reactions

Reaction	E_a (kcal/mole)	$\log(A/T^{1/2})_{exp}$[a]	d_{12} (Å)	p	Reference
$H_2 + I_2 \rightarrow 2HI$	40.7	9.78	3.5	0.28	11
$2HI \rightarrow H_2 + I_2$	43.7	8.97	3.5	0.44	11
$2NOCl \rightarrow 2NO + Cl_2$	25.8	9.51	3.5	1.1	12
$2CH_3 \rightarrow C_2H_6$	~ 0	9.24	3.5	0.5	13
$NO + Cl_2 \rightarrow NOCl + Cl$	19.6	8.00	3.5	0.014	14
$Cl + NOCl \rightarrow Cl_2 + NO$	0.75	8.59	4.0	0.042	17
$Br + CH_4 \rightarrow HBr + CH_3$	17.8	9.17	3.0	0.21	15
$Br + CHCl_3 \rightarrow HBr + CCl_3$	8.86	7.82	5.0	10^{-3}	16
2[1,3-pentadiene] → dimer	25.4	5.92	6.0	1×10^{-4}	18
$C_2F_4 + C_2F_3Cl \rightarrow$ cyclo C_4F_7Cl	25.6	6.30	5.0	4×10^{-4}	19

[a] A is in $M^{-1}\ sec^{-1}$.

Ternary Collisions

A number of the gas-phase association reactions involving the formation of a single product molecule from two reacting atoms or from simple molecules do not occur in a bimolecular-collision process because there is no likely process by which the energy of the reaction can be removed from the excited product. In such cases, dissociation into the original reactants occurs, usually within the period of a single vibration, about 10^{-13} sec, unless a third atom or molecule can enter into the collision and stabilize the product molecule by removing its excitation energy as kinetic energy. Since very little energy of activation, if any, should be needed in the three-body encounters, the overall rate of the association reaction is usually equal to the frequency of the termolecular collisions.

A ternary collision may be conveniently pictured as a very rapid succession of two binary collisions: one to form the unstable product, and the second, occurring within a period of about 10^{-13} sec or less, to stabilize the product. It is immediately obvious that it is not possible to use the elastic-hard-sphere molecular model to represent ternary collisions: since two such spheres would be in collision contact for zero time, the probability of a third molecule making contact with the colliding pair would be strictly zero. It is therefore necessary to assume a potential model involving forces which are exerted over an extended range. One such model is that of point centers having either inverse-power repulsive or inverse-power attractive central forces. This potential, shown in Fig. 2-1f, is represented by $U(r) = \pm K/r^s$. For the sake of convenience, we shall make several additional assumptions: first, at the interaction distances of interest the intermolecular forces are weak, that is, $U(r) < kT$; second, when the reactants A and B approach each other, they form an unstable product molecule $A \cdot\cdot B$ when their internuclear separations are in the range $d_{AB} \le r \le d_{A \cdot\cdot B}$; third, the unstable product is in essential equilibrium with the reactants so that

$$A + B \underset{k_r}{\overset{k_f}{\rightleftharpoons}} A \cdot\cdot B$$

where $k_f = Z_{AB}/c_A c_B$, $k_r = k_f/K$, and the mean lifetime of $A \cdot\cdot B$ is $1/k_r$. In the light of the preceding assumptions, the equilibrium concentration of $A \cdot\cdot B$ complexes is given by

$$c_{A \cdot\cdot B} = \kappa \int_V c_A c_B e^{-U(r)/kT}\, dV$$

$$\cong \kappa c_A c_B \int_0^{2\pi} \int_0^{\pi} \int_{d_{AB}}^{d_{A \cdot\cdot B}} \left[1 - \frac{U(r)}{kT} \cdots \right] r^2 \sin\theta \, d\theta \, d\phi \, dr \qquad (2\text{-}36)$$

where V is the volume within which A and B must lie in order to form an $\mathrm{A} \cdot\cdot \mathrm{B}$ complex. Equation (2-36) may be integrated directly by setting $U(r) = \pm K/r^s$, and the following result is obtained:

$$\begin{aligned} c_{\mathrm{A}\cdot\cdot\mathrm{B}} &= \kappa \frac{4\pi}{3} c_{\mathrm{A}} c_{\mathrm{B}} (d^3_{\mathrm{A}\cdot\cdot\mathrm{B}} - d^3_{\mathrm{AB}}) \left[1 \pm \frac{\mathrm{K}}{(s-3)kT} \left(\frac{1}{d^s_{\mathrm{A}\cdot\cdot\mathrm{B}}} \right) - \frac{1}{d^s_{\mathrm{AB}}} \right) \right] \\ &= \kappa \frac{4\pi}{3} c_{\mathrm{A}} c_{\mathrm{B}} (d^3_{\mathrm{A}\cdot\cdot\mathrm{B}} - d^3_{\mathrm{AB}}) \left[1 + \frac{U(d_{\mathrm{A}\cdot\cdot\mathrm{B}}) - U(d_{\mathrm{AB}})}{(s-3)kT} \right] \\ &= \kappa \frac{4\pi}{3} c_{\mathrm{A}} c_{\mathrm{B}} (d^3_{\mathrm{A}\cdot\cdot\mathrm{B}} - d^3_{\mathrm{AB}})^* \end{aligned} \tag{2-37}$$

We shall designate the third body as X, where X may be A or B or any other stable molecule. The ternary collision rate per unit volume is equal to the binary collision rate per unit volume between $\mathrm{A} \cdot\cdot \mathrm{B}$ complexes and X molecules. By analogy with Eq. (2-26), the desired collision rate is

$$\begin{aligned} Z_{\mathrm{ABX}} &= \left(\frac{m_{\mathrm{A}\cdot\cdot\mathrm{B}} m_{\mathrm{X}}}{m_{\mathrm{A}\cdot\cdot\mathrm{B}} + m_{\mathrm{X}}} \frac{1}{kT} \right)^{3/2} \left(\frac{2}{\pi} \right)^{1/2} c_{\mathrm{A}\cdot\cdot\mathrm{B}} c_{\mathrm{X}} \\ &\quad \times \int_0^\infty \exp\left(- \frac{m_{\mathrm{A}\cdot\cdot\mathrm{B}} m_{\mathrm{X}}}{m_{\mathrm{AB}} + m_{\mathrm{X}}} \frac{v_r{}^2}{2kT} \right) S_{\mathrm{ABX}}(v_r) v_r{}^3 \, dv_r \end{aligned} \tag{2-38}$$

where $S_{\mathrm{ABX}}(v_r)$ is the collision cross section for the encounters between the $\mathrm{A} \cdot\cdot \mathrm{B}$ complexes and the X molecules of diameter d_{X}. If we are willing to regard these particular binary encounters as equivalent to those between elastic hard spheres, with a collision cross section equal to $\pi(d_{\mathrm{A}\cdot\cdot\mathrm{B}} + d_{\mathrm{X}}/2)^2$, Eq. (2-38) may be integrated directly, and the result after substituting the right-hand side of Eq. (2-37) for $c_{\mathrm{A}\cdot\cdot\mathrm{B}}$ becomes

$$Z_{\mathrm{ABX}} = 4\kappa c_{\mathrm{A}} c_{\mathrm{B}} c_{\mathrm{X}} \frac{\pi^2}{3} (d^3_{\mathrm{A}\cdot\cdot\mathrm{B}} - d^3_{\mathrm{AB}})^* \left(d_{\mathrm{A}\cdot\cdot\mathrm{B}} + \frac{d_{\mathrm{X}}}{2} \right)^2 \left[\frac{8kT}{\pi} \left(\frac{1}{m_{\mathrm{A}\cdot\cdot\mathrm{B}}} + \frac{1}{m_{\mathrm{X}}} \right) \right]^{1/2} \tag{2-39}$$

If the encounters between $\mathrm{A} \cdot\cdot \mathrm{B}$ complexes and X molecules are not assumed to be those between elastic hard spheres, the functional form of $S_{\mathrm{ABX}}(v_r)$ appropriate to the assumed interaction potential between $\mathrm{A} \cdot\cdot \mathrm{B}$ and X must be known before Eq. (2-38) can be integrated. The arbitrary cutoffs introduced by defining an $\mathrm{A} \cdot\cdot \mathrm{B}$ complex as an A and a B molecule with internuclear separations in the range $d_{\mathrm{AB}} \leq r \leq d_{\mathrm{A}\cdot\cdot\mathrm{B}}$ reduces the substitution of a realistic cross section $S_{\mathrm{ABX}}(v_r)$ for $\pi(d_{\mathrm{A}\cdot\cdot\mathrm{B}} + d_{\mathrm{X}}/2)^2$ to an exercise in elegant formalism.

As an example of a termolecular reaction consider the recombination of atomic hydrogen:

$$H + H + H \rightarrow H_2 + H$$

The rate constant at 303°K is 1.42×10^{16} cc^2 $mole^{-2}$ sec^{-1} [20]; this is consistent with Eq. (2-39) if we take $d_{A\cdot\cdot B} + d_X/2$ equal to 3 Å and $(d^3_{A\cdot\cdot B} + d^3_{AB})^*$ equal to 2.1×10^{-23} cm^3. These numbers are in reasonable agreement with similar parameters from other sources. Actually, termolecular reactions are often more complex than indicated here; they are discussed in greater detail in Chapter 6.

The simplified-kinetic-theory treatment of reaction rates must be regarded as relatively crude for several reasons. Numerical calculations are usually made in terms of either elastic hard spheres or hard spheres with superposed central attractions or repulsions, although such models of molecular interaction are better known for their mathematical tractability than for their realism. No account is taken of the internal motions of the reactants. The fact that every combination of initial and final states must be characterized by a different reaction cross section is not considered. In fact, the simplified-kinetic-theory treatment is based entirely on classical mechanics. Finally, although reaction cross sections are complicated averages of many inelastic cross sections associated with all possible processes by which reactants in a wide variety of initial states are converted to products in a wide variety of final states, the simplified kinetic theory approximates such cross sections by elastic cross sections appropriate to various transport properties, by cross sections deduced from crystal spacings or thermodynamic properties, or by order-of-magnitude estimates based on scientific experience and intuition. It is apparent, therefore, that the usual collision theory of reaction rates must be considered at best an order-of-magnitude approximation; at worst it is an oversimplification that may be in error in principle as well as in detail.

2-3 COLLISION DYNAMICS BY COMPUTER SIMULATION

A straightforward approach toward the problem of calculating reaction rates is to divide the process into three distinct parts. First, the multidimensional potential energy surface reflecting the energy of interaction between all atoms must be determined. The many complexities involved in such a determination have already been discussed; it is sufficient to say here that an exact surface is not yet known for any chemical reaction. However, given a potential energy surface, the second part of the problem is to solve the quantum-mechanical or classical equations of motion as a function of all initial states of reactants and final states of products. The solution of these equations gives the cross sections for reaction. Finally, the rate of the chemical reaction can be calculated by integrating the cross section over appropriate initial state distributions. The kinetic theory already presented represents a simple approach to this more general problem. Although solutions of the equations of motion for all possible initial and final states obviously cannot be determined, fast digital computers can be used to obtain approximate solutions of the equations of motion for randomly selected initial and final states (consistent with specified distribution functions). A chemical kinetics experiment is then carried out using the computer: the equations of motion are solved many times for a number of different initial energy conditions and the trajectories of the atoms are examined to see if a chemical reaction occurs. If a sufficient number of trajectories are observed, reaction cross sections can be obtained by comparing the number of reactions occurring with the total number of trajectories. The integration of the cross sections obtained over given initial energy distributions allows calculation of total reaction rates. Because of the random selection processes used to generate initial conditions, procedures such as already outlined are frequently called Monte Carlo calculations.

The first chemical kinetics experiments carried out with a fast digital computer were done by Wall *et al.* [21]. Although their experiments must be regarded as fairly primitive by present-day standards because of the rapid development of computers in recent years, we nevertheless consider their procedure and results in some detail since they do embody the basic principles of computer experiments. As might be expected, the reaction selected for study was the exchange reaction between H_2 and H. The simple potential-energy surface used was that due to London, Eyring, and Polanyi discussed earlier. Furthermore, the assumption was made that only colinear collisions occur. This assumption is reasonable in that the activation energy is lowest for a colinear collision; on the other hand, the probability of a colinear collision is rather small. The use of this assumption considerably reduces the difficulty of the calculation. Finally, the assumption was made that the classical equations of motion can be employed; this assumption is quite reasonable in the temperature range of interest.

The easiest way to formulate the equations of motion is to make use of Hamiltonian mechanics. The Hamiltonian H, or total energy of the system in this case, can be written as

$$H = T + U \tag{2-40}$$

$$H = \tfrac{1}{2}m(\dot{x}_x{}^2 + \dot{x}_y{}^2 + \dot{x}_z{}^2) + U(r_{xy}, r_{xz}, r_{yz}) \tag{2-41}$$

where T is the kinetic energy of the system, U is the potential energy of interaction between atoms, m is the mass of the hydrogen atom, the x's are the coordinates of the atoms ($\dot{x}$ denotes the time derivative of x), and the r's are the distances between the three atoms. The equations of motion can now be found by the relationships [22]

$$\frac{\partial H}{\partial p_i} = \dot{q}_i \tag{2-42}$$

$$\frac{\partial H}{\partial q_i} = -\dot{p}_i \tag{2-43}$$

where the p's and q's are conjugate momenta and coordinates. In order to make use of Eqs. (2-42) and (2-43), Eq. (2-41) must be expressed in terms of an appropriate set of p's and q's. Since only the relative motion of the atoms is of interest, a convenient set of generalized coordinates is

$$q_1 = x_z - x_x, \qquad q_2 = x_z - x_y, \qquad q_3 = x_z \tag{2-44}$$

The conjugate momenta can be easily generated as

$$\begin{aligned} p_1 &= \frac{\partial T}{\partial \dot{q}_1} = m(\dot{q}_1 - \dot{q}_3) = -m\dot{x}_x \\ p_2 &= \frac{\partial T}{\partial \dot{q}_2} = m(\dot{q}_2 - \dot{q}_3) = -m\dot{x}_y \\ p_3 &= \frac{\partial T}{\partial \dot{q}_3} = m(3\dot{q}_3 - \dot{q}_1 - \dot{q}_2) = m(\dot{x}_x + \dot{x}_y + \dot{x}_z) \end{aligned} \tag{2-45}$$

The Hamiltonian can now be expressed in terms of the p's and q's as

$$H = (1/m)(p_1{}^2 + p_2{}^2 + p_1 p_2 + \tfrac{1}{2}p_3{}^2 + p_2 p_3 + p_1 p_3) + U(q_1, q_2) \tag{2-46}$$

where $r_{xy} = q_1 - q_2$, $r_{xz} = q_1$, and $r_{yz} = q_2$. Only the relative motion of the particles is of interest so that p_3 can be set equal to zero; this corresponds to ignoring the motion of the center of mass. The Hamiltonian is then a function of only four variables, and the four equations of motion for the system of three colinear atoms can be found by use of Eqs. (2-42) and (2-43). These equations were integrated on a computer using time increments, δt, small compared to the harmonic oscillation of the three atoms. (The time increment used was 2×10^{-16} sec.) The initial conditions were such that the forces between the atom and molecule were negligible ($r_{xz} = 5$ Å). The changes in p's and q's were then calculated over δt; a new set of variables was obtained, and the cycle was repeated. This was continued either until the atom and molecule were greater than 5 Å apart, or until 1024 cycles were performed. The total energy was varied from values below the expected activation energy to values greatly in excess of the minimum energy for activation. Two initial states of the hydrogen molecule were considered, one

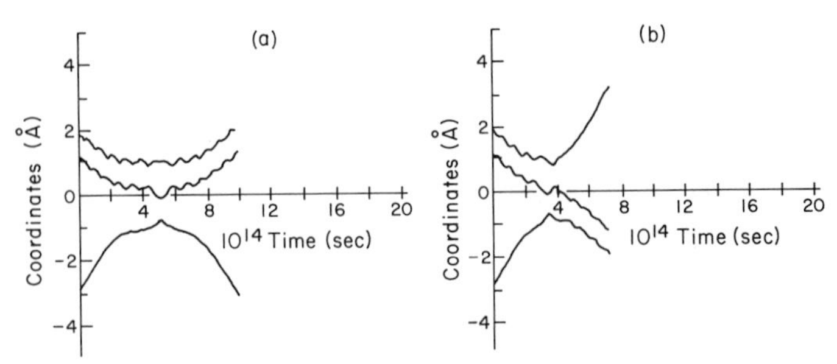

Fig. 2-6 Calculated trajectories for collisions of H and H_2. (a) Coordinates of atoms relative to center of gravity plotted versus time for collision of vibrating molecule (vibrational energy equal to the zero point energy) and atom with total initial kinetic energy of translation equal to 10.71 kcal/mole. Reaction did not occur. (b) Same as (a) except with the total initial kinetic energy of translation equal to 12.26 kcal/mole. Reaction occurred. (Adapted from Wall *et al.* [21].)

with no vibrational energy, the other with vibrational energy equal to the quantum-mechanical zero-point energy.

Several hundred calculations were performed and the trajectories observed. Two typical cases are shown in Fig. 2-6. In Fig. 2-6a, reaction did not occur, while in Fig. 2-6b reaction quickly occurred. No long-lived complex is observed in the latter case, but the vibrational energy is quickly transferred to the new molecule as indicated by the oscillating lines. In general, the probability of reaction is less than 0.5 even if the total energy exceeds the activation energy; in fact, the highest probability of reaction is found when the total energy is approximately equal to the activation energy. Putting energy into vibration helps somewhat, but vibrational energy is only about one-sixth as effective as translational energy in promoting reaction. Finally, some fairly long-lived complexes are observed; however, later calculations suggest this is an artifact of the simplified system studied.

Wall and co-workers also investigated the three-dimensional case where the atoms are not colinear [23]. The Hamiltonian for this situation is

$$H = \frac{1}{2m} \sum_{i=1}^{3} (p_{ix}^2 + p_{iy}^2 + p_{iz}^2) + U(r_{xy}, r_{xz}, r_{yz}) \tag{2-47}$$

By use of Eqs. (2-42) and (2-43), eighteen differential equations of motion can be obtained for the system. Since the equations of motion of the center of mass are not of interest, this number can be reduced to twelve. In principle, the number of equations of motion could be reduced further since the total energy and three angular momenta are constants of the motion. However, it proved more convenient to solve the full set of twelve equations and to use the constants of motion as checks on the computer results. The initial conditions were chosen by a pseudorandom process; no attempt was made to use a proper Monte Carlo method whereby correct statistical weights would be applied to each set of initial conditions. Seven hundred calculations were carried out over a range of total energies. The total energy was distributed in translational, rotational, and vibrational modes and a range of impact parameters was probed. Only six reactions occurred; far too small a number to draw any statistically significant conclusions about the requirements necessary for a chemical reaction to occur. At that time, it was concluded that a proper Monte Carlo approach would be prohibitively time consuming.

Some years later, aided by considerably more rapid computers than available to Wall and co-workers, Karplus, Porter, and Sharma reinvestigated the exchange reaction between H_2 and H [24]. As with the earlier work, the twelve classical equations of motion were solved. In addition, discrete quantum-mechanical vibrational and rotation states were included in the total energy so that the trajectories were examined as a function of the initial relative velocity v_r of the atom and molecule and the rotational and vibrational quantum numbers j and v of the molecule. The more sophisticated potential energy surface of Porter and Karplus was used [7], and the impact parameter, orientation and momentum of the reactants, and vibration phase were selected at random from appropriate distribution functions. This Monte Carlo approach was used to examine 200–400 trajectories for each set of v_r, j, and v. The reaction probability P can be written as

$$P(v_r, j, \mathrm{v}, b) = \lim_{N \to \infty} \frac{N_r(v_r, j, \mathrm{v}, b)}{N(v_r, j, \mathrm{v}, b)} \tag{2-48}$$

where N_r is the number of reactions resulting from a set of N trajectories with initial conditions chosen as previously described. The cross section for reaction, S, is given by

$$S_R(v_r, j, \mathrm{v}) = 2\pi \int_0^{b\max} P(v_r, j, \mathrm{v}, b) b \, db \tag{2-49}$$

The impact parameter was included as a Monte Carlo variable, with the distribution of values being uniform in b^2. Thus average values of P can be obtained which are independent of the impact parameter, and Eq. (2-49) can be written as

$$S_R(v_r, j, \mathrm{v}) = \pi b_{\max}^2 \lim_{N \to \infty} \frac{N_r(v_r, j, \mathrm{v})}{N(v_r, j, \mathrm{v})} \tag{2-50}$$

Finally, the overall rate constant can be evaluated by averaging the cross section over the appropriate energy distributions. Thus the translation averaged rate constant is

$$k_{j,\mathrm{v}} = \int_0^\infty dv_{\mathrm{H}} \int_0^\infty dv_{\mathrm{H}_2} f_{\mathrm{H}}(v_{\mathrm{H}}) f_{\mathrm{H}_2}(v_{\mathrm{H}_2}) v_r S_R(v_r, j, \mathrm{v}) \tag{2-51}$$

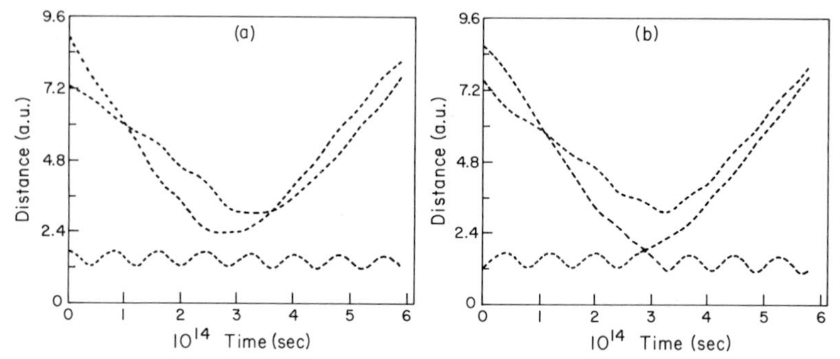

Fig. 2-7 Calculated trajectories for collisions of $H + H_2$. (a) Typical nonreactive $H + H_2$ collision trajectory. (b) Typical reactive $H + H_2$ collision trajectory. $j = 5$, $v = 0$, $v_r = 1.18 \times 10^6$ cm/sec. (Adapted from Karplus *et al.* [24].)

where the f's are the normalized velocity distributions $[(m_i/2\pi kT)^{3/2} \exp(-m_i v_i^2/2kT)]$. The overall rate constant is

$$k = \sum_{j,\mathrm{v}} k_{j,\mathrm{v}} F_{\mathrm{H}_2}(j, \mathrm{v}) \tag{2-52}$$

The rotational–vibrational distribution function F_{H_2} is given by

$$F_{\mathrm{H}_2} = g_j(2j + 1)e^{-\epsilon_{\mathrm{v},j}/kT}/f_{\mathrm{RV}} \tag{2-53}$$

where the g_j are the rotational degeneracies, $\epsilon_{\mathrm{v},j}$ the rotational–vibrational energy of state v, j, and f_{RV} the rotational–vibrational partition function.

Typical trajectories for reactive and nonreactive collisions are shown in Fig. 2-7. Distances are given in atomic units, 1 atomic unit = 0.529 Å. The crossing of lines is caused by molecular rotation. Very little difference appears to exist between reactive and nonreactive collisions. In both cases, the total interaction time between reactants is about 10^{-14} sec. No evidence for a collision complex was found in any case; the longest interaction times are about 3×10^{-14} sec. Obviously the interaction time is too short to permit equilibration between the various degrees of freedom of the three-atom system and almost no energy exchange occurs. A typical plot of the reaction probability as a function of the impact parameter is shown in Fig. 2-8. A smooth curve is drawn through the histogram actually obtained. Even for zero impact parameter collisions, the reaction probability does not approach unity. This can be thought of as a simple steric effect whereby the relative orientation of reactants become less important at small values of the impact parameter, but can never be completely neglected.

A typical plot of the cross section versus the relative velocity is shown in Fig. 2-9. The dependence of the cross section on j is not very large, although the threshold energy for reaction is a slowly increasing function of j. This

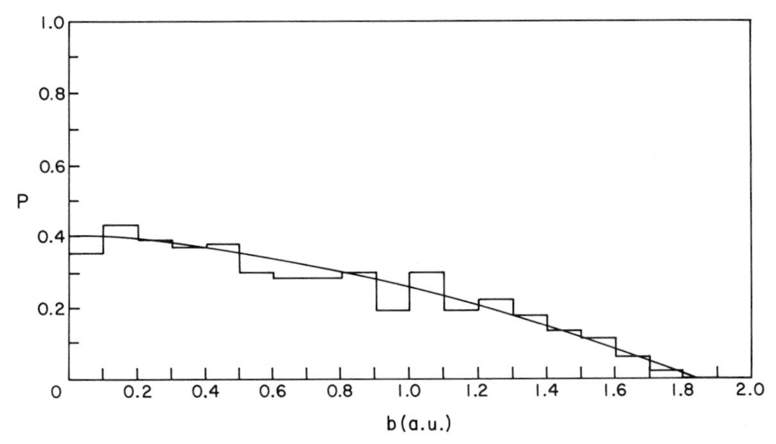

Fig. 2-8 Reaction probability P as a function of the impact parameter b for $H + H_2$ (v = 0, $j = 0$, $v_r = 1.174 \times 10^6$ cm/sec). The bar graph is from the computer experiments. The smooth curve corresponds to the function $P = a\cos(\pi b/2b_{max})$, $b \leq b_{max}$; $P = 0$, $b > b_{max}$. (Adapted from Karplus *et al.* [24].)

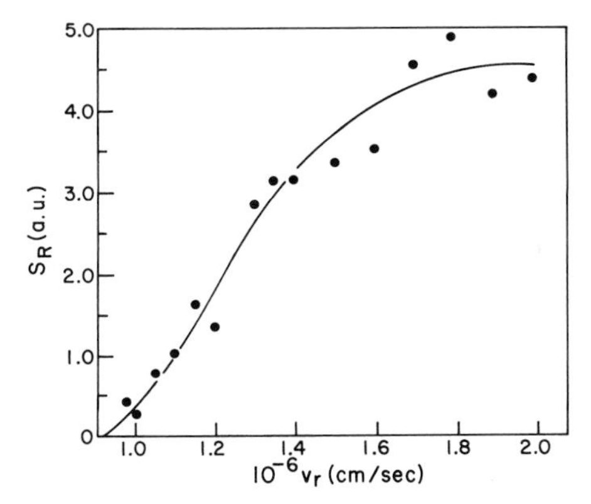

Fig. 2-9 Example of a plot of the reaction cross section $S_R(v_r, j, \text{v})$ versus v_r for $H + H_2$ (v = 0, $j = 1$) as obtained from computer experiments. (Adapted from Karplus *et al.* [24].)

latter effect is probably due to the difficulty in properly orienting a rapidly rotating molecule. The dependence of the cross section on v was not investigated in detail because vibrational excitation is not of great importance in the temperature region of interest. However, preliminary results suggest the reaction cross section is significantly dependent on v. In Fig. 2-10, the dependence of the reaction cross section on the relative energy of motion is shown for three cases: the hard-sphere model where $S_R = 0$ if $E < E^*$ and $S_R = \pi d_{12}^2$ for $E \geq E^*$; the model for elastic hard spheres with superposed

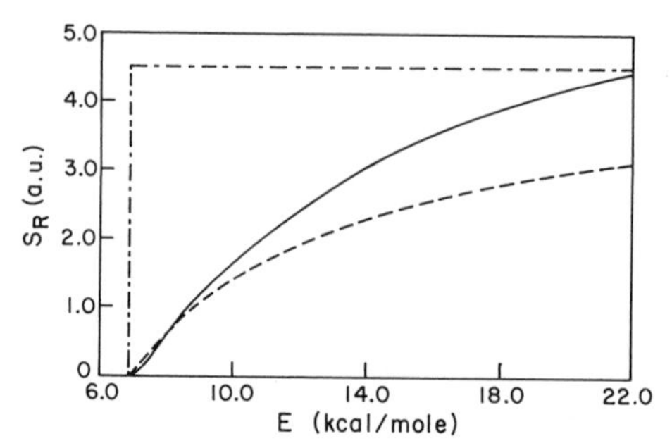

Fig. 2-10 Reaction cross section $S_R(E, j, v)$ as a function of E: ---, $S_R(E, 0, 0)$ from computer experiments; -·-, hard sphere model; —, $S_R = \pi d_{12}^2(1 - E^*/E)$ for $E \geq E^*$. (Adapted from Karplus *et al.* [24].)

TABLE 2-2

Trajectory Calculations and Experimental Values of the Rate Constant for the Reaction of $H_2 + H$

T (°K)	$10^{-8}k^a$ (M^{-1} sec^{-1})	$10^{-8}k^b$ (M^{-1} sec^{-1})
300	0.000185	0.0014–0.0020 [36]
400	0.0356	0.017–0.061 [37]
500	0.221	—
600	0.780	—
700	1.97	2.49–4.99 [37, 38]
800	4.05	4.66–9.32 [37, 38]
900	7.20	7.6–15.2 [37–39]
1000	11.5	11.0–22.0 [38, 39]

[a] Computer values, Karplus *et al.* [24].
[b] Experimental values.

repulsive forces where $S_R = 0$ for $E < E^*$ and $S_R = \pi d_{12}^2(1 - E^*/E)$ for $E \geq E^*$; and the computer experiment case. The curve for the computer results lies between the two models which are commonly used. The minimum energy for reaction, E^*, is always significantly larger than the potential energy barrier height, but is not dependent in any simple way on the barrier height and total energy of the system.

Finally, the overall rate constants were obtained by use of Eqs. (2-51)–(2-53). The values obtained are assembled in Table 2-2, along with the experimental rate constants. The precision in the measured values is not very good, but in general they agree quite well with those obtained from the computer analysis. The temperature dependence of the computer rate constants is consistent with the simple Arrhenius equation with an activation energy of 7.4 kcal/mole and a preexponential factor of 4.3×10^{10} M^{-1} sec^{-1};

the corresponding values consistent with the experimental data are 7.5 kcal/mole and $5.4 \times 10^{10}\ M^{-1}\ \text{sec}^{-1}$. If elementary kinetic theory is used to analyze the data [Eq. (2-33)], the apparent hard-sphere cross section is about 3.3 Å^2 for the experimental rate constants and about 4.0 to 4.8 Å^2 for the computer values.

In assessing the validity of the computer experiments, two factors must be considered. First, is the potential-energy surface correct? The surface used was the best available, but it is still not known how favorably this compares with reality. Second, given the potential-energy surface, are the solutions of the equations of motion correct? In order to be certain, the quantum-mechanical equations of motion should be solved; this has been done for some special cases of the reaction of $H + H_2$, and the results do not differ greatly from the classical calculations [25]. Also, even in the studies of Karplus and co-workers [24], only a limited number of trajectories were examined. Equilibrium energy distributions were used in the averaging process, although this may not be entirely correct. However, in spite of the questions remaining, these computer experiments represent an interesting and significant step forward. The results obtained from other investigations of this type will be utilized in later discussions.

2-4 TRANSITION-STATE THEORY

The use of collision dynamics in developing theories of chemical kinetics requires specific models for molecular interactions and quite dilute systems. An alternative approach is to consider the statistical properties of the reacting system without worrying about the details of the molecular collisions. Only the potential energy of interaction of the reactants must be known, and the properties of the reactants as they proceed along the reaction pathway of the multidimensional potential-energy surface (see Fig. 2-3) are utilized to calculate the reaction rate. This theory was first proposed by Pelzer and Wigner [26] in 1932 and was extensively developed by Eyring and collaborators [27]. It is particularly useful in condensed phases where collision theories are not useful. Known as the transition-state theory or, occasionally, as the theory of absolute reaction rates, it can be conveniently divided into three stages: the construction of a multidimensional potential-energy surface, which already has been discussed and which can in principle be utilized to obtain the energy of activation from a priori theoretical considerations; the formulation of the rate equation using statistical mechanics to describe the equilibrium between reactants and the activated complex; and formulation of the rate equation in terms of thermodynamic state functions characterizing the transition state complex and the reactants.

Statistical-Mechanical Derivation of the Rate Equation

Consider the general reaction

$$aA + bB + \cdots \rightleftharpoons M^{\ddagger} \rightarrow \text{products} \tag{2-54}$$

in terms of Fig. 2-3, showing the variation of potential energy of the reacting system with changing distance along the reaction coordinate. We shall assume that $M^{\ddagger}$, the activated complex located at the potential-energy maximum, is in virtual equilibrium with A and B and that the order and molecularity of the reaction are equal, namely, $a + b + \cdots$. We endow $M^{\ddagger}$ with all the properties of a stable molecule with two exceptions: one of its vibrational degrees of freedom is considered "frozen," and no account will be taken of it in formulating statistical-mechanical partition functions; the missing vibration is replaced by a fourth translation, which the system uses as it approaches the top of the potential barrier along the reaction coordinate, crosses over the top, and finally rearranges or dissociates into products by the time it reaches the region of relative stability associated with its final state. The rate of the reaction is given by the equilibrium concentration of transition-state complexes at the top of the barrier $M_e^{\ddagger}$ multiplied by the frequency with which the complex crosses the barrier. The transition state is associated with a distance δ along the reaction coordinate, where δ may be considered to be of the order of magnitude of 1 Å, although this quantity cancels out when the final expression for the rate constant is obtained. Half of the transition complexes will be moving to the right for the forward reaction and half to the left for the reverse reaction. Therefore, the frequency with which the activated complex crosses the barrier is simply $\bar{v}/2\delta$, where $\bar{v}$ is the average speed of the molecule along the reaction coordinate. Thus, for a reaction in any phase,

$$\text{rate of reaction} = k(A_e)^a(B_e)^b \cdots = \frac{\varkappa(M_e^{\ddagger})\bar{v}}{2\delta} \tag{2-55}$$

where (A_e) and (B_e) have been written instead of (A) and (B) to emphasize that $(M_e^{\ddagger})$ is taken to be an equilibrium concentration with respect to the reactants. The constant $\varkappa$, usually called the transmission coefficient, represents the fraction of transition-state complexes which cross the top of the barrier to form products; the possibility exists that because of some unusual feature of the potential-energy surface the complex will not lead to product. The transmission coefficient will be assumed to have a value of 1, although there are a number of reactions for which this assumption is not valid. The average speed $\bar{v}$ is obtained by the usual process of averaging with a distribution function, in this case, a one-dimensional function, which describes the equilibrium distribution of speeds for particles of mass $m^{\ddagger}$ $(= am_A + bm_B + \cdots)$ moving along a single position coordinate, the reaction coordinate.

The appropriate relation is

$$\bar{v} = (2kT/\pi m^{\ddagger})^{1/2} \tag{2-56}$$

In addition,

$$\frac{(\mathrm{M_e}^{\ddagger})}{(\mathrm{A_e})^a(\mathrm{B_e})^b \cdots} = K^{\ddagger} \frac{\delta(2\pi m^{\ddagger}kT)^{1/2}}{h} = K \tag{2-57}$$

where $K^{\ddagger}$ is the equilibrium constant involving the reactants and a transition-state complex which is missing one vibrational degree of freedom. The multiplication of $K^{\ddagger}$ by the partition function for a single (fourth) translation, $\delta(2\pi m^{\ddagger}kT)^{1/2}/h$, gives the equilibrium constant for the postulated type of transition-state complex capable of dissociating into products. Strictly speaking, $K^{\ddagger}$ is not a proper equilibrium constant, although it is usually treated as one. By combining Eqs. (2-55) to (2-57) with $\varkappa = 1$, we find

$$k = (kT/h)K^{\ddagger} \tag{2-58}$$

for the rate constant of a reaction of any order.

If we express the "equilibrium" constant $K^{\ddagger}$ in Eq. (2-58) in terms of the appropriate partition functions, we obtain

$$K^{\ddagger} = \frac{f_{\mathrm{M}}^{\ddagger}}{(f_{\mathrm{A}})^a(f_{\mathrm{B}})^b \cdots} \exp\left(-\frac{\Delta E_0^{\ddagger}}{RT}\right) \tag{2-59}$$

where f_i represent appropriate products of partition functions for translation, rotation, vibration, etc., and $\Delta E_0^{\ddagger}$ is the difference in internal energy between the transition-state complex and the reactants at 0°K.

The assumption of a special translational degree of freedom in the theory just outlined is difficult to justify theoretically, especially if we consider that the activated complex is assumed to be in equilibrium with the reactants while it is moving quite rapidly along the reaction coordinate (or equivalently that the rate of passage of transition states over the barrier is the same as at equilibrium). An alternative derivation of Eq. (2-58) assumes that the frequency of passage across the barrier is given by the frequency $\nu^{\ddagger}$, which characterizes the decomposition. In this case,

$$\text{rate of reaction} = \nu^{\ddagger}(\mathrm{M_e}^{\ddagger}) = \nu^{\ddagger}K(\mathrm{A_e})^a(\mathrm{B_e})^b \cdots \tag{2-60}$$

and therefore, using Eq. (2-57),

$$k = \nu^{\ddagger}K \tag{2-61}$$

If it is now assumed that the "reaction vibration" corresponds to the motion of one of the normal coordinates of the activated complex with frequency $\nu^{\ddagger\prime}$ and further that $\nu^{\ddagger\prime} \ll kT/h$, then the associated vibrational partition function $[1 - \exp(-h\nu^{\ddagger\prime}/kT)]^{-1}$ is equal to $kT/h\nu^{\ddagger\prime}$, and the following equation is obtained:

$$k = (v^\ddagger/v^{\ddagger\prime})(kT/h)K^{\ddagger\prime} \tag{2-62}$$

If $v^\ddagger = v^{\ddagger\prime}$, an equation of the same form as (2-58) is obtained. This alternative derivation also involves many assumptions, and there is no reason to prefer one development of the theory over the other.

For the simple illustrative example of a second-order reaction between structureless reactants, which is also bimolecular, namely

$$A + B \leftrightharpoons M^\ddagger \rightarrow \text{products}$$

we obtain, on combining Eqs. (2-58) and (2-59),

$$k_2 = \frac{kT}{h}\frac{(2\pi m^\ddagger kT)^{3/2}h^{-3}8\pi^2 IkTh^{-2}\exp(-\Delta E_0{}^\ddagger/RT)}{(2\pi m_A kT)^{3/2}(2\pi m_B kT)^{3/2}h^{-6}} \tag{2-63}$$

where $(2\pi m_i kT)^{1/2}h^{-1}$ is the partition function assuming unit length for a single degree of translation, and $8\pi^2 IkTh^{-2}$ is the classical partition function for a rigid diatomic molecule of moment of inertia I. [No partition function for vibration appears in Eq. (2-63) since the single vibration in $m^\ddagger$ has been replaced by a fourth translation, whose partition function has been combined with the expression for $\bar{v}$, Eq. (2-56), to yield the term kT/h, as was previously shown in Eq. (2-58).]

If $r_A + r_B$, the sum of the "radii" of A and B, is taken to be the internuclear distance in $M^\ddagger$, then

$$I = m^*(r_A + r_B)^2 = (m_A m_B/m^\ddagger)(r_A + r_B)^2 \tag{2-64}$$

and Eq. (2-63) reduces to

$$k_2 = \pi(r_A + r_B)^2(8kT/\pi m^*)^{1/2}\exp(-\Delta E_0{}^\ddagger/RT) \tag{2-65}$$

This is identical with the expression previously obtained (Section 2-2) for the bimolecular rate constant for a reaction involving dissimilar elastic hard spheres with a steric factor of unity, since $r_A + r_B = d_{12}$ and $\Delta E_0{}^\ddagger/RT = E_a/RT = \epsilon_a/kT$.

If, instead of being structureless reactants, A and B are nonlinear polyatomic molecules containing n_A and n_B atoms, respectively, the partition functions for the transition-state complex and reactants may be written

$$f_\ddagger = (f_T{}^3 f_R{}^3 f_V{}^{3n_A + 3n_B - 7})_\ddagger$$
$$f_A = (f_T{}^3 f_R{}^3 f_V{}^{3n_A - 6})_A$$
$$f_B = (f_T{}^3 f_R{}^3 f_V{}^{3n_B - 6})_B$$

where f_T, f_R, and f_V are the partition functions for translation, rotation, and vibration per degree of freedom. If we assume that the translational partition functions for A and B and the transition-state complex are approximately equal and that this approximate equality also holds for the rotational and vibrational partition functions, we obtain

$$k_2' \cong \frac{kT}{h} \frac{f_V^{\,5}}{f_T^{\,3} f_R^{\,3}} \exp\left(-\frac{\Delta E_0^{\ddagger}}{RT} \right) \cong k_2 \left(\frac{f_V}{f_R} \right)^5 \tag{2-66}$$

where k_2' and k_2 are the respective bimolecular rate constants for the systems of structured and structureless reactants. Thus the steric factor p, which was introduced into the kinetic-theory expression for the collision frequency per unit volume leading to reaction [Eq. (2-31)], may be identified with $(f_V/f_R)^5$. Since f_R is usually larger than f_V, bimolecular reactions involving polyatomic reactants would be expected to have rather small steric factors. This predicted behavior is shown by several of the systems in Table 2-1. Similar considerations and conclusions are applicable to reactions of higher order.

The fundamental equation for the rate constant, Eq. (2-58), contains the assumption that the reactants (and the activated complex as well) are ideal, namely, that all components have activity coefficients of unity at all concentrations. The assumption of ideality is implicit in Eq. (2-57), in which the equilibrium constant K is related to concentrations rather than activities. For reactions in the gas phase the restriction to ideality is not, in general, a serious limitation, since departure from ideal behavior will be slight at the moderate densities commonly used. In solution, however, ionized solutes are far from ideal even at low concentrations, and even nonionized solutes exhibit significant nonideal behavior at concentrations frequently used in rate studies.

Equation (2-58) may be modified to take account of nonideality in the following simple manner. For the generalized reaction of any order in any phase represented by Eq. (2-54),

$$\text{rate of reaction} = k(\mathrm{A_e})^a(\mathrm{B_e})^b \cdots = \frac{\varkappa(\mathrm{M_e}^{\ddagger})\bar{v}}{2\delta} \tag{2-67}$$

where $(\mathrm{A_e})$, $(\mathrm{B_e})$, and $(\mathrm{M_e}^{\ddagger})$ represent *concentrations*. However, since we are treating the reactants and activated complex as the components of a nonideal system in equilibrium, we have at once

$$K = \frac{(\mathrm{M_e}^{\ddagger})}{(\mathrm{A_e})^a(\mathrm{B_e})^b \cdots} \frac{\gamma_{\mathrm{M}}^{\ddagger}}{\gamma_{\mathrm{A}}^{\,a}\gamma_{\mathrm{B}}^{\,b} \cdots} = \frac{[\mathrm{M_e}^{\ddagger}]}{[\mathrm{A_e}]^a[\mathrm{B_e}]^b \cdots} \tag{2-68}$$

where the bracketed symbols are activities and the γ_i the corresponding activity coefficients. Substitution of $(\mathrm{M_e}^{\ddagger})$ from Eq. (2-68) into Eq. (2-67) leads to the fundamental relation for the rate constant for a nonideal system:

$$k = \frac{kT}{h} K^{\ddagger} \frac{\gamma_{\mathrm{A}}^{\,a}\gamma_{\mathrm{B}}^{\,b} \cdots}{\gamma_{\mathrm{M}}^{\ddagger}} \tag{2-69}$$

Thermodynamic Formulation of the Rate Equation

The equilibrium constant between the transition-state complex and the reactants is related to the thermodynamic state functions by

$$\Delta G^{0\ddagger} = -RT \ln K^{\ddagger} = \Delta H^{0\ddagger} - T\,\Delta S^{0\ddagger} \tag{2-70}$$

where $\Delta G^{0\ddagger}$, $\Delta H^{0\ddagger}$, and $\Delta S^{0\ddagger}$ are referred to as the standard free energy of activation, the standard enthalpy of activation, and the standard entropy of activation, respectively. In each case they represent differences between the state functions of the activated complex in a particular standard state and the state functions of the reactants referred to the same standard state. (We are now endowing the constant $K^{\ddagger}$ with all the characteristics of a thermodynamic equilibrium constant, although strictly speaking it should be multiplied by a translational partition function. This in no way invalidates the analysis presented.) For ideal systems the magnitude of $\Delta H^{0\ddagger}$ does not depend upon the choice of standard state, and for most of the nonideal systems which are encountered the dependence is slight. For all systems, however, the magnitudes of $\Delta G^{0\ddagger}$ and $\Delta S^{0\ddagger}$ depend strongly on the choice of standard state, so that it is not useful to say that a particular reaction is characterized by specified numerical values of $\Delta G^{0\ddagger}$ and $\Delta S^{0\ddagger}$ unless the standard states associated with these values are clearly identified.

For the generalized reaction represented by Eq. (2-54), we recall that

$$\text{rate of reaction} = \mathscr{k}(\mathrm{A_e})^a(\mathrm{B_e})^b \cdots = (kT/h)K^{\ddagger}(\mathrm{A_e})^a(\mathrm{B_e})^b \cdots \tag{2-71}$$

Since the dimensions of $\mathscr{k}(\mathrm{A_e})^a(\mathrm{B_e})^b \cdots$ will always be concentration $\times$ time^{-1} and the dimension of the term kT/h will always be time^{-1}, $K^{\ddagger}$ will have precisely the same concentration dependence as $\mathscr{k}$, namely,

$$K^{\ddagger} \propto \text{concentration}^{-(a+b+\cdots-1)} \tag{2-72}$$

and the units chosen for the concentrations of A, B, . . . will be the concentration units common to $\mathscr{k}$ and $K^{\ddagger}$, subject to the dependence expressed by Eq. (2-72).

Now for the nonequilibrium change of state,

$$a\mathrm{A}(c_1) + b\mathrm{B}(c_2) \cdots = \mathrm{M}^{\ddagger}(c_3) \qquad (P,\ T \text{ constant}) \tag{2-73}$$

where c_1, c_2, and c_3 are arbitrary concentrations, the free-energy change is

$$\Delta G^{\ddagger} = RT \ln Q^{\ddagger} - RT \ln K^{\ddagger} = RT \ln(Q^{\ddagger}/K^{\ddagger}) \tag{2-74}$$

where $Q^{\ddagger} = (\mathrm{M}^{\ddagger}_{c_3})/(\mathrm{A}_{c_1})^a(\mathrm{B}_{c_2})^b \cdots$. Note that $\Delta G^{\ddagger}$, being a function of the ratio $Q^{\ddagger}/K^{\ddagger}$, does not change its numerical value as we select different concentration units to specify c_1, c_2, and c_3 and the equilibrium concentrations contained in $K^{\ddagger}$. Since we have stated that c_1, c_2, and c_3 are arbitrary, we may set each of them equal to a standard-state concentration whose magnitude is unity and whose unit we may choose to suit our convenience. In this case, $\Delta G^{\ddagger}$ becomes $\Delta G^{0\ddagger}$, $RT \ln Q^{\ddagger}_{\text{std}}$ is zero, and we have at once the first two members of Eq. (2-70). It is apparent, however, that the particular unit which was chosen for the standard state must influence the numerical magnitude of $\Delta G^{0\ddagger}$ (and $K^{\ddagger}$) in Eq. (2-70), since otherwise $\Delta G^{\ddagger}$ in Eq. (2-74) could not be invariant to the choice of concentration units. It follows finally that the units chosen for the concentrations of the reactants (A), (B), . . . , which are also the concentration units of these same substances in their standard states, determine both the magnitude and units of $K^{\ddagger}$ and, through $K^{\ddagger}$, the magnitudes of $\Delta G^{0\ddagger}$ and $\Delta S^{0\ddagger}$.

Having analyzed the role of the standard state with reference to Eqs. (2-70) and (2-71), we continue the thermodynamic formulation of the transition-state theory by considering the temperature dependence of the rate constants in terms of the parameters of absolute rate theory. For reactions in the gas phase, rate constants are normally expressed in terms of concentration units so that the equilibrium constant $K^{\ddagger}$ in Eq. (2-71) also is in concentration units. However, the standard state normally employed for gases is 1 atm. The relationship between the equilibrium constant expressed in terms of concentration, $K_c^{\ddagger}$, and the equilibrium constant expressed in terms of pressures, $K_p^{\ddagger}$, for ideal gases is

$$K_c^{\ddagger} = K_p^{\ddagger}(RT)^{-\Delta n\ddagger} \tag{2-75}$$

where $\Delta n^{\ddagger} = 1 - (a + b + \cdots)$. (Note that R must have dimensions consistent with those used in defining $K_c^{\ddagger}$ and $K_p^{\ddagger}$.) Combining

$$\frac{d \ln K_c^{\ddagger}}{dT} = \frac{\Delta E^{0\ddagger}}{RT^2} \tag{2-76}$$

with Eq. (2-58) gives

$$\frac{d \ln k}{dT} = \frac{\Delta E^{0\ddagger} + RT}{RT^2} \tag{2-77}$$

Here $\Delta E^{0\ddagger}$ is the standard internal energy of activation. The Arrhenius activation energy E_a is by definition given by

$$\frac{d \ln k}{dT} = \frac{E_a}{RT^2} \tag{2-78}$$

Comparison of Eqs. (2-77) and (2-78) gives

$$E_a = \Delta E^{0\ddagger} + RT \tag{2-79}$$

The relationship between $\Delta H^{0\ddagger}$ and $\Delta E^{0\ddagger}$ is

$$\Delta H^{0\ddagger} = \Delta E^{0\ddagger} + \Delta(PV)^{\ddagger} \tag{2-80}$$

For ideal gases, $\Delta(PV)^{\ddagger} = \Delta n^{\ddagger} RT$; therefore,

$$E_a = \Delta H^{0\ddagger} - (\Delta n^{\ddagger} - 1)RT \tag{2-81}$$

For condensed phases, $\Delta(PV)^{\ddagger} \approx 0$ so that $\Delta H^{0\ddagger} = \Delta E^{0\ddagger}$. Also, in liquids the standard state is normally taken as 1 M. Therefore for condensed phases

$$E_a = \Delta H^{0\ddagger} + RT \tag{2-82}$$

A procedure for calculating the thermodynamic quantities characterizing the activation process is to plot $\ln(k/T)$ versus $1/T$. The slope of such a plot is $-\Delta H^{0\ddagger}/R$ in condensed phases or $-\Delta E^{0\ddagger}/R$ for ideal gases; $\Delta G^{0\ddagger}$ and $\Delta S^{0\ddagger}$ can be readily obtained from Eqs. (2-58) and (2-70). Alternatively, if the Arrhenius energy of activation is of interest, it may be obtained in the usual manner from a plot of $\ln k$ versus $1/T$, and can be related to the transition-state thermodynamic parameters through Eqs. (2-58), (2-70), (2-81), and (2-82). In carrying out these calculations, the role of the choice of standard state in relation to the numerical values of $\Delta S^{0\ddagger}$ and $\Delta G^{0\ddagger}$ should be firmly kept in mind. Since $\Delta H^{0\ddagger}$ and $\Delta E^{0\ddagger}$ are independent of the standard state for ideal systems, the standard state is automatically determined by the units used for the rate constant [Eq. (2-72)] (unless a conversion factor such as that for going from $K_c^{\ddagger}$ to $K_p^{\ddagger}$ is used). For nonideal systems, no additional problems are involved, since, as shown previously, the effects of nonideality may be restricted to the ratio of activity coefficients which appear in Eq. (2-69). However, both $\Delta H^{0\ddagger}$ and $\Delta E^{0\ddagger}$ will be functions of temperature if the heat capacity of activation is not equal to zero.

In applying the concepts of the transition-state theory, we assume that at any time t we may characterize a kinetic system (one that changes to a known extent in a specified time from reactants to products) by a free energy which is a function of the composition of the system at a specified temperature and pressure. Strictly, the only justification for this assumption is the hope that its consequences will be useful, since the only systems for which we can be certain that the concept of a free-energy function is valid are those whose state variables do not change with time. This situation is usually of no kinetic interest, however. Our use of a free-energy function in a time-dependent system implies that the time scale for the reaction is long compared with that required for the molecules to come into statistical equilibrium with respect to the degrees of freedom determining the thermodynamic state. We expect this to be true for all reactions except a few which may occur at extraordinary speeds.

For a simple one-step reaction the concept of a reaction coordinate is readily understandable, and the course of a reaction is often depicted as a

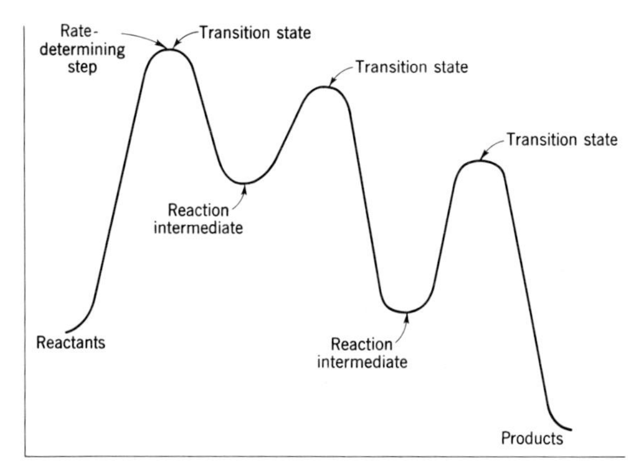

Fig. 2-11 Diagram for a reaction involving a series of activated complexes and reaction intermediates.

plot of the energy versus the reaction coordinate as in Fig. 2-3. For complex reactions involving several steps, the reaction progress is frequently depicted as in Fig. 2-11 where the nature of the ordinate and abscissa is considered shortly. This resembles a plot of the energy versus the reaction coordinate, and indicates the various states the reactants go through to reach products. The maxima indicate different transition states, and the minima indicate reaction intermediates. The highest maximum is identified with the rate-determining step in the mechanism. The reaction intermediates are experimentally detectable in principle, and occasionally so in practice. The activated complexes, however, are unlikely to be detectable because of the relatively small magnitude of K and hence of $(M_e^{\ddagger})$. We recall that

$$K = \mathscr{k} \frac{h}{kT} \frac{\delta(2\pi m^{\ddagger} kT)^{1/2}}{h} \approx 10^{-13} \mathscr{k}$$

(at room temperature and with seconds as the time unit) since the translational partition function is approximately unity when $\delta \sim 10^{-8}$ cm. Thus K, and consequently $(M_e^{\ddagger})$, are of appreciable magnitude only if the rate constant $\mathscr{k}$ is extraordinarily large. If this were to occur, the reaction would be so rapid that the thermodynamic formulation would be of doubtful validity.

We now consider more carefully exactly what Fig. 2-11 represents. The most obvious possibility is that this is a plot of the potential energy (or standard-state enthalpy) versus reaction coordinate. This is reasonable providing all reactions have the same stoichiometry; that is, the products of a given reaction in the overall mechanism must have the same standard-state energy or enthalpy as the reactants of the next step in the mechanism. For example, consider the decomposition of N_2O_5 to NO_2 and O_2 which proceeds by the mechanism

$$N_2O_5 \rightarrow NO_2 + NO_3$$
$$NO_2 + NO_3 \rightarrow NO + O_2 + NO_2$$
$$NO_3 + NO \rightarrow 2\,NO_2$$

The standard-state enthalpy of formation of the products of the second step is 29.7 kcal/mole, while the corresponding enthalpy for the reactants of the third step is 38.7 kcal/mole. Clearly these states cannot be connected on an energy profile. The standard-state enthalpies of the three transition states are 25.2, 29.7 and 40 kcal/mole. This does not imply, however, the rate determining step in the mechanism is the third reaction: for N_2O_5 alone as the starting reactant, the rate-determining step is the second reaction; if a large amount of NO is initially added, the rate-determining step is the first reaction; finally, if a large excess of NO_2 is added to a mixture of N_2O_5 and NO, the rate-determining step becomes the third reaction. Thus the rate-determining step is determined both by the initial concentrations and the energy barriers.

Diagrams such as Fig. 2-11 are often referred to as free-energy profiles, but the meaning of this designation is obscure. Obviously the actual free energy is not meant since the transition state and reactants have equal free energies at equilibrium. The use of standard-state free energies involves the same problems as the use of standard-state energies and enthalpies, namely, the stoichiometry must be preserved throughout the reaction to make meaningful comparisons. Again the magnitudes of the standard-state free energies for the transition states of the above mechanism do not determine what is the rate-determining step. Furthermore, the standard-state free energies are only defined for the reactants, products, and transition state, so that connecting these quantities by curved lines with maxima and minima is not meaningful; instead, bar graphs should be used with the abscissa undefined as shown schematically in Fig. 2-12.

A diagram such as that in Fig. 2-11 may be regarded as a qualitative picture of the reaction progress which arbitrarily connects the products and reactants of consecutive steps and in which the rate-determining step has the most extreme maximum. For the decomposition of N_2O_5, three different diagrams are necessary since, as already discussed, which step is rate determining depends on the initial conditions. The ordinates and abscissa are best left unlabeled. (A more extensive discussion of this problem in terms of the mechanism for N_2O_5 decomposition is given by Johnston [28].) For some complex reactions, where the stoichiometry is preserved throughout the reaction sequence, plots of standard-state energy or free energy may be useful, but each case should be examined with extreme caution.

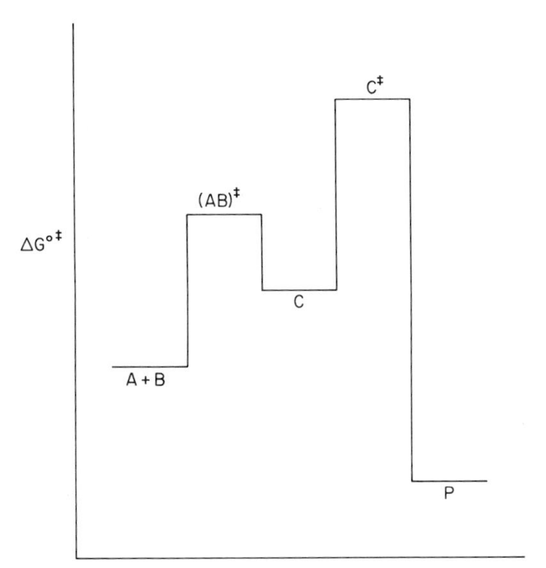

Fig. 2-12 Schematic standard free energy diagram for the reaction $A + B \rightleftharpoons C \rightleftharpoons P$.

From the preceding discussion it is clear that the concept of a rate-determining step must be used with care. The rate-determining step cannot be inferred only by comparing values of rate constants since in general the rate of each step in the reaction mechanism is a product of concentrations and a rate constant. For some complex reactions, a rate-determining step may be hard to identify (e.g., complex chain reactions with complex rate laws). If a rate-determining step can be identified, the experimental rate law will yield information about the activated complex associated with that particular step. Thus, if it is possible to characterize such a complex reaction with a well-defined order, the number of reactant molecules which combine to form the activated complex associated with the rate-determining step corresponds to this order. It is not possible from the experimentally observed rate law to obtain information from the transition-state theory about activated complexes in those steps of the overall reaction which are not rate-determining. When the rate of individual steps in a complex reaction can be obtained directly, the preceding limitations apply to each step.

In view of the nature of the assumptions in the various parts of the transition-state theory, it is well to remember that the transition-state theory should be used with caution and discretion. Because the theory is useful in describing and correlating rate studies, particularly in solution, it does not follow that its microscopic (molecular) details correspond to the actual situation in nature. Until theories are verified by experiment, one should be prepared to reject today a theory which appeared to be satisfactory yesterday. In any event, the transition-state theory provides an extremely useful conceptual framework for reaction kinetics.

2-5 PHENOMENOLOGICAL THEORY OF REACTION RATES

Kinetic theory and transition-state theory try to calculate the rates of chemical reactions starting from a model of molecular interactions. A less ambitious task is to correlate reaction rates with phenomenological laws of various macroscopic processes which have been established experimentally. This type of theory can be termed a phenomenological theory of reaction rates. For the purpose of calculating theoretical reaction rates, chemical reactions are divided into three categories: bimolecular associations, unimolecular dissociations, and intramolecular transformations.

Let us first consider the assessment of the upper limit for the rate constant of a bimolecular reaction: this is analogous to finding the collision frequency in kinetic-theory calculations. The rate with which two systems of molecules can come together is governed by the phenomenological laws of diffusion (Fick's laws). The first of these laws states that the flux density J (the number of molecules flowing through a square centimeter per second) is proportional to the concentration gradient of the substance under consideration:

$$J_A = -D_A \nabla C_A \tag{2-83}$$

Here D_A is by definition the diffusion coefficient and has the dimensions of $cm^2\ sec^{-1}$ if J is expressed in the conventional units given above, and C_A is the concentration, in molecules/cm^3. Strictly speaking, this equation is correct only for a two-component system; it is applicable in multicomponent systems only if the interaction between flows can be neglected [29]. If the flow of substance A is affected by an external potential (for example, electrostatic interactions greatly influence ionic reactions), Eq. (2-83) can be modified to give

$$J_A = -D_A\left(\nabla C_A + \frac{C_A}{kT}\nabla U\right) \tag{2-84}$$

where U is the potential energy of interaction. If matter is to be conserved in a constant-volume system, the change in concentration with time must equal the divergence of J_A. This result is expressed by Fick's second law of diffusion

$$\frac{\partial C_A}{\partial t} = \nabla \cdot \left[D_A\left(\nabla C_A + \frac{C_A}{kT}\nabla U\right)\right] \tag{2-85}$$

In order to calculate the maximum rate of the bimolecular reaction, the assumption is made that every time two molecules collide, a chemical reaction occurs. Consider the diffusion of a system of A molecules into stationary B molecules. If every time A collides with B a reaction occurs, the concentration of A at the surface of B must equal zero, while the concentration at a large distance from B is equal to the bulk concentration C_{A0}. In order to simplify the problem mathematically, the molecules are assumed to be spherical, so that the diffusion process is spherically symmetric, and the potential energy U is assumed to be a function of r only. This model is depicted in Fig. 2-13. A general solution of Eq. (2-85) is still not possible, but if a steady state is assumed (that is, $\partial C_A/\partial t = 0$), the total flux through the surface of a sphere of radius r around B is constant for all values of r and is

$$I_A = \int_0^{2\pi} \int_0^{\pi} J_A r^2 \sin\theta \, d\theta \, d\phi = \text{const}$$

$$= 4\pi r^2 D_A \left(\frac{\partial C_A}{\partial r} + \frac{C_A}{kT} \frac{\partial U}{\partial r} \right) = \text{const} \qquad (2\text{-}86)$$

This equation can be integrated easily if the assumption is made that the diffusion constant is independent of concentration. The boundary conditions are when $r = \infty$, $C_A = C_{A0}$, and $U = 0$, and when $r = a$ (a being the distance

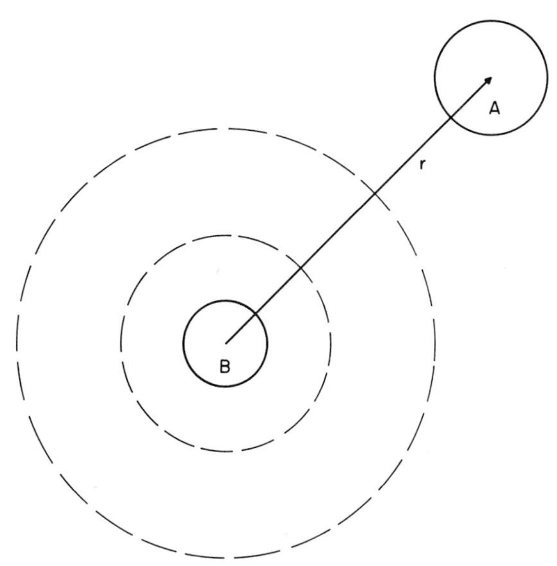

Fig. 2-13 An illustration of the spherically symmetric model used for deriving the rates of diffusion controlled reactions. The total flux through the surfaces of spheres concentric with B (---) is constant in the steady state. The concentration of A increases with increasing r for the association reaction and decreases with increasing r for the dissociation reaction.

of closest approach, i.e., the sum of the radii of A and B), $C_A = 0$, and $U = U(a)$. Under these conditions, the solution of Eq. (2-86) is

$$I_A \int_a^\infty \frac{e^{U/kT}}{r^2}\, dr = 4\pi D_A \int_0^{C_{A0}} d(C_A e^{U/kT})$$

or

$$I_A = \frac{4\pi D_A a C_{A0}}{a \int_a^\infty e^{U/kT}\, dr/r^2} \tag{2-87}$$

The fact that the B molecule is actually not stationary but is also undergoing Brownian motion can be taken into account by replacing D_A with $D_A + D_B$. The preceding expression is for a single B molecule, so that the total rate of reaction R_a is

$$R_a = \kappa 4\pi a(D_A + D_B) f C_{A0} C_{B0} \tag{2-88}$$

where

$$f = \left(a \int_a^\infty e^{U/kT} \frac{dr}{r^2} \right)^{-1} \tag{2-89}$$

The factor κ is introduced here exactly as in the previous discussion of the kinetic theory of gases. On the other hand, the conventional kinetic rate equation for a bimolecular association is

$$R_a = k_a C_{A0} C_{B0}$$

so that the diffusion-controlled bimolecular rate constant in the conventional units of M^{-1} sec^{-1} is

$$k_a = \kappa 4\pi a(D_A + D_B) f(N_0/1000) \tag{2-90}$$

where N_0 is the Avogadro number, a is in centimeters, and $(D_A + D_B)$ has the units of cm^2 sec^{-1}. Equation (2-90) was first derived by Debye [30], but a similar result for the case of no potential energy of interaction was obtained many years earlier by Smoluchowski [31]. The rate constant k_a is the largest possible bimolecular rate constant to be expected and can be readily calculated with the equation above using only experimentally determinable parameters.

Before presenting some calculations of k_a for specific cases, let us examine the assumptions made in obtaining Eq. (2-90). First, the variation of the diffusion coefficient with concentration has been neglected. Since the concentration of the reactants is often quite low and a constant environment is usually maintained by some inert substance, e.g., a salt or buffer in aqueous solution, the diffusion coefficient does remain constant, to a good approximation, throughout the course of the reaction. The correct diffusion constants to use for calculations are the mutual diffusion coefficients determined in the appropriate environment. Second, a steady state has been assumed in order to obtain an analytic solution to the diffusion equation. General solutions have been obtained for some special forms of U, for example, $U = 0$, $U = z_A z_B e^2/\epsilon r$ [30, 32], and the results obtained are identical to Eq. (2-90) except for an additional time-dependent term, which is usually negligible at times greater than about 10^{-7} sec. For shorter times, the rate constant is time-dependent, but this time-dependent period is unimportant in most cases. Finally, spherical symmetry has been assumed. In all but very simple reactions, this is not quite true. The general effect of nonspherical symmetry is to reduce the solid angle from 4π to a smaller value. The size of this steric effect can be easily estimated for a given reaction and taken into account in the calculation of k_a. A comprehensive discussion of diffusion-controlled reactions has been given by Noyes [33].

In liquid solutions, the interaction potential is predominantly electrostatic in nature. For reactions between neutral molecules, the potential energy of interaction can be taken as zero, and f is equal to unity. In aqueous solution, typical values of $(D_A + D_B)$ and a are 10^{-5} cm^2 sec^{-1} and 5 Å, respectively. The resultant value of k_a is 4×10^9 M^{-1} sec^{-1}, which is the "typical" upper limit for the rate constant of a bimolecular reaction between neutral molecules in aqueous solution. This rate constant is not the same as the collision frequency. This is a consequence of the "cage effect" in liquids; that is, because of the liquid structure, if two molecules collide, the probability of their colliding again is higher than the original random probability. Thus the collision frequency at room temperature in liquids is about 10^{11} M^{-1} sec^{-1}, exactly as in gases, while the upper bounds for the bimolecular rate constants are about 10^9 and 10^{11} M^{-1} sec^{-1}, respectively, in the two phases. The simplest form of U for ionic reactions is a coulombic potential $U = z_A z_B e^2/\epsilon r$, where ϵ is the dielectric constant of the medium, z_A and z_B are the valences of substances A and B, and e is the charge on an electron. The electrostatic factor f can be obtained in a closed form from Eq. (2-89) and is

$$f = \frac{z_A z_B e^2}{\epsilon k T a}\left[\exp\left(\frac{z_A z_B e^2}{\epsilon k T a}\right) - 1\right]^{-1} \tag{2-91}$$

As expected, ions of opposite charge react more rapidly than neutral molecules, while ions of the same charge react more slowly. If the ionic strength of the solution is sufficiently high ($> \sim 10^{-4}$ M), a simple coulombic potential is no longer adequate, and some type of Debye–Hückel potential should be used. In this case, f must be evaluated by numerical integration.

Making use of a similar model, let us now derive the expression for the maximum rate of a unimolecular dissociation. This rate, of course, is unimolecular only with respect to substances other than the solvent; one or more solvent molecules can be associated with the actual process. The physical process involved is obviously the separation by diffusion of two molecules, so that Eq. (2-86) is applicable, although different boundary conditions are necessary. The appropriate boundary conditions are when $r = \infty$, $c_A = 0$, and $U = 0$, and when $r = a$, $C_A = 1/\Delta V_A$, and $U = U(a)$. Here ΔV_A is the volume containing the molecule AB, i.e., approximately $4/3\pi a^3$, so that $1/\Delta V_A$ is the concentration of A (or B) in this volume. Following the procedure used previously, the maximum rate constant for dissociation, k_d, is

$$k_d = \frac{4\pi(D_A + D_B)af}{\Delta V_A}\left[\exp\frac{U(a)}{kT}\right] \tag{2-92}$$

This equation was first proposed by Eigen [34] for the case where U is a coulombic potential. For $U = 0$ and the same values of the other parameters as used in the calculation of k_a, k_d is 1.2×10^{10} sec^{-1}. Note that the equilibrium constant

$$K = \frac{k_a}{k_d} = \frac{C_{AB}}{C_A C_B} = \frac{\kappa 4\pi a^3 N_0}{3000}\exp\left[-\frac{U(a)}{kT}\right] \quad M^{-1} \tag{2-93}$$

is independent of the diffusion constants, as it should be. In fact, this equation is valid for any complex not involving true chemical bonds. An identical expression was obtained in a quite different manner for the special case of ion-pair formation in liquids by Fuoss [35].

Finally, let us consider intramolecular transformations. The term intramolecular actually means that no solvent molecules or only several can take part in the reaction. In this case, a quantitative theory cannot be formulated; however, some useful results can be obtained with a crude model. A very simple model assumes that a chemical transformation occurs due to a molecular vibration of unusually large amplitude. In particular, a chemical reaction occurs whenever a critical bond (or bonds) reaches a critical length or, alternatively, whenever the necessary amount of energy is accumulated in a critical mode of vibration. This model is similar to those proposed for unimolecular dissociations in the gas phase, where quantitative theories have been developed (cf. Chapter 4), but for liquids and solids such a theory does not exist at present. The form of the result obtained, however, would probably be the same. The simplest type of behavior to be expected is that

$$k_I = \nu_0 \exp(-\epsilon_a/kT) \tag{2-94}$$

where ν_0 is a frequency which is a particular average of molecular vibrational frequencies, and ϵ_a is the activation energy per molecule necessary for the process. Since ν_0 is a vibrational frequency, the maximum value of k_I is of the order of magnitude of 10^{13} to 10^{14} sec^{-1}.

For many reactions involving ions in liquids, the rate of ion-pair formation is so fast compared to the rate controlling intramolecular transformation that the initial step is always at equilibrium. For example, consider the mechanism

$$A + B \underset{k_d}{\overset{k_a}{\rightleftharpoons}} AB \xrightarrow{k_I} AB^*$$

If the first step is assumed to be at equilibrium and the concentration of AB is small, the experimental overall rate constant is

$$k_f = Kk_I = \frac{\kappa 4\pi N a^3 \nu_0}{3000} \exp\left(-\frac{\epsilon_A}{kT}\right)\exp\left[-\frac{U(a)}{kT}\right] \tag{2-95}$$

Only if the rate of chemical transformation is rapid compared to the rate at which reaction partners diffuse apart will the reaction rate be diffusion-controlled. This can be seen by assuming that AB is in a steady state and solving for k_f. The result is

$$k_f = k_a/(1 + k_d/k_I)$$

Thus if $k_I \gg k_d$, k_f is equal to k_a. The extension of these considerations to multistep mechanisms is possible, and the results are of essentially the same form.

The use of a phenomenological theory of reaction rates has the important advantage that it is based on experimentally determined facts. Therefore, it can be used with considerable confidence. The great disadvantage, of course, is that such a theory is not based on first principles of physics and chemistry. Although most of the material presented here applies equally well in all phases, it is probably most useful in liquids, where the lack of a satisfactory microscopic theory of liquids makes the possibility of an adequate theory of kinetics in liquid solutions seem quite remote.

Problems

2-1 A useful approximation to the intermolecular potential energy $U(r)$ of a pair of spherically symmetric nonpolar molecules is given by the Lennard–Jones formula $U(r) = Ar^{-12} - Cr^{-6}$ where A and C are constants and r is the internuclear separation.

(a) Find an expression in terms of A and C for the distance r_m at which $U(r)$ passes through its minimum value.

(b) Denoting the minimum value of $U(r)$ by $-\epsilon$, evaluate the constants A and C in terms of r_m and ϵ, and then express $U(r)$ in terms of ϵ, r_m, and r.

(c) Find the relation between r_m and the distance σ at which $U(r)$ passes through the value zero.

(d) Find the relation between r_m and the distance at which the force between the two molecules has its minimum value.

2-2 Assume that two reactive molecules A (mol wt = 40) and B (mol wt = 60) can be characterized by an intermolecular-pair potential corresponding to repelling hard spheres, for which

$$U(r) = \infty, \qquad r < d_{12} \qquad (= 3 \times 10^{-8} \text{ cm})$$

$$U(r) = \frac{2.74 \times 10^{-80}}{r^9} \text{ erg/molecule}, \qquad r > d_{12}$$

(a) If A and B react according to a simple bimolecular mechanism

$$A + B \rightarrow C + D$$

calculate the initial rate of reaction, in M sec^{-1}, when A and B are introduced into a constant-volume system each at a partial pressure of 1 atm at 300°K.

(b) If, on the other hand, A and B are sufficiently simple in structure (atoms or simple free radicals) so that association proceeds according to the highly exothermic reaction

$$A + B + X \rightarrow AB + X$$

where X is an inert molecule (mol wt = 100), estimate the corresponding initial rate of reaction when the initial partial pressures of A, B, and X are each 1 atm at 300°K.

Hints: (1) Give your final answer in units of M sec^{-1}; (2) assume that the intermolecular potential is negligible at a value of r for which $U(r)$ has fallen to 1% of $U(d_{12})$; (3) assume a collision distance of 6×10^{-8} cm for encounters between X molecules and A · · B activated complexes; (4) note that $U(d_{12}) \gg kT$.

2-3 Show that the binary collision rate per unit volume between elastic-hard-sphere molecules, when the energy of the component of the initial relative motion along the line of centers exceeds a specified value ϵ_a $(= \frac{1}{2}m^* v_a^2)$, is given by Eq. (2-31) with p equal to unity.

Hint: Recall that in Eq. (2-25), $S(v_r) = 2\pi \int_0^{b_{\max}} b\,db$, where, for elastic hard spheres, $b = d_{12} \sin\theta$, and that the component of $\mathbf{v}_r$ along the line of centers at impact satisfies the condition $v_r \cos\theta \geq v_a$. The limits of θ are specified by $1 \geq \cos\theta \geq v_a/v_r$.

2-4 The specific rate constant k for the homogeneous gas-phase reaction

$$C_2H_4 + H_2 \rightarrow C_2H_6$$

has the values

$$6.98 \times 10^{-9} \quad M^{-1}\text{ sec}^{-1} \quad \text{at } 500°\text{K}$$

$$4.16 \times 10^{-6} \quad M^{-1}\text{ sec}^{-1} \quad \text{at } 600°\text{K}$$

Find:

(a) The experimental energy of activation

(b) The kinetic-theory activation energy (taking into account the temperature dependence of the preexponential factor)

(c) The enthalpy and entropy of activation

(d) The concentration of transition-state complexes in equilibrium with C_2H_4 and H_2 each at a concentration of 0.05 M.

(e) The "effective" reaction cross section for the association reaction

Assume $T = 500°K$.

2-5 In terms of the transition-state theory, formulate the rate constant for the exchange reaction

$$H + H_2 \rightleftharpoons H_3^{\ddagger} \rightarrow H_2 + H$$

in terms of the appropriate partition functions. Assume $H_3^{\ddagger}$ has a linear symmetric configuration with a ground-state electronic degeneracy of 2 and a symmetry number of 2. The electronic degeneracies for ground-state H_2 and H are 1 and 2, respectively.

The experimental rate constants can be represented by [39]

$$k = 10^{8.94} T^{1/2} e^{-5500/RT} \quad M^{-1}\ \text{sec}^{-1}$$

Theoretical calculations suggest that the moment of inertia of $H_3^{\ddagger}$ is 3.34×10^{-40} g cm^2 and that the fundamental frequencies of the transition state are 3650, 670, and 670 cm^{-1} [1]. The moment of inertia of H_2 is 0.459×10^{-40} g cm^2, and its fundamental vibrational frequency is 4395.2 cm^{-1}. Compare the theoretical and experimental values of the frequency factor at 300°K.

2-6 (a) Using transition-state terminology and introducing the concept of a volume of activation $\Delta V^{0\ddagger}$, derive an equation for the pressure dependence of the rate constant.

(b) The following rate constants have been determined for the hydrolysis of acetamide by hydroxyl ion as a function of pressure at 25°C [40].

Pressure (lb/in.^{2}):	14.7	4000	10,000	15,000
$k(M^{-1}\ \text{sec}^{-1})$:	3.77×10^{-5}	4.44×10^{-5}	5.47×10^{-5}	6.80×10^{-5}

Calculate the volume of activation.

(c) Discuss any correlation you might expect to find between $\Delta V^{0\ddagger}$ and $\Delta S^{0\ddagger}$ for a series of related reactions.

2-7 (a) Assuming that the enthalpy and entropy of activation are temperature dependent and that the heat capacity of activation $\Delta C_P^{0\ddagger}$ is independent of temperature, derive an expression for the rate constant in terms of transition-state formalism.

(b) The following equation has been found to fit the experimental data for the hydrolysis of ethyl benzenesulfonate over a temperature range of 9 to 75°C:

$$\log k = -(7063/T) - 16.38 \log T + 59.22$$

Calculate $\Delta H_0^{0\ddagger}$, $\Delta S_0^{0\ddagger}$, and $\Delta C_P^{0\ddagger}$.

2-8 The kinetics of the reaction

$$H^+ + B^- \underset{k_{-1}}{\overset{k_1}{\rightleftharpoons}} HB$$

in aqueous medium has been studied for several different bases B^-, and the following values of k_1 have been obtained at 25°C [41]:

B^-:	F^-	HS^-	$O_2NC_6H_4O^-$ (meta)
$k_1\ (M^{-1}\ \text{sec}^{-1})$:	1×10^{11}	7.5×10^{10}	4.2×10^{10}

(a) Calculate the theoretical values of the rate constant k_1 assuming that the reaction rate is diffusion-controlled. To a very good approximation, the sum of the diffusion constants can be taken as 13×10^{-5} cm^2 sec^{-1} and the distance of closest approach as 8 Å in all cases. The dielectric constant of water can be taken as 80.

(b) Account for the deviations between the experimental and calculated values.

2-9 The second-order rate constant for the reaction

$$H^+(aq) + OH^- \rightarrow H_2O$$

is 1.4×10^{11} M^{-1} sec^{-1} in pure water at 25°C. The diffusion coefficient of H^+ is 9.31×10^{-5} cm^2 sec^{-1}, and that of OH^- is 5.26×10^{-5} cm^2 sec^{-1}. Assuming a dielectric constant of 80 for liquid water, calculate the reaction radius.

The effective radius of a water molecule is about 2.8 Å. On the basis of these data, suggest a plausible structure for H^+ (aq).

References

1. H. Eyring and M. Polanyi, *Z. Phys. Chem. (Leipzig)* **B12**, 279 (1931).
2. F. London, *Z. Elektrochem.* **35**, 552 (1929).
3. A. Farkas, *Z. Phys. Chem. (Leipzig)* **B10**, 419 (1930).
4. S. Sato, *J. Chem. Phys.* **23**, 592, 2465 (1955); *Bull. Chem. Soc. Japan* **28**, 450 (1955).
5. R. E. Weston, Jr., *J. Chem. Phys.* **31**, 892 (1959).
6. T. E. Sharp and H. S. Johnston, *J. Chem. Phys.* **37**, 1541 (1962).
7. R. N. Porter and M. Karplus, *J. Chem. Phys.* **40**, 1105 (1954).
8. R. D. Present, "Kinetic Theory of Gases." McGraw-Hill, New York, 1958; *Proc. Nat. Acad. Sci. USA* **41**, 415 (1955).
9. R. H. Fowler, "Statistical Mechanics." Cambridge Univ. Press, London and New York, 1936.
10. S. W. Benson, "Foundations of Chemical Kinetics." McGraw-Hill, New York, 1960.
11. J. H. Sullivan, *J. Chem. Phys.* **31**, 1292, 1577 (1959); **36**, 1925 (1962); **39**, 3001 (1963).
12. T. Welinsky and H. A. Taylor, *J. Chem. Phys.* **6**, 466 (1938).
13. G. B. Kistiakowsky and E. K. Roberts, *J. Chem. Phys.* **21**, 1637 (1953).
14. P. G. Ashmore and J. Cahnmugam, *Trans. Faraday Soc.* **49**, 270 (1953).
15. G. B. Kistiakowsky and E. R. van Artsdalen, *J. Chem. Phys.* **10**, 305 (1942); **12**, 469 (1944).
16. J. H. Sullivan and N. Davidson, *J. Chem. Phys.* **19**, 143 (1951); **17**, 176 (1949); A. A. Miller and J. E. Willard, *ibid.* **17**, 168 (1949).
17. W. G. Burns and F. S. Dainton, *Trans. Faraday Soc.* **48**, 39 (1952).
18. J. B. Harkness, G. B. Kistiakowsky, and W. H. Mears, *J. Chem. Phys.* **5**, 682 (1937).
19. J. R. Lacker, G. W. Tompkin, and J. D. Park, *J. Am. Chem. Soc.* **74**, 1693 (1952); B. Atkinson and A. B. Trenwith, *J. Chem. Phys.* **20**, 754 (1952).
20. I. Amdur, *J. Am. Chem. Soc.* **60**, 2347 (1938).
21. F. T. Wall, L. A. Hiller, Jr., and J. Mazur, *J. Chem. Phys.* **29**, 255 (1958).
22. H. Goldstein, "Classical Mechanics." Addison-Wesley, Reading, Massachusetts, 1950.
23. F. T. Wall, L. A. Hiller, Jr., and J. Mazur, *J. Chem. Phys.* **35**, 1284 (1961).
24. M. Karplus, R. N. Porter, and R. D. Sharma, *J. Chem. Phys.* **43**, 3259 (1965).
25. M. Karplus and K. T. Tang, *Discuss. Faraday Soc.* **44**, 56 (1967).
26. H. Pelzer and E. Wigner, *Z. Phys. Chem. (Leipzig)* **B15**, 445 (1932).
27. S. Glasstone, K. J. Laidler, and H. Eyring, "Theory of Rate Processes." McGraw-Hill, New York, 1940.

28. H. S. Johnston, "Gas Phase Reaction Rate Theory," pp. 298–320. Ronald Press, New York, 1966.
29. L. Onsager, *Ann. NY Acad. Sci.* **46**, 241 (1945).
30. P. Debye, *Trans. Electrochem. Soc.* **82**, 265 (1942).
31. M. V. Smoluchowski, *Z. Phys. Chem. (Leipzig)* **92**, 129 (1917).
32. E. W. Montroll, *J. Chem. Phys.* **14**, 202 (1946).
33. R. M. Noyes, *Prog. React. Kinet.* **1**, 129 (1961).
34. M. Eigen, *Z. Phys. Chem. (Leipzig)* **NF1**, 176 (1954).
35. R. M. Fuoss, *J. Am. Chem. Soc.* **80**, 5059 (1958).
36. K. Geib and P. Harteck, *Z. Phys. Chem. Bodenstein Festband* p. 849 (1931).
37. M. Van Meersche, *Bull. Soc. Chim. Belg.* **60**, 99 (1951).
38. G. Boato, G. Careri, A. Cimino, E. Molinari, and G. G. Volpi, *J. Chem. Phys.* **24**, 783 (1956).
39. A. Farkas and L. Farkas, *Proc. Roy. Soc. London Ser. A* **152**, 124 (1935).
40. K. J. Laidler and D. Chen, *Trans. Faraday Soc.* **54**, 1026 (1958).
41. M. Eigen and K. Kustin, *J. Am. Chem. Soc.* **82**, 5952 (1960).

CHAPTER 3

REACTIONS IN THE GAS PHASE

3-1 INTRODUCTION

This chapter discusses in detail the experimental results of a class of reactions in the gas phase for the purpose of illustrating dynamical features of theories previously presented. In the course of the development, it will be found that while in some cases it is possible to make a quantitative comparison with theory, in others conclusions concerning the dynamical features must assume a more qualitative, albeit specific, character. The reason for selecting gas-phase reactions should now be obvious; at present it is only in the gas phase that we may hope to find an environment sufficiently simple to justify comparison with a dynamical theory. Accordingly, we shall discuss in some detail the series of reactions

$$H_2 + X_2 \rightleftharpoons 2HX \tag{3-1}$$

where X_2 is a halogen molecule. This is a particularly useful series, since the mechanism of the reaction is not the same for each of the halogens. There are interesting transitions from apparent simplicity to complexity as X_2 changes from I_2 to F_2, which can be explained in relatively logical fashion in terms of the thermodynamic and kinetic parameters of postulated elementary reactions in the mechanism schemes. In addition, the mechanism for a well-studied branching chain reaction

$$2H_2 + O_2 \rightarrow 2H_2O$$

is discussed.

3-2 $H_2 + I_2 \rightleftharpoons 2HI$

This reaction, which has been very extensively studied, has been shown to have two unusual features. At sufficiently low temperatures (below about 600°K) the reaction is second order in both directions [1, 2]. In addition, the equilibrium constant K has a value whose magnitude is such that it is convenient to study either the rate of the reversible formation of HI from H_2 and I_2 or the rate of the reversible decomposition of HI into H_2 and I_2.

Because the rates of the reverse reactions cannot be neglected, this system must be treated as a kinetically reversible second-order reaction:

$$H_2 + I_2 \underset{k_r}{\overset{k_f}{\rightleftharpoons}} 2HI \tag{3-2}$$

Thus, if we start with a constant-volume system at a fixed temperature with an initial H_2 concentration a, an initial I_2 concentration b, and no initial HI concentration,

$$\begin{aligned}\frac{1}{2}\frac{d}{dt}(\text{HI}) = \frac{1}{2}\frac{dx}{dt} &= k_f\left(a - \frac{x}{2}\right)\left(b - \frac{x}{2}\right) - k_r x^2 \\ &= k_f\left[ab - (a+b)\frac{x}{2} + \frac{x^2}{4}\right] - k_r x^2\end{aligned} \tag{3-3}$$

where x is the concentration of HI at time t. Since the equilibrium constant $K = k_f/k_r$, Eq. (3-3) may be rewritten to eliminate k_r:

$$\frac{1}{2}\frac{dx}{dt} = k_f\left[\left(\frac{1}{4} - \frac{1}{K}\right)x^2 - (a+b)\frac{x}{2} + ab\right] \tag{3-4}$$

and Eq. (3-4) may be integrated between 0 and x and 0 and t to give

$$k_f = \frac{1}{mt}\left\{\ln\frac{[(a+b+m)/(1-4/K)] - x}{[(a+b-m)/(1-4/K)] - x} - \ln\frac{a+b+m}{a+b-m}\right\} \tag{3-5}$$

where $m = [(a+b)^2 - 4ab(1-4/K)]^{1/2}$.

If we now start with a constant-volume system at the same fixed temperature with an initial HI concentration c and no initial concentrations of H_2 or I_2, then

$$-\frac{1}{2}\frac{d}{dt}(\text{HI}) = \frac{1}{2}\frac{dy}{dt} = k_r(c-y)^2 - k_f\left(\frac{y}{2}\right)^2 \tag{3-6}$$

where y is the concentration of HI that has reacted at time t. Integration of Eq. (3-6) between 0 and y and 0 and t gives

$$k_r = \frac{1}{2cKt}\left\{\ln\frac{[c(1+\sqrt{K/4})/(1-K/4)] - y}{[c(1-\sqrt{K/4})/(1-K/4)] - y} - \ln\frac{1+\sqrt{K/4}}{1-\sqrt{K/4}}\right\} \tag{3-7}$$

where, as in Eq. (3-4), $K = k_f/k_r$. Thus, if equilibrium measurements are made to obtain K directly at a given temperature, Eqs. (3-5) and (3-7) can then be used to obtain independent values of k_f and k_r from rate measurements. A comparison of the ratio k_f/k_r with K can be used to judge the self-consistency of the rate measurements.

Bodenstein originally pointed out [1] that two possible mechanisms for this reaction are either a one-step bimolecular mechanism in both directions

$$H_2 + I_2 \overset{k_1}{\rightleftharpoons} HI \tag{3-8}$$

or a mechanism in which molecular iodine rapidly dissociates to atomic iodine followed by two atoms of iodine reacting with H_2

$$I_2 \overset{K_D}{\rightleftharpoons} 2I, \qquad 2I + H_2 \overset{k_2}{\rightleftharpoons} 2HI \tag{3-9}$$

The rate laws for these two mechanisms are identical: in the former case, $k_f = k_1$, while in the latter case, assuming the first step is at equilibrium and the concentration of iodine atoms is small, $k_f = K_D k_2$.

If Eq. (3-8) is the mechanism, the measured rate constant can be readily analyzed in terms of kinetic theory. By measuring K, k_f, and k_r over a range of temperature (below about 600°K), we can obtain the energies of activation from plots of $\ln(k/T^{1/2})$ versus $1/T$ if we assume that the rate constant may be written in the form

$$k = A'T^{1/2}e^{-E_a/RT}$$

Table 3-1 gives a comparison of the experimental results with the predictions of the collision theory. Since the entries in the third column are the elastic-hard-sphere results for the preexponential factor $A/T^{1/2}$, it would appear that the simple collision theory is not inconsistent with the experimental results on the reversible decomposition of HI. However, this does not necessarily prove that the collision theory is correct, since the transition-state theory is also in agreement with theory when reasonable parameters are selected for evaluation of the various partition functions.

TABLE 3-1

Experimental and Theoretical Kinetic Parameters[a]

Reaction	A' (M^{-1} sec^{-1} deg$^{-1/2}$)	E_a (kcal/mole)	$\kappa\pi d^2\left(\frac{8k}{\pi m^*}\right)\frac{N_0}{1000}$ (M^{-1} sec^{-1} deg$^{-1/2}$)	Steric factor
$H_2 + I_2 \rightarrow 2HI$	$10^{9.78}$	40.7	$10^{10.38}$	0.28
$2HI \rightarrow H_2 + I_2$	$10^{8.97}$	43.7	$10^{9.33}$	0.44

[a] d has been arbitrarily taken as 3.5×10^{-8} cm.

The formation of HI from H_2 and I_2 has been analyzed in terms of transition-state theory, assuming the activated complex is a planar molecule H_2I_2, with an axis of symmetry [3–5].

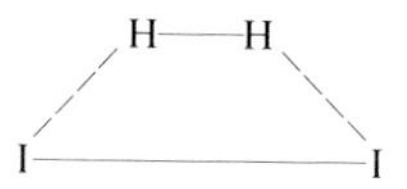

The rate constant can be written in terms of the appropriate partition functions. After simplification we obtain

$$k = \frac{h^3}{\pi^2 kT 2^{3/2}} \frac{(ABC)^{1/2}}{m^{*3/2} I_{H_2} I_{I_2}}$$

$$\times \frac{[1 - \exp(-h\nu_{H_2}/kT)][1 - \exp(-h\nu_{I_2}/kT)] \exp(-\Delta E_0^{\ddagger}/RT)}{\prod_{i=1}^{5} [1 - \exp(-h\nu_i^{\ddagger}/kT)]} \tag{3-10}$$

where a symmetry number of 2 has been assigned to the activated complex. Reasonable bond distances in the activated complex have been assumed for estimating the three moments of inertia A, B, and C. These distances lead to values of $A = 921.5$, $B = 6.9$, and $C = 928.4 \times 10^{-40}$ g cm^2; also it is known that $I_{H_2} = 0.456$ and $I_{I_2} = 748.5 \times 10^{-40}$ g cm^2. Finally, with the aid of various empirical rules, the vibrational frequencies of the transition state have been estimated as 994, 86, 1280, 1400, and 1730 cm^{-1}. For H_2, ν is 4395.2 cm^{-1}, and for I_2, ν is 214.57 cm^{-1}. Equation (3-10) can now be evaluated at any desired temperature. At 600°K, after setting $\Delta E_0^{\ddagger} \approx E_a$, we obtain $k = 10^{10.3} e^{-E_a/RT}$. The experimental results (Table 3-1) give $k = 10^{11.2} e^{-E_a/RT}$. In view of the many assumptions which were made, the order-of-magnitude agreement is considered satisfactory. Note, however, that the transition state analysis is equally valid for both mechanisms!

The simple bimolecular mechanism was widely accepted until 1967 when Sullivan [6] carried out an experiment in which the reaction rate was measured under conditions where the temperature was so low that the reaction rate would be immeasurably small except for the fact that iodine atoms were continuously produced at high concentrations by photodissociation of I_2. The activation energy of the measured rate constant in this experiment, k^{ph}, was identical with that of the rate constant measured at higher temperatures without photodissociation, k^{th}, divided by K_D, i.e.

$$k^{th} = K_D k^{ph} \tag{3-11}$$

If both mechanisms are operative,

$$k^{th} = k_1 + K_D k_2 \tag{3-12}$$

At the low temperatures utilized by Sullivan $k_1 = 0$, and if it is assumed that $k^{ph} = k_2$, then it follows that $k^{th} = K_D k_2$. In other words the mechanism by which HI is formed from H_2 and I_2 would be Eq. (3-9) rather than Eq. (3-8).

Unfortunately this simple analysis is not correct, and in fact the two mechanisms cannot be experimentally distinguished [7]. The identification of k^{ph} with k_2 is based on the assumption that the only effect of the photostationary state is to raise greatly the concentration of iodine atoms without increasing the concentration of the most reactive of the molecular I_2, I_2^*, which reacts with H_2. If the ratio $(I_2^*)/(I_2)$ is enhanced by the factor x in the photostationary state, then the net rate of the photochemical reaction is

$$R = xk_1(I_2) + k_2(I)^2(H_2) = k^{ph}(H_2)(I)^2 \tag{3-13}$$

Now if the enhancement of the concentration of iodine atoms is identical with the enhancement of the ratio $(I_2^*)/(I_2)$, that is, $(I)^2/(I_2) = xK_D$, then from Eq. (3-13)

$$k^{ph} = K_D^{-1}k_1 + k_2 \tag{3-14}$$

Thus the result found by Sullivan, Eq. (3-11), is predicted from Eqs. (3-12) and (3-14) of this analysis.

Of course, the preceding analysis assumes that the enhancement of the concentration of reactive molecular iodine is identical with the enhancement of the concentration of atomic iodine. In molecular terms, this requires that the redissociation of I_2^* to iodine atoms be faster than its deexcitation since if deexcitation were much faster no enhancement at all of the concentration of I_2^* would be possible. In other words, I_2^* and 2I are so rapidly equilibrated that they act as a single kinetically undifferentiated species. Whenever two (or more) possible reactant species are rapidly equilibrated with respect to the rate of the overall reaction, it is not possible to ascertain which of the potential reactants actually is the predominant reactant by conventional kinetics. In the present case, it might actually be possible to determine experimentally whether redissociation or deexcitation of highly excited vibration–rotation states of molecular iodine is occurring, but until this information is available the mechanism of the supposedly simple reaction Eq. (3-2) must be considered unresolved.

3-3 $H_2 + Br_2 \rightarrow 2HBr$

The homogeneous reaction between H_2 and Br_2 to form HBr was investigated in 1906 by Bodenstein and Lind [8], who studied it in the range 230–300°C. Their empirical representation of the rate was given by

$$\frac{1}{2}\frac{d}{dt}(HBr) = \frac{k(H_2)(Br_2)^{1/2}}{1 + k'(HBr)/(Br_2)} \tag{3-15}$$

The denominator gives an immediate clue to part of the mechanism, namely, that as the reaction proceeds, the product HBr tends to inhibit the rate of reaction. The dimensionless inhibition constant k' was found [9] to be practically constant between 25 and 300°C, increasing slowly from 0.116 to 0.122. Thus, during the initial stages of the reaction, when $(HBr) \ll (Br_2)$, the inhibition has not set in to an appreciable extent, and

$$\frac{1}{2}\frac{d}{dt}(HBr) \cong k(H_2)(Br_2)^{1/2} \tag{3-16}$$

Equation (3-16) indicates that Br atoms rather than Br_2 molecules are involved in the rate-determining step and that the mechanism will therefore be complex rather than simple, as in the case of HI formation. The mechanism independently proposed by Christiansen, Herzfeld, and Polanyi [10–12] about thirteen years later is

$$Br_2 + M \underset{k_2}{\overset{k_1}{\rightleftharpoons}} 2Br + M, \qquad \Delta H^0_{1,2} = 46.1 \quad \text{kcal/mole}$$

$$Br + H_2 \underset{k_4}{\overset{k_3}{\rightleftharpoons}} HBr + H, \qquad \Delta H^0_{3,4} = 16.6 \quad \text{kcal/mole}$$

$$H + Br_2 \xrightarrow{k_5} HBr + Br, \qquad \Delta H^0_{5,6} = -41.4 \quad \text{kcal/mole}$$

where M refers to any gaseous species and the ΔH^0 are for standard states at 25°C and 1 atm.

On the basis of the above mechanism we have

$$\frac{d}{dt}(HBr) = k_3(Br)(H_2) + k_5(H)(Br_2) - k_4(HBr)(H) \tag{3-17}$$

while for the intermediate species Br and H we have

$$\frac{d}{dt}(Br) = 2k_1(Br_2)(M) + k_4(HBr)(H) + k_5(H)(Br_2) - 2k_2(Br)^2(M) - k_3(Br)(H_2) \tag{3-18}$$

and

$$\frac{d}{dt}(H) = k_3(Br)(H_2) - k_4(H)(HBr) - k_5(H)(Br_2) \tag{3-19}$$

If we ignore the induction period and make use of the stationary-state hypothesis, namely, that $d(Br)/dt = 0$ and $d(H)/dt = 0$, we can solve for (Br) and (H) at the steady state as follows.

From Eq. (3-19) with $d(H)/dt = 0$

$$k_3(Br)(H_2) = k_4(H)(HBr) + k_5(H)(Br_2) \tag{3-20}$$

which, when substituted into Eq. (3-18) with $d(\mathrm{Br})/dt = 0$, gives

$$(\mathrm{Br}) = \left(\frac{k_1}{k_2}\right)^{1/2} (\mathrm{Br}_2)^{1/2} = K_{1,2}^{1/2} (\mathrm{Br}_2)^{1/2} \tag{3-21}$$

Similarly, from Eq. (3-19),

$$k_4(\mathrm{H})(\mathrm{HBr}) + k_5(\mathrm{H})(\mathrm{Br}_2) = k_3(\mathrm{Br})(\mathrm{H}_2) = k_3 K_{1,2}^{1/2} (\mathrm{Br}_2)^{1/2}(\mathrm{H}_2)$$

or

$$(\mathrm{H}) = \frac{k_3 K_{1,2}^{1/2} (\mathrm{Br}_2)^{1/2} (\mathrm{H}_2)}{k_4(\mathrm{HBr}) + k_5(\mathrm{Br}_2)} \tag{3-22}$$

Substitution of Eqs. (3-20) and (3-22) into Eq. (3-17) gives

$$\frac{1}{2}\frac{d}{dt}(\mathrm{HBr}) = k_5(\mathrm{H})(\mathrm{Br}_2) = \frac{k_3 K_{1,2}^{1/2} (\mathrm{H}_2)(\mathrm{Br}_2)^{1/2}}{1 + k_4(\mathrm{HBr})/k_5(\mathrm{Br}_2)} \tag{3-23}$$

which has the same form as the empirical rate equation and leads to the following relations for the empirical rate constants:

$$k = k_3 K_{1,2}^{1/2}, \qquad k' = k_4/k_5$$

The suggested mechanism represents a fairly simple chain reaction, wherein the chains are initiated by thermal dissociation of bromine molecules into atoms. The chain-propagating steps involve the conversion of the reactant H_2 into HBr by reaction with Br atoms and, since H atoms are produced in the process, the conversion of the reactant Br_2 into HBr by reaction with H atoms. The destruction of HBr by H atoms in step 4, which is characterized by k_4, accounts for the observed inhibition by the product of the reaction. Finally, the chains are terminated by the termolecular association of Br atoms to form Br_2 molecules.

Referring to the relations for k and k', we can calculate k_3 from the experimental value of k since $K_{1,2}$ is independently known [13]. The relation for k_3 has been given in the form

$$\log k_3 = 12.308 + \log T^{1/2} - \frac{17{,}600}{2.303RT}$$

(for k_3 in cc mole^{-1} sec^{-1}) from which the activation energy E_{a3} is 17.6 kcal/mole. Since

$$E_{a3} - E_{a4} = \Delta E_{3,4}^0 = \Delta H_{3,4}^0 = 16.6 \quad \text{kcal/mole}$$

$E_{a4} = 1.0$ kcal/mole, which is the activation energy associated with k_4. Furthermore, $k' = k_4/k_5$ has virtually no temperature dependence, so that E_{a5} (associated with k_5) must also be about 1 kcal/mole. If the termolecular association of Br atoms to form Br_2 molecules is assumed to have zero activation energy, the bimolecular dissociation of Br_2 into 2Br has an activation energy of about 46.1 kcal/mole. We therefore know E_a for every step in the mechanism. Further, we can proceed to assume reasonable diameters in order to calculate steric factors for reactions 3, 4, and 5, whose rate constants k_3, k_4, and k_5 appear in k and k'. These turn out to be quite reasonable, between about 0.03 and 0.1, and lend further support to the likelihood of the proposed mechanism.

The mechanism is further supported by showing on the basis of magnitudes of energies of activation and the absence of certain terms in the empirical rate equation that, under ordinary conditions, the following possible reactions contribute negligibly to the mechanism:

$$H_2 + Br_2 \rightarrow 2HBr, \qquad H_2 + M \rightleftharpoons 2H + M$$

$$Br + HBr \rightarrow H + Br_2, \qquad M + H + Br \rightleftharpoons M + HBr$$

3-4 THE GENERAL MECHANISM

We are now in position to make certain generalized correlations concerning the kinetics of the reaction $X_2 + H_2 \rightarrow 2HX$, where X_2 is a halogen. If we assume that the chain mechanism deduced for $Br_2 + H_2$ is of general validity, then

$$X_2 + M \underset{k_2}{\overset{k_1}{\rightleftharpoons}} 2X + M, \qquad X + H_2 \underset{k_4}{\overset{k_3}{\rightleftharpoons}} HX + H, \qquad H + X_2 \xrightarrow{k_5} HX + X$$

The relevant thermodynamic data for the various halogens are summarized in Table 3-2.

We recall that for $H_2 + Br_2$ the overall chain rate constant k was given by $k = k_3 K_{1,2}$ so that the overall activation energy associated with k would be $E_a = E_{a3} + \Delta E^0_{1,2}/2 \cong E_{a3} + \Delta H^0_{1,2}/2$. Since $E_{a3} - E_{a4} = \Delta E^0_{3,4} = \Delta H^0_{3,4}$, the

TABLE 3-2

Thermochemical Data for Halogen Reactions[a]

Reaction no.	Reaction type	F ΔH^0	F ΔS^0	Cl ΔH^0	Cl ΔS^0	Br ΔH^0	Br ΔS^0	I ΔH^0	I ΔS^0
1, 2	$X_2 \rightleftharpoons 2X$	38.6	27.2	58.0	25.6	46.1	25.0	35.4	24.0
3, 4	$X + H_2 \rightleftharpoons HX + H$	−30.4	−0.3	1.0	1.4	16.6	1.8	32.9	2.3
5, (6)	$H + X_2 (\rightleftharpoons) HX + X$	−98.0	2.8	−45.1	3.3	−41.4	3.2	−36.0	2.8

[a] ΔH^0 in kcal/mole, ΔS^0 in cal deg^{-1} mole^{-1} for the reactions as written. The pairs of numbers in the first column identify the forward and reverse reactions in terms of the subscripts associated with the appropriate rate constants. Since there is no step (or reaction) characterized by k_6, a parenthetical notation has been used in the final entries of the first two columns.

TABLE 3-3

Minimum Activation Energies for the Reaction $H_2 + X_2 \rightarrow 2HX$

Reaction	$E_{a,\min}$
$F_2 + H_2 \rightarrow 2HF$	—
$Cl_2 + H_2 \rightarrow 2HCl$	30.0
$Br_2 + H_2 \rightarrow 2HBr$	39.7
$I_2 + H_2 \rightarrow 2HI$	50.6

minimum value of E_a is obtained by letting $E_{a3} \cong \Delta H^0_{3,4}$ or $E_{a,\min} \cong \Delta H^0_{3,4} + \Delta H^0_{1,2}/2$. In this way we find the values of $E_{a,\min}$ given in Table 3-3.

For $I_2 + H_2 \rightarrow 2HI$ the activation energy for the single-step bimolecular recombination or its kinetic equivalent Eq. (3-9), about 41 kcal/mole, is sufficiently less than the $E_{a,\min}$, about 51 kcal/mole, to show that at least at low temperatures we should not expect a chain reaction.

We may sharpen the comparison by removing the approximation $E_{a3} \cong \Delta H^0_{3,4}$. Table 3-4 lists experimental activation energies for all the steps in the generalized mechanism except the initial dissociation of X_2 into atoms, for which E_{a1} is approximated by $\Delta H^0_{1,2}$. Thus, for $H_2 + Cl_2 \rightarrow 2HCl$, E_a is very nearly 34 kcal/mole compared with about 41 kcal/mole for $H_2 + Br_2 \rightarrow 2HBr$, and this corresponds to a factor of about 2000 at the very low temperature of 180°C. In the photolytic reactions the $\Delta H^0_{1,2}$ term is not present in E_a, which is essentially just E_{a3}. In such cases the difference in activation energies for the Cl_2 and Br_2 reactions is about 13 kcal/mole.

In the case of $F_2 + H_2 \rightarrow 2HF$, complications arise due to the extreme reactivity of the F atoms, which is comparable to that of H atoms. We must therefore consider H + H + M, H + F + M, and F + F + M all as possible

TABLE 3-4

Activation Energies for X_2–H_2 Reactions

Reaction no.	Reaction	Activation energy		
		Cl	Br	I
2	$X + X + M \rightarrow X_2 + M$	0[a]	0[a]	0[a]
3	$X + H_2 \rightarrow HX + H$	4.6–6.1	17.6	33.4
4	$H + HX \rightarrow H_2 + X$	3.6–5.1	0.9	0.5
5	$H + X_2 \rightarrow HX + X$	2–4	0.9	0

[a] These activation energies are actually slightly negative (cf. Chapter 6).

chain-terminating steps. Although the overall reaction is difficult to work with, some of the individual steps in the chain reaction have been extensively studied (cf. Chapter 6 and Gross and Bott [14]).

While the thermal reaction of Cl_2 and H_2 has been extensively studied [15–17], the results are not nearly so satisfactorily correlated as the reactions of I_2 and H_2 or of Br_2 and H_2. This is because the chain-propagation steps proceed very much more rapidly than in the case of $H_2 + Br_2$; heterogeneous reactions involving termination of chains at the wall are important (in such cases diffusion is as important as chemical factors in determining the rate of the reaction); and the reaction is extremely sensitive to small traces of impurity. In connection with the rapid rate at which chains are propagated, the chain cycle

$$Cl + H_2 \rightarrow HCl + H, \qquad H + Cl_2 \rightarrow HCl + Cl$$

is actually too fast to be balanced by the homogeneous termination step

$$2Cl + M \rightarrow Cl_2 + M$$

and the reaction therefore does not reach a stationary state in the gas phase. If the temperature could be lowered sufficiently to slow down the chain cycle relative to the termination step, which is virtually independent of temperature, we might expect a stationary state to be reached. This can be shown to be what one would expect below about 200°C. Although the rate of the thermal reaction becomes too slow to test this, that of the photochemical reaction is not. Experimentally, the photochemical reaction, whose mechanism would be expected to be the same as that of the thermal reaction except for the step producing chlorine atoms by dissociation of Cl_2, is found to be well behaved below 172°C but not above. It is of interest that at sufficiently high temperatures, between 1100 and 1600°K, the reaction of H_2 and Br_2 shows behavior similar to that of H_2 and Cl_2 in that the reaction rate is too fast to permit the active chain propagators to attain steady-state concentrations [18–21].

In rather general terms, the reactions of the halogens with molecular hydrogen may in principle be regarded as possible chain reactions with a common mechanism scheme similar to that for $H_2 + Br_2$. In cases where suitable experimental information exists for the homogeneous reactions, it has been shown that these experimental results can be well correlated with the thermal data (ΔH^0's and ΔS^0's) and with activation energies for the individual steps in the mechanism. The correlation includes an explanation of the simple behavior of the reversible reaction of $H_2 + I_2$ below about 600°K, where it was shown that the chain behavior is less probable than the direct reaction of molecular iodine or iodine atoms, Eqs. (3-8) and (3-9). However, Sullivan has found that above 600°K the chain mechanism for the reaction of either H_2 or D_2 with I_2 becomes increasingly important [2]. At

800°K, for example, about 95% of all the HI formed is produced by reaction of atomic iodine with molecular hydrogen (reaction 3 of the general mechanisms). The increased importance of this step in the mechanism is a consequence of the rapid increase with temperature of the rate of dissociation of molecular iodine in the initiation step.

3-5 REACTION OF H_2 AND O_2

The combination of H_2 and O_2 has been studied extensively for many years by a variety of other methods, but the complete mechanism is probably not yet unambiguously established. Under appropriate conditions, however, the reaction proceeds by a chain mechanism and, specifically, by steps involving branching chains, in which the loss of a single propagator is compensated by the simultaneous production of two other propagators. Extensive discussion of this reaction occurs in the literature, particularly in sources which treat chain reactions in gases more fully [22–27].

A mechanism consistent with results for systems with constant temperature during the course of the reaction involves an initiation step, in which chain propagators are produced at a rate R_0. The character of this step depends upon such experimental factors as temperature, pressure, and composition. If partial dissociation of either H_2 or O_2 occurs, the initial propagator is H or O. If H_2 and O_2 interact in the initiation step, the free radical OH is the propagator, so that a reaction scheme which takes all these possibilities into account continues, after the initiation step, according to

$$
\begin{array}{rll}
H + O_2 \rightarrow OH + O & & \\
O + H_2 \rightarrow OH + H & & \text{Propagation} \\
OH + H_2 \rightarrow H_2O + H & & \\
\left.\begin{array}{r} H + O_2 + X \rightarrow HO_2 + X \\ H + H + X \rightarrow H_2 + X \\ H + OH + X \rightarrow H_2O + X \end{array}\right\} & & \text{Termination (gas phase)} \\
\text{H, OH, or } HO_2 + \text{wall} \rightarrow \text{stable species} & & \text{Termination (wall)}
\end{array}
\qquad (3\text{-}24)
$$

A description of some of the experimental results obtained when H_2 and O_2 react in a constant-volume closed system is helpful in understanding the scheme represented in Eq. (3-24). At temperatures around 550°C, a stoichiometric mixture ($2H_2 : O_2$) reacts slowly and smoothly if the total pressure is sufficiently low, of the order of a fraction of a millimeter. As the total pressure is increased, the reaction rate increases slowly, until a pressure of a few millimeters is reached. The rate then suddenly becomes too rapid to follow, and a nonthermal explosion occurs. As the total pressure is further increased, the reaction continues to be explosive until a pressure of about 100 mm is reached. The rate then abruptly decreases to a conveniently measurable value

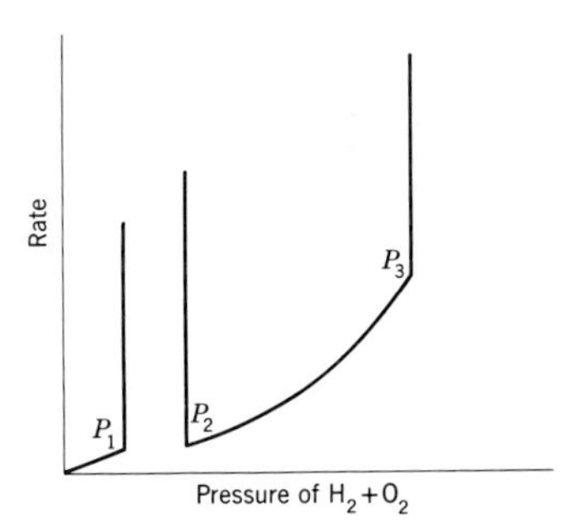

Fig. 3-1 Schematic drawing of explosion limits of a mixture of H_2 and O_2.

and continues to proceed at a measurable, increasing rate as the total pressure is still further increased. The behavior described above is shown schematically in Fig. 3-1, which shows the first and second nonthermal explosion limits P_1 and P_2, as well as a third limit P_3, which is observed at sufficiently high pressures. P_3 represents the onset of a true thermal explosion produced by the liberation of heat in the exothermic combination of H_2 and O_2 more rapidly than it can be dissipated. Isothermal conditions can no longer be maintained, and the rate of reaction increases exponentially with increasing temperature until the reactants are consumed with explosive violence.

The effect of temperature on the reaction in general, as well as on the three explosion limits, is illustrated in Fig. 3-2. The vertical line at 550°C represents the behavior described above and shown in Fig. 3-1. In addition, Fig. 3-2 shows that no explosions are observed below about 400°C. In this low-temperature region the reaction, although of measurable rate, is predominantly heterogeneous and can be markedly retarded by poisoning of the wall, e.g., with a coating of KCl. Above about 600°C the reaction is explosive at

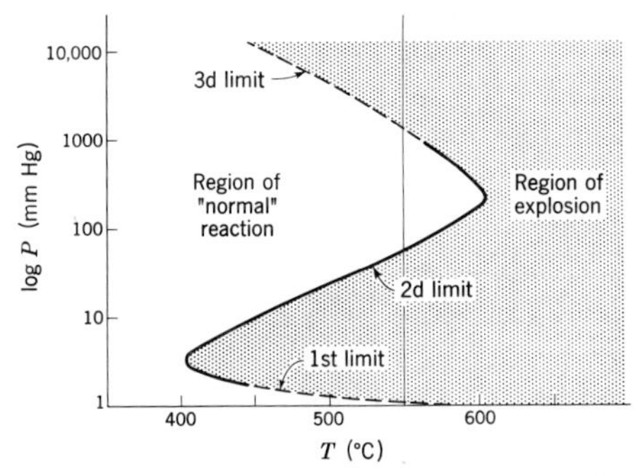

Fig. 3-2 Explosion limits of a stoichiometric mixture of $H_2 + O_2$ in a KCl-coated spherical vessel (7.4 cm diameter). [From S. W. Benson, "Foundations of Chemical Kinetics," Copyright 1960, McGraw-Hill, New York. Used with permission of McGraw-Hill Book Company.]

all pressures. Between these two temperatures the first explosion limit shows relatively little dependence on temperature compared to the second limit.

The observations summarized in Figs. 3-1 and 3-2 and the effects on P_1 and P_2 of certain experimental variations provide a basis for the mechanism of Eq. (3-24). The occurrence of sudden explosions as the pressure is varied under isothermal conditions is a consequence of the presence of branching chains, as shown in the first two propagation steps. The third propagation step, which utilizes OH radicals produced in the second step or in the initiation step, does not interfere with chain branching, since the destruction of OH is compensated by production of H, a propagator which does produce chain branching. The existence of two nonthermal explosion limits suggests two different mechanisms for chain termination, one occurring at the wall, the other in the gas phase. The suggestion is supported by several experimental observations. For example, if the size of the vessel is increased, or if, in a given vessel, the partial pressure of H_2 is increased, or if inert gas is added, the pressure at which the first limit appears is lowered. This behavior is characteristic of chain termination at the wall, since the time required for propagators to diffuse to the wall increases with total pressure or with increasing diameter of the reaction vessel. The rate at which branching chains are produced, however, increases with increasing total pressure. These opposing effects are not offset by differences in the concentration dependence of chain destruction at the wall and chain production in the gas phase, since both processes are first order with respect to the concentration of the propagators. Thus, at sufficiently low pressures destruction of propagators at the wall counteracts their production in the gas, and the reaction proceeds slowly. As the total pressure is increased, however, the efficiency of termination at the wall decreases relative to the increase in the rate of production of branching chains, since diffusion is inversely proportional to the first power of the pressure and chain branching is proportional to the square of the pressure. The result is a nonthermal explosion. It is therefore necessary to include in the mechanism a step for termination of chains at the wall, such as the last step in Eq. (3-24).

The second nonthermal explosion limit is virtually independent of the vessel diameter. If the partial pressure of H_2 is decreased, restoration of the original total pressure by addition of an inert gas such as argon will prevent a lowering of the limit. Behavior of this nature is characteristic of termination of chains in the gas phase. Since production of branching chains involves binary encounters, whereas destruction of propagators in the gas phase depends, in almost all cases, upon ternary collisions, the efficiency of chain destruction relative to that of chain production increases with increasing total pressure. A pressure P_2 is thus reached, where the two processes are near balance, and the rate of reaction is quickly reduced to the relatively low value above P_2. It is therefore necessary to include in the mechanism termolecular steps for termination of branching chains in the gas phase. The particular termination step in Eq. (3-24), in which HO_2 is formed, is chosen to permit continuation of the relatively slow reaction between P_2 and P_3 by a nonbranching chain reaction which utilizes HO_2 as a propagator [24, 26].

The concentration of OH radicals as a function of time has been observed directly in a shock tube (see Chapter 6 for details of this technique) under conditions where the reaction between H_2 and O_2 proceeds smoothly (1100–2600°K, in the presence of excess argon) [28–30]. An induction period approximately 10 to 100 μsec long is followed by a rapid rise of the OH concentration to a maximum value and finally a slow approach to equilibrium. The rise of the OH concentration over its equilibrium value (a phenomenon also observed for H and O) results from a partial approach to equilibrium in the three propagation steps in the mechanism before a significant amount of termolecular reaction occurs in the gas phase termination step. A detailed analysis shows that this overshoot can occur only if there is a decrease in the number of moles in the overall reaction, since it is only in this case that the termolecular termination step may become rate limiting.

The interpretation of the shock tube results is that during the induction period the branching chain reaction predominates until significant amounts of H_2 and O_2 have reacted and back reactions have become important. After the induction period, the concentration of free radical propagators go through a maximum and then slowly approach equilibrium values until the higher-order termination steps limit chain propagation, and the back reactions become important. All of the individual rate constants in the mechanism could be evaluated and are consistent with simple collision theory.

Although this discussion of gas-phase reactions can hardly be considered comprehensive, many important and general features of gas-phase mechanisms have been considered. Unimolecular decompositions are discussed in some detail in the next chapter, and some approaches to elementary reactions in the gas phase are considered in Chapters 5 and 6.

Problems

3-1 The following mechanism has been proposed for the thermal decomposition of acetone:

$$CH_3COCH_3 \xrightarrow{k_1} CH_3 + CH_3CO, \qquad E_a = 84 \text{ kcal/mole}$$
$$CH_3CO \xrightarrow{k_2} CH_3 + CO, \qquad E_a = 10 \text{ kcal/mole}$$
$$CH_3 + CH_3COCH_3 \xrightarrow{k_3} CH_4 + CH_2COCH_3, \qquad E_a = 15 \text{ kcal/mole}$$
$$CH_2COCH_3 \xrightarrow{k_4} CH_3 + CH_2CO, \qquad E_a = 48 \text{ kcal/mole}$$
$$CH_3 + CH_2COCH_3 \xrightarrow{k_5} C_2H_5COCH_3, \qquad E_a = 5 \text{ kcal/mole}$$

(a) Express the overall rate in terms of the individual rate constants and concentrations of stable chemical species.
(b) Calculate the overall energy of activation.

3-2 The formation of gaseous phosgene $COCl_2(g)$ from $CO(g)$ and $Cl_2(g)$ and its decomposition into $CO(g)$ and $Cl_2(g)$ have been explained by two alternative mechanisms:

$$\text{I.} \quad Cl_2 \rightleftharpoons 2Cl, \text{ equilibrium, } K_1$$
$$Cl + CO \rightleftharpoons COCl, \text{ equilibrium, } K_2$$
$$COCl + Cl_2 \xrightarrow{k_3} COCl_2 + Cl$$
$$Cl + COCl_2 \xrightarrow{k_4} COCl + Cl_2$$
$$\text{II.} \quad Cl_2 \rightleftharpoons 2Cl, \text{ equilibrium, } K_1$$
$$Cl + Cl_2 \rightleftharpoons Cl_3, \text{ equilibrium, } K_3$$
$$Cl_3 + CO \xrightarrow{k_5} COCl_2 + Cl$$
$$Cl + COCl_2 \xrightarrow{k_6} CO + Cl_3$$

Derive the expression for $d(COCl_2)/dt$ for both mechanisms and indicate whether or not kinetic measurements can determine which mechanism is correct.

3-3 Pure parahydrogen can be prepared at low temperatures, and the kinetics of the conversion of pure parahydrogen to an equilibrium (3:1) mixture of ortho- and parahydrogen has been studied at 923°K [31]. The rate of conversion is first order with respect to parahydrogen in a given run at constant pressure, but the observed first-order rate constant is proportional to the square root of the total hydrogen pressure. Postulate a mechanism that is consistent with the observed rate law.

3-4 The decomposition of ozone in ozone–oxygen mixtures has been extensively studied (cf. Benson and Axworthy [32]), and proceeds by a relatively simple mechanism.

In pure ozone, during the early stages of the reaction

$$-\frac{d}{dt}(O_3) = k(O_3)^2$$

where k is associated with an activation energy of 24 kcal/mole. When $O_2 \gg O_3$,

$$-\frac{d}{dt}(O_3) \approx k' \frac{(O_3)^2}{(O_2)}$$

and k' is associated with an activation energy of 30 kcal/mole. At low pressures of O_3, O_2 is an inhibitor of the ozone decomposition, whereas at high pressures of O_3, it is an accelerator. Inert gases can also accelerate the reaction rate under certain conditions.

Postulate a mechanism and show that it is consistent with the behavior described above. Derive the rate law and determine the activation energy of as many of the individual steps in the mechanism as possible. The following thermodynamic information may prove useful:

$$K = \frac{(O_2)(O)}{(O_3)} = 7.7 \times 10^4 e^{-24,600/RT} \quad M, \qquad K' = \frac{(O)^2}{(O_2)} = 1.1 \times 10^5 e^{-119,100/RT} \quad M$$

3-5 The decomposition of di-*t*-butyl peroxide has been studied in both the vapor and liquid phases [33]. The solvents used were isopropylbenzene, isobutylbenzene, and tri-

n-butylamine. In all cases, the reaction was first order, the frequency factor for the rate constant was about 10^{16} sec^{-1}, and the experimental activation energy ranged from 37 to 39 kcal/mole. In the vapor phase the stoichiometry of the reaction is

$$(CH_3)_3COOC(CH_3)_3 \rightarrow 2(CH_3)_2CO + C_2H_6$$

In the liquid phase t-butyl alcohol as well as acetone was found, but the sum of the moles of t-butyl alcohol and acetone formed per mole of di-t-butyl peroxide was always 2. (Actually about 5% of the peroxide was found to give other products, but this fact can be neglected for our purposes.)

The addition of nitric oxide and propylene to the reaction mixture produced no change in the rate of decomposition. (These substances are known to be inhibitors of chain reactions.) However, when nitric oxide was added, formaldoxime ($CH_2{=}NOH$) was formed, and when propylene was added, appreciable amounts of isopentane, n-butane, and similar substances were found. The energy of the O—O bond in the peroxide has been estimated as 38 kcal/mole.

Postulate a reasonable mechanism for the di-t-butyl peroxide decomposition. Indicate how all the above facts are consistent with the proposed mechanism.

3-6 The system CO–O_2 displays two explosion limits. The proposed mechanism involves chain branching by excited CO_2 formed via the reaction $CO + O \rightarrow CO_2^*$. The mechanism can be written as

$$O_2 \rightleftharpoons 2O \qquad \text{Initiation}$$

$$O + CO \underset{k_{-1}}{\overset{k_1}{\rightleftharpoons}} CO_2^* \qquad \text{Transfer}$$

$$CO_2^* + O_2 \xrightarrow{k_2} CO_2 + 2O \qquad \text{Branching}$$

(Transfer and Branching: Propagation)

$$CO_2^* + X \xrightarrow{k_3} CO_2 + X$$

$$CO_2^* \xrightarrow[\text{(at walls)}]{k_4} CO_2$$

(Termination)

Assuming that the pressure of X (any molecule or atom present, that is, $\sum P_i$) is equal to the total pressure, derive an expression for the pressure of the second explosion limit in terms of the above mechanism. Discuss the validity of the assumption.

References

1. M. Bodenstein, *Z. Phys. Chem. (Leipzig)* **13**, 56 (1894); **22**, 1 (1897); **29**, 295 (1898).
2. J. H. Sullivan, *J. Chem. Phys.* **30**, 1292, 1577 (1959); **36**, 1925 (1962); **39**, 3001 (1963).
3. A. Wheeler, B. Topley, and H. Eyring, *J. Chem. Phys.* **4**, 178 (1936).
4. W. Altar and H. Eyring, *J. Chem. Phys.* **4**, 661 (1936).
5. S. Glasstone, K. J. Laidler, and H. Eyring, "The Theory of Rate Processes." McGraw-Hill, New York, 1941.
6. J. H. Sullivan, *J. Chem. Phys.* **46**, 73 (1967).
7. G. G. Hammes and B. Widom, *J. Am. Chem. Soc.* **96**, 7621 (1974).
8. M. Bodenstein and S. C. Lind, *Z. Phys. Chem. (Leipzig)* **57**, 168 (1906).

9. M. Bodenstein and G. Jung, *Z. Phys. Chem.* (*Leipzig*) **121**, 127 (1926).
10. J. A. Christiansen, *K. Dan. Vidensk. Selsk. Mat. Fys. Medd.* **1**, 1, 14 (1919).
11. K. F. Herzfeld, *Z. Elektrochem.* **25**, 301 (1919); *Ann. Phys.* (*Leipzig*) **59**, 635 (1919).
12. M. Polanyi, *Z. Elektrochem.* **26**, 50 (1920).
13. M. Bodenstein and F. Cramer, *Z. Elektrochem.* **22**, 327 (1916).
14. R. W. F. Gross and J. F. Bott, eds., "Handbook of Chemical Lasers." Wiley (Interscience), 1976.
15. J. C. Morris and R. N. Pease, *J. Am. Chem. Soc.* **61**, 394, 396 (1939).
16. W. J. Kramers and L. A. Moignard, *Trans. Faraday Soc.* **45**, 903 (1943).
17. P. G. Ashmore and J. Chanmugam, *Trans. Faraday Soc.* **49**, 254 (1953).
18. D. Britton and N. Davidson, *J. Chem. Phys.* **23**, 2461 (1955).
19. M. Gilbert and D. Altman, *Symp. Combust. 6th Yale Univ. 1956* p. 222 (1957).
20. J. C. Giddings and H. Shin, *Trans. Faraday Soc.* **57**, 468 (1961).
21. H. Shin, *J. Chem. Phys.* **39**, 2937 (1963).
22. N. Semenov, "Chemical Kinetics and Chain Reactions." Oxford Univ. Press, London and New York, 1935.
23. B. Lewis and G. von Elbe, "Combustion, Flames and Explosions." Academic Press, New York, 1951.
24. S. W. Benson, "Foundations of Chemical Kinetics." McGraw-Hill, New York, 1960.
25. C. N. Hinshelwood and A. T. Williamson, "The Reaction between Hydrogen and Oxygen." Oxford Univ. Press, London and New York, 1934.
26. C. N. Hinshelwood, "Kinetics of Chemical Change." Oxford University Press, London and New York, 1940.
27. G. L. Schott and W. Getzinger, "Physical Chemistry of Fast Reactions" (B. Levitt, ed.), Vol. 1, pp. 81–160. Plenum, New York, 1973.
28. G. L. Schott and J. L. Kinsey, *J. Chem. Phys.* **29**, 1177 (1958).
29. G. L. Schott, *J. Chem. Phys.* **32**, 710 (1960).
30. T. Asaba, W. C. Gardiner, Jr., and R. F. Stubbeman, *Int. Symp. Combust. 10th* p. 295. Combustion Inst. Pittsburgh, Pennsylvania, 1965.
31. A. Farkas, *Z. Phys. Chem.* (*Leipzig*) **B10**, 419 (1930).
32. S. W. Benson and A. E. Axworthy, Jr., *J. Chem. Phys*, **26**, 1718 (1957).
33. J. H. Raley, F. F. Rust, and W. E. Vaughan, *J. Am. Chem. Soc.* **70**, 88, 1336, 2767 (1948).

CHAPTER 4

UNIMOLECULAR DECOMPOSITIONS IN THE GAS PHASE

4-1 INTRODUCTION

Attempts to find a satisfactory microscopic theory for unimolecular reactions in the gas phase started in 1919, when Perrin [1] suggested that infrared radiation must be absorbed by the reactants in the activation step of such reactions. He argued that since the phenomenological rate law

$$-\frac{1}{c}\frac{dc}{dt} = k$$

showed that the fractional rate of reaction was a constant independent of pressure, collisional activation could not be involved, because collisions would tend to vanish as the pressure approached zero, whereas $(-1/c)\,dc/dt$ would retain its constant value. He accordingly postulated that activation energy was supplied by blackbody radiation characteristic of the temperature of the system, as described by Planck's law for the dependence of radiation density on frequency at a given temperature. By assuming that the specific rate was proportional to the radiation density at a frequency ν, where this frequency was related to the activation energy by $N_0 h\nu = E_a$, he obtained a rate expression of the correct form

$$k = \text{const}\, e^{-E_a/RT}$$

Attempts to find experimental justification for Perrin's theory were unsuccessful. For example, the experimental activation energy for the first-order thermal decomposition of N_2O_5 corresponded to infrared radiation with a wavelength of 1.16 μm as the source of activation. Irradiation of N_2O_5 with high-intensity light of this wavelength at temperatures below that where thermal decomposition occurred produced no decomposition [2]. Further, examination of the spectrum of N_2O_5 showed that the molecule was unable to absorb radiation at or near this wavelength. By 1922 it was clearly established that Perrin's basic premise, namely, that $(-1/c)\,dc/dt = k$ even at vanishingly low pressure, was incorrect. It was found that although the first-order rate law was obeyed at relatively high pressures, as the pressure was decreased, k became pressure-dependent. Accordingly, collisional activation could no longer be excluded, and in 1922 Lindemann [3] proposed that activated molecules were produced by binary collisions but that there was a time lag, during which a small fraction of the activated molecules underwent unimolecular decomposition. Most of the activated molecules, however, would be deactivated by collision before decomposition occurred. The process can be represented by the scheme (4-1).

$$\text{Reactant molecules} \underset{k_2}{\overset{k_1}{\rightleftharpoons}} \text{activated molecules} \xrightarrow[\text{(slow, first-order decomposition)}]{k_3} \text{products} \tag{4-1}$$

In terms of a decomposition whose stoichiometry is represented by

$$A = B + C \tag{4-2}$$

the mechanism of the Lindemann scheme leads to the following phenomenological analysis.

The activation and deactivation steps are

$$A + A \underset{k_2}{\overset{k_1}{\rightleftharpoons}} A^* + A$$

and the decomposition step is

$$A^* \xrightarrow{k_3} B + C$$

where A* represents an activated reactant molecule. Assuming the concentration of A* to be small compared to that of A, B, and C, we can write

$$-\frac{d}{dt}(A) = \frac{d}{dt}(B) = k_3(A^*) \tag{4-3}$$

and application of the steady-state assumption to A* leads to

$$(A^*) = \frac{k_1(A)^2}{k_2(A) + k_3} \tag{4-4}$$

so that

$$-\frac{d}{dt}(A) = \frac{k_3 k_1(A)^2}{k_2(A) + k_3} = k_{exp}(A) \tag{4-5}$$

Equation (4-5) states that at pressures sufficiently high so that $k_2(A) \gg k_3$ the reaction follows a first-order rate law with

$$k_{exp} = k_3 k_1 / k_2$$

As the pressure is decreased until k_3 is no longer negligible, k_{exp} begins to decrease in value and becomes a function of pressure, namely, $k_{exp} = k_3 k_1(A)/[k_2(A) + k_3]$. At sufficiently low pressures, where $k_2(A) \ll k_3$, the reaction should appear to be second order, since $k_{exp} = k_1(A)$ and $-d(A)/dt = k_1(A)^2$. If we refer to the Lindemann scheme, the high-pressure constancy of k corresponds to the rate-determining slow decomposition preceded by a rapid activation–deactivation step. As the pressure is decreased, the activation–deactivation step becomes slower as the binary-collision frequency decreases quadratically with pressure. Thus the rate of this activation–deactivation step tends to approach that of the decomposition step, which decreases linearly with pressure. In principle, the rate of the unimolecular decomposition of A* may become rapid relative to the activation–deactivation step, and the overall reaction is then characterized by the slow bimolecular step and follows a second-order rate law.

4-2 CLASSICAL MICROSCOPIC THEORIES

The first microscopic theory of unimolecular reactions was developed by Hinshelwood [4], who used a rather simple molecular model. His assumptions were as follows:

(1) the problem could be treated classically;

(2) the energy of activation was supplied by binary collisions in which the vibrational energy in the bonds was available for activation;

(3) for activation it was sufficient to have the minimum energy ϵ_a in the molecule as a whole; it was not necessary to localize this energy in the bond where dissociation would eventually occur, nor was the probability of reaction increased if the energy in the molecule as a whole exceeded ϵ_a; and

(4) the fraction W of reactant molecules which contain energy per mole in excess of E_a is given by the classical equilibrium distribution for a system of molecules having $n/2$ vibrational modes which contribute to the activation process.

For $E_a \gg (n/2 - 1)RT$, this distribution is [5]

$$W = \frac{e^{-E_a/RT}(E_a/RT)^{n/2-1}}{\Gamma(n/2)} + \cdots \tag{4-6}$$

where $\Gamma(n/2)$ is the gamma function and the higher-order terms are negligible for the values of E_a/RT which are usually involved. (For a system of classical oscillators, $n/2$ has a maximum value of $3s - 6$ for a nonlinear molecule of s atoms and $3s - 5$ for a linear molecule.)

The expression for W is obtained by recalling that the fraction of molecules possessing energy between E and $E + dE$ for one degree of vibrational freedom is $(1/RT)e^{-E/RT}\,dE$. Therefore, for $n/2$ vibrational degrees of freedom, the fraction of molecules possessing energy between E_1 and $E_1 + dE_1$ in the first vibrational mode, between E_2 and $E_2 + dE_2$ in the second vibrational mode, etc., is

$$\frac{1}{(RT)^{n/2}}\,e^{-E_1/RT}e^{-E_2/RT}\cdots e^{-E_{n/2}/RT}\,dE_1\,dE_2\cdots dE_{n/2}$$

or

$$\frac{1}{(RT)^{n/2}}\,e^{-E/RT}\,dE_1\,dE_2\cdots dE_{n/2} \tag{4-7}$$

where use has been made of the fact that the total energy is $E = \sum_{i=1}^{n/2} E_i$. We wish to know the fraction of molecules with total energy between E and $E + dE$. To obtain this we first integrate Eq. (4-7) over the energy range 0 to E. The multiple integral involved can be written as

$$\underset{\sum E_i \leq E}{\int\cdots\int} dE_1\,dE_2\cdots dE_{n/2} = \int_0^E dE_1 \int_0^{E-E_1} dE_2 \cdots \int_0^{E-(E_1+E_2+\cdots+E_{n/2-1})} dE_{n/2}$$

$$= \frac{E^{n/2}}{\Gamma(n/2+1)}$$

[Recall that $\Gamma(n/2 + 1) = (n/2)!$ when $n/2$ is an integer.] Differentiation of this expression gives the corresponding relationship for the range of energies between E and $E + dE$:

$$\frac{E^{n/2-1}\,dE}{\Gamma(n/2)}$$

Combining this with Eq. (4-7) gives the equilibrium fraction of molecules with total energy between E and $E + dE$ distributed over $n/2$ vibrational degrees of freedom:

$$dW = \frac{1}{\Gamma(n/2)RT}\left(\frac{E}{RT}\right)^{(n-2)/2} e^{-E/RT}\,dE \tag{4-8}$$

The fraction of molecules which contains more than energy E_a is

$$W = \int_{E_a}^{\infty} dW = \int_{E_a}^{\infty} \frac{1}{\Gamma(n/2)RT}\left(\frac{E}{RT}\right)^{(n-2)/2} e^{-E/RT}\,dE$$

This can be integrated by parts to give

$$W = \left[\frac{1}{\Gamma(n/2)}\left(\frac{E_a}{RT}\right)^{n/2-1} + \frac{1}{\Gamma(n/2-1)}\left(\frac{E_a}{RT}\right)^{n/2-2} + \cdots\right] e^{-E_a/RT}$$

which yields Eq. (4-6) if $E_a \gg (n/2 - 1)RT$.

On the basis of the preceding assumptions, Hinshelwood's theory proceeds as follows. Let c be the concentration of reactant, W the equilibrium fraction of activated molecules, i.e., that at high pressures, where the rate of reaction is extremely small relative to the rate of deactivation, and c^* the concentration of activated molecules at any pressure. The formal expression for the number of deactivating collisions at high pressures, namely, $accW$, where a was found from elastic-hard-sphere collision theory to be equal to $2\pi d^2(kT/\pi m)^{1/2}$ $(= Z_{11}/c^2)$, represents the rate of activation at all pressures. Since at the steady state at any pressure this rate of activation will be equal to the rate of deactivation plus the rate of reaction,

$$ac^2 W = acc^* + k(E)c^* \tag{4-9}$$

where $k(E)$ is the microscopic rate constant, which, according to the assumptions of the theory, has a finite constant value for all energies greater than E_a and which is also independent of concentration and temperature. Since the phenomenological and microscopic rate equations must agree, we have at once that

$$k_{\text{exp}} = k(E)c^*/c \tag{4-10}$$

where, from Eq. (4-9),

$$c^* = \frac{cW}{1 + k(E)/ac} \tag{4-11}$$

Thus

$$k_{\text{exp}} = \frac{k(E)W}{1 + k(E)/ac} \tag{4-12}$$

and since at sufficiently high pressure $c^* = cW$ and

$$k_{\text{exp}} = k_\infty = k(E)W \tag{4-13}$$

we have from Eq. (4-12)

$$k_{\text{exp}} = \frac{k_\infty}{1 + k_\infty/aWc} = \frac{k_\infty}{1 + k_\infty RT/aWN_0P} \tag{4-14}$$

As $P \to 0$, we find

$$k_{\text{exp}} = aWc \tag{4-15}$$

and $-dc/dt = k_{\text{exp}}c = ac^2W$, a result previously obtained from the collision theory of bimolecular reactions with $n = 2$ in Eq. (4-6).

The Hinshelwood theory predicts a straight-line plot for $1/k_{\text{exp}}$ versus $1/P$ with an intercept $1/k_\infty$ and slope

$$\frac{RT}{aN_0W} = \frac{\Gamma(n/2)(M/\pi)^{1/2}e^{Ea/RT}(RT)^{(n-1)/2}}{2N_0\, d^2 E_a^{n/2-1}}$$

The theoretical curve defined by the reciprocal of Eq. (4-14) is calculated as follows. E_a is obtained from a plot of $\ln k_\infty$ versus $1/T$; the high-pressure rate constant k_∞ is measured directly or obtained from extrapolation of the plot of $1/k_{\text{exp}}$ versus $1/P$; the effective number of oscillators $n/2$ is obtained by locating the pressure at which k_∞ begins to fall off; it is assumed that at this pressure the rate of activation is equal to the first-order rate of reaction, that is, $ac^2 W = k_{\text{exp}} c$, a relation which will yield a value of $n/2$ after insertion of the experimental value of E_a and a reasonable value for the elastic-hard-sphere diameter d.

Although the Hinshelwood theory of unimolecular reactions represents an analytical formulation of the time-lag theory of Lindemann, it does not in general account satisfactorily for the observed variation of the rate constant with pressure. This failure may in large measure be attributed to the rather unrealistic assumptions that the activation energy per molecule may be distributed among all or many of the vibrational bonds of the molecule and that excess of energy in these bonds above the activation energy does not increase the probability of reaction. To overcome this difficulty, Kassel [6] proposed a classical theory which is a modification of similar theories previously proposed by Rice and Ramsperger [7]. It is assumed that a molecule must possess at least the activation energy in a single vibrational normal mode as a necessary condition for decomposition and, further, that the excess energy in the molecule above the activation energy enhances the probability of reaction by increasing the chance that the bond in question will possess the necessary threshold energy increment.

If a molecule contains total vibrational energy of magnitude E/N_0 distributed among n quadratic terms associated with $n/2$ vibrational bonds, the temperature-independent probability per unit time for decomposition can be shown to be [6, 7]

$$k(E) = A\left(\frac{E - E_a}{E}\right)^{(n-2)/2} \tag{4-16}$$

where A is a proportionality constant. Of those molecules which have energy E/N_0 per molecule associated with n quadratic terms representing vibrational energy, $[(E - E_a)/E]^{(n-2)/2}$ is the fraction which have at least E_a/N_0 per molecule in two quadratic terms (one vibrational degree of freedom).

The proportionality constant A in Eq. (4-16) is obtained from the following relation, which is valid in the limit of high pressure, where an equilibrium distribution of vibrational energy among the $n/2$ bonds may be assumed:

$$\begin{aligned} c\,dk_\infty &= c k(E)\,dW \\ &= cA\left(\frac{E - E_a}{E}\right)^{(n-2)/2}\left[\left(\frac{E}{N_0}\right)^{(n-2)/2} \frac{e^{-E/RT}}{\Gamma(n/2)(RT/N_0)^{n/2}}\right]\frac{dE}{N_0} \end{aligned} \tag{4-17}$$

Equation (4-17) is the rate of unimolecular reaction at high pressures for those molecules which possess vibrational energy between E/N_0 and $E/N_0 + dE/N_0$. Integration of Eq. (4-17) yields

$$c \int_0^{k_\infty} dk_\infty = k_\infty c = c \int_{E_a}^{\infty} k(E)\,dW = cAe^{-E_a/RT} \tag{4-18}$$

and Eq. (4-16) may therefore be written

$$k(E) = k_\infty e^{E_a/RT} \left(\frac{E - E_a}{E} \right)^{(n-2)/2} \tag{4-19}$$

The derivation of a theoretical relation for k/k_∞ is similar to that in the Hinshelwood theory except that the starting relations are differential equations because $k(E)$ is no longer a constant but a function of energy. The balance at any pressure between activation rate and deactivation rate plus reaction rate is

$$ac^2\,dW = ac\,dc^* + k(E)\,dc^* \tag{4-20}$$

and for molecules with energies between E/N_0 and $E/N_0 + dE/N_0$,

$$dk_{\text{exp}} = \frac{k(E)\,dc^*}{c} \tag{4-21}$$

On substituting dc^* from Eq. (4-20) into Eq. (4-21), the relation for dk_{exp} becomes

$$dk_{\text{exp}} = \frac{k(E)\,dW}{1 + k(E)RT/aN_0P} \tag{4-22}$$

where $P/kT = N_0P/RT$ has replaced c and where dW has been previously given in Eq. (4-8). After the relation in Eq. (4-19) for $k(E)$ and that for dW contained in Eq. (4-8) are introduced into Eq. (4-22), integration leads to the following relation for k_{exp}/k_∞:

$$\frac{k_{\text{exp}}}{k_\infty} = \frac{1}{\Gamma(n/2)(RT)^{n/2}} \int_{E_a}^{\infty} \frac{(E - E_a)^{(n-2)/2} e^{-(E-E_a)/RT}\,dE}{1 + [k_\infty e^{E_a/RT} RT/(aN_0P)][(E - E_a)/E]^{(n-2)/2}} \tag{4-23}$$

Experimental values of k/k_∞ can be compared with Eq. (4-23) by obtaining k_∞ at a given temperature directly from experiment and E_a from $(d \ln k_\infty)/[d(1/T)]$. A preliminary value of n is then selected which is less than the maximum possible value of $6(s - 2)$, where s is the number of atoms in the (nonlinear) decomposing molecule. The hard-sphere collision diameter is estimated from viscosity measurements or similar sources. Now if, in Eq.

(4-23), $E - E_a$ is set equal to a new variable X, then

$$\frac{k_{\exp}}{k_\infty} = D \int_0^\infty \frac{X^{(n-2)/2} e^{-X/RT}\, dX}{1 + [k_\infty e^{E_a/RT} RT/(aN_0 P)][X/(X + E_a)]^{(n-2)/2}}$$

$$= D \int_0^\infty \psi(X, P)\, dX \tag{4-24}$$

where D is a constant which does not depend on X or P. For a selected series of pressures from $P = \infty$ to values such that k/k_∞ is about 0.1, values of the integrand $\psi(X, P)$ are plotted against X as a running variable. The areas under the various curves are normalized to that under the curve for $P = \infty$, which is scaled to a value of unity; that is, $k/k_\infty = 1$ when $P = \infty$. The normalized areas under the other curves then yield directly k/k_∞ at the parametric pressures associated with the curves. As $P \to 0$, it can be shown that Eq. (4-24) reduces to Eq. (4-15). If *reasonable* variations in the initially selected values of n and d can bring the experimental and theoretical curves of k/k_∞ into agreement, it is assumed that the Ramsperger–Rice–Kassel classical model is a possible one for the reaction.

In most cases it is found that the RRK theory provides satisfactory agreement with experiment, so that the extra computational labor associated with its use is not without reward. It is interesting that Hinshelwood's theory would be physically plausible only if a single vibrational bond in the molecule were capable of containing energy E/N_0, where $E/N_0 > E_a/N_0$; i.e., when $n = 2$. If n is set equal to 2 in Eq. (4-23), the integration can be carried out analytically, and the Hinshelwood expression for k/k_∞ is obtained as a special case of the RRK theory!

4-3 QUANTUM THEORY

Kassel's quantum-theory treatment of unimolecular decomposition [6] is probably more realistic than the corresponding classical theory. It has the disadvantage, however, of having an additional adjustable parameter.

For a molecule of s atoms, Kassel assumes that the S oscillators [where $S = 3(s - 2)$] have vibrational energy equivalent to j quanta with a single common frequency v associated with every quantum. The probability that at least m of these quanta will be found in a single selected oscillator is given by

$$\frac{(j - m + S - 1)!\,j!}{(j + S - 1)!\,(j - m)!} \tag{4-25}$$

(This is simply the ratio of the number of ways to distribute j quanta in S oscillators with m quanta in a single oscillator divided by the number of ways to distribute j quanta in S oscillators.) The chance that a molecule in the jth quantum state will undergo unimolecular decomposition in unit time is

$$k_j = A\frac{(j-m+S-1)!j!}{(j+S-1)!(j-m)!} \tag{4-26}$$

By analogy with the classical case, it can be shown that

$$A = k_\infty \exp(E_a/RT) = k_\infty \exp(N_0 mh\nu/RT) \tag{4-27}$$

Similarly, the form of the relation for the rate constant k_{exp} parallels that of the classical model and may be written

$$k_{\text{exp}} = \sum_m^\infty \frac{k_j W_j}{1 + k_j RT/aN_0 P} \tag{4-28}$$

where W_j is the Boltzmann relation for the fraction of molecules in the jth quantum state, namely,

$$W_j = \frac{g_j \exp(-N_0\epsilon_j/RT)}{\sum g_i \exp(-N_0\epsilon_i/RT)}, \qquad \epsilon_j = jh\nu, \quad \epsilon_i = ih\nu \tag{4-29}$$

and where the quantum weights are given by

$$g_j = \frac{(j+S-1)!}{j!(S-1)!}, \qquad g_i = \frac{(i+S-1)!}{i!(S-1)!} \tag{4-30}$$

The summation in Eq. (4-29) yields

$$W_j = \frac{(j+S-1)!}{j!(S-1)!}\exp\left(-\frac{N_0 jh\nu}{RT}\right)\left[1-\exp\left(-\frac{N_0 h\nu}{RT}\right)\right]^S \tag{4-31}$$

If Eqs. (4-26), (4-27), and (4-31) are substituted into Eq. (4-28), we obtain the Kassel quantum-theory expression for k_{exp}/k_∞:

$$\frac{k_{\text{exp}}}{k_\infty} = \left[1-\exp\left(-\frac{N_0 h\nu}{RT}\right)\right]^S \sum_{p=0}^\infty \frac{\dfrac{(p+S-1)!}{p!(S-1)!}\exp\left(-\dfrac{N_0 ph\nu}{RT}\right)}{1+\dfrac{ART}{aN_0 P}\dfrac{(p+m)!(p+S-1)!}{p!(p+m+S-1)!}} \tag{4-32}$$

where p is a running index equal to $j - m$.

The comparison between the prediction of Eq. (4-32) and experimental values of k_{exp}/k_∞ is carried out in a manner similar to that used for the classical case except that additional flexibility in the choice of parameters is afforded by the relation

$$E_a = N_0 mh\nu \tag{4-33}$$

which fixes the value of the product $m\nu$ and not the individual values m and ν. Equation (4-32), like its classical analog, is in agreement with much of the experimental work on unimolecular decompositions.

4-4 TRANSITION-STATE THEORY

Marcus and Rice [8] have formulated a theory of unimolecular decompositions which is based on a transition-state model. The physical assumptions of the theory are essentially the same as those of the RRK theory, but the reaction mechanism is postulated as

$$A + A \rightleftharpoons A^* + A, \qquad A^* \rightleftharpoons A^\ddagger \rightarrow \text{products} \tag{4-34}$$

This approach takes into account the change from energized molecules to molecules in the transition state. The internal rearrangement may be accompanied by large changes in vibrational frequencies. All internal degrees of freedom are included in calculating the rate constants, and quantum and symmetry effects can readily be taken into account.

A nonrigorous semiclassical approach to the theory is adopted here. Consider first an activated complex with a specified energy distribution. If N_ϵ is the energy-level density (number of energy levels per unit energy) of the active molecules (*not* activated complexes), with energy ϵ in the internal degrees of freedom, then $1/N_\epsilon$ can be considered as the average energy between energy levels. The average lifetime of an active molecule in the energy level ϵ is τ_ϵ; that is, $1/\tau_\epsilon$ is the rate constant for decomposition of the activated molecule. We can now invoke the uncertainty principle in the form

$$\tau_\epsilon / N_\epsilon = h \tag{4-35}$$

(A rather detailed argument indicates that h rather than $h/2\pi$ should be used in this statement of the uncertainty principle [9].) In order for reaction to occur, a certain amount of energy ϵ_m must be localized in the reaction coordinate. The number of ways a transition state of total internal energy ϵ can be set up is simply the sum of all the degeneracies of the internal energy states with a positive energy less than or equal to $\epsilon - \epsilon_m$; that is, $\sum_0^{\epsilon-\epsilon_m} g_{\epsilon_i}$, where g_{ϵ_i} is the degeneracy of the energy state ϵ_i. A schematic energy diagram for this model is shown in Fig. 4-1. The probability of decomposition of an activated molecule of energy ϵ in unit time is

$$k(\epsilon) = \sum_0^{\epsilon-\epsilon_m} \frac{g_{\epsilon i}}{hN_\epsilon} \tag{4-36}$$

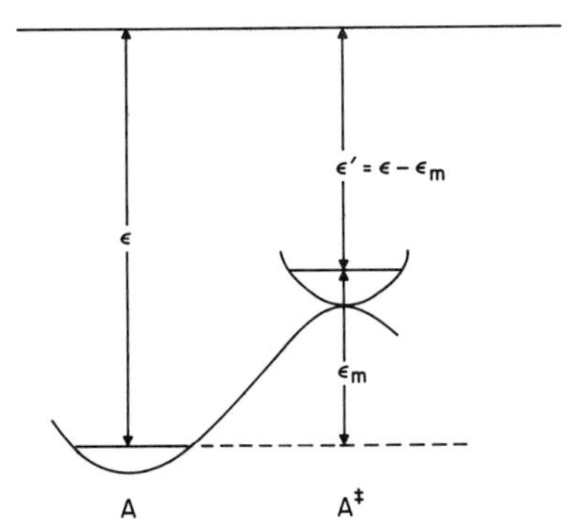

Fig. 4-1 Schematic energy diagram for the RRKM theory of unimolecular decompositions. Here ϵ is the energy of the active molecule, ϵ_m is the minimum energy necessary to reach the ground state of the activated complex, and ϵ' is the energy available for distribution in the internal degrees of freedom of the activated complex.

The fraction of activated molecules with energy between ϵ and $\epsilon + d\epsilon$ is

$$dW_\epsilon = N_\epsilon e^{-\epsilon/kT}(d\epsilon/f) \qquad (4\text{-}37)$$

where f is the partition function of the internal energy states of the reacting molecules. Applying the usual steady-state treatment to molecules with energy in the range $d\epsilon$, relations analogous to Eqs. (4-20) and (4-21) can be used to obtain

$$dk_{\text{exp}} = \frac{1}{hf} \frac{(\sum_0^{\epsilon-\epsilon_m} g_{\epsilon_i})e^{-\epsilon/kT}\, d\epsilon}{1 + [(\sum_0^{\epsilon-\epsilon_m} g_{\epsilon_i})/(achN_\epsilon)]} \qquad (4\text{-}38)$$

which is the analog of Eq. (4-22). Finally, if the minimal energy necessary for reaction is ϵ_m, the experimental rate constant can be written as

$$k_{\text{exp}} = \frac{1}{hf} \int_{\epsilon_m}^{\infty} \frac{(\sum_0^{\epsilon-\epsilon_m} g_{\epsilon_i})e^{-\epsilon/kT}\, d\epsilon}{1 + [(\sum_0^{\epsilon-\epsilon_m} g_{\epsilon_i})/(achN_\epsilon)]} \qquad (4\text{-}39)$$

or

$$k_{\text{exp}} = \frac{e^{-\epsilon_m/kT}}{hf} \int_0^{\infty} \frac{(\sum_0^{\epsilon'} g_{\epsilon_i})e^{-\epsilon'/kT}\, d\epsilon'}{1 + [(\sum_0^{\epsilon'} g_{\epsilon_i})/(achN_{\epsilon_m+\epsilon'})]} \qquad (4\text{-}40)$$

in terms of a new variable ϵ', which is equal to $\epsilon - \epsilon_m$. Equation (4-40), which is the analog of Eq. (4-24), is quite general. However, evaluation of $\sum_0^{\epsilon'} g_{\epsilon_i}$ and N_ϵ is difficult and will not be considered here. A number of approximations have been used, and these are discussed in detail elsewhere [8–10].

It is interesting to examine the high- and low-pressure limits of Eq. (4-40). The high-pressure limit ($c \to \infty$) may be obtained by integrating Eq. (4-40) in the following manner:

$$k_\infty = \frac{e^{-\epsilon_m/kT}}{hf}\left[\int_0^{\epsilon_1} g_{\epsilon_0} e^{-\epsilon'/kT}\,d\epsilon' + \int_{\epsilon_1}^{\epsilon_2} (g_{\epsilon_0} + g_{\epsilon_1}) e^{-\epsilon'/kT}\,d\epsilon' + \int_{\epsilon_2}^{\epsilon_3} (g_{\epsilon_0} + g_{\epsilon_1} + g_{\epsilon_2}) e^{-\epsilon'/kT}\,d\epsilon' + \cdots\right]$$

$$= \frac{e^{-\epsilon_m/kT}}{hf}\left(\int_0^{\infty} g_{\epsilon_0} e^{-\epsilon'/kT}\,d\epsilon' + \int_{\epsilon_1}^{\infty} g_{\epsilon_1} e^{-\epsilon'/kT}\,d\epsilon' + \int_{\epsilon_2}^{\infty} g_{\epsilon_2} e^{-\epsilon'/kT}\,d\epsilon' + \cdots\right)$$

$$= \frac{e^{-\epsilon_m/kT}}{hf}\sum_0^\infty g_{\epsilon_i}\int_{\epsilon_i}^{\infty} e^{-\epsilon'/kT}\,d\epsilon'$$

$$= \frac{kTe^{-\epsilon_m/kT}}{hf}\sum g_{\epsilon_i} e^{-\epsilon_i/kT} = \frac{kTf^\ddagger}{hf} e^{-\epsilon_m/kT} \tag{4-41}$$

since $\sum_0^\infty g_{\epsilon_i} e^{-\epsilon_i/kT}$ is the partition function $f^\ddagger$ of the internal-energy states of the activated complex. Equation (4-41), which is exactly what simple transition-state theory would predict, provides an explicit formulation of the preexponential factor. In the limit of very low pressures, Eq. (4-40) reduces to

$$k = \frac{ac}{f}\int_{\epsilon_m}^{\infty} N_\epsilon e^{-\epsilon/kT}\,d\epsilon \tag{4-42}$$

which is formally the same as the result obtained from simple collision theory, since

$$f^{-1}\int_{\epsilon_m}^{\infty} N_\epsilon e^{-\epsilon/kT}\,d\epsilon$$

is the equilibrium fraction of molecules with energy greater than ϵ_m.

In most cases, in order to conserve angular momentum, it is assumed that the rotational energy levels do not exchange energy with the vibrational degrees of freedom, and the rotational degrees of freedom are then said to be "adiabatic." In this situation, Eq. (4-40) is slightly modified

$$k_{\text{exp}} = \frac{I_r e^{-\epsilon_m/kT}}{f_v h}\int_0^\infty \frac{(\sum_0^{\epsilon'} g_{\epsilon_i}) e^{-\epsilon'/kT}\,d\epsilon'}{1 + [I_r(\sum_0^{\epsilon'} g_{\epsilon_i})/(achN_{\epsilon_m+\epsilon'})]} \tag{4-43}$$

Here I_r is the ratio of the rotational partition function of the activated complex to that of the reactant molecule; f_v is the vibrational partition function of the reactant; and $\sum_0^{\epsilon'} g_{\epsilon_i}$ involves vibrational-energy states only. This theory, called the RRKM theory, satisfactorily accounts for available experimental results (cf. [10–12]).

4-5 CRITIQUE OF THEORIES

Slater [13] has criticized the RRKM theory, mainly on the grounds that the proposed model assumes that the intramolecular transfer of energy between vibrational degrees of freedom is very rapid compared to the rate of reaction. In fact, all theories of unimolecular reactions, aside from Slater's, involve this particular assumption. Slater assumes that the rate of intramolecular vibrational-energy transfer is very slow. He then performs a normal-mode analysis of the reacting molecule, assuming that reaction occurs when the bond to be broken reaches a certain critical length. A formally correct expression for k_∞ is obtained, namely,

$$k_\infty = Ae^{-E_a/RT} \tag{4-44}$$

where $A = (\sum_i^s \alpha_i^2 v_i^2 / \sum_i^s \alpha_i^2)$ and the α_i are amplitude factors characteristic of the vibrational coordinate associated with the frequency v_i. The detailed theory of the pressure dependence of k_{exp} is in reasonable agreement with experiment. Slater's general approach, including quantum-mechanical and transition-state treatments of the problem, has been extensively developed.

Several recent studies suggest that the basic assumption of the RRKM theory of rapid intramolecular vibrational-energy transfer is, in fact, essentially correct. For example, several cases have been well documented where the experimental results are contrary to those predicted from physical assumptions of Slater's theory [14]. Butler and Kistiakowsky [15] studied the isomerization of vibrationally excited methylcyclopropane molecules formed by two different chemical reactions (I).

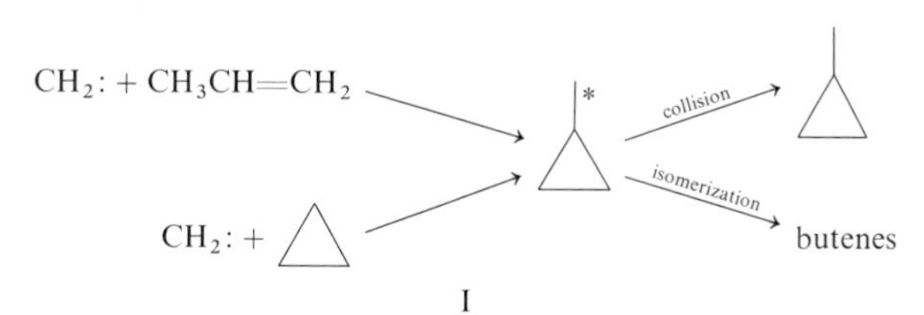

I

In both cases, the methylcyclopropane is produced with more than enough energy for isomerization so that the isomerization occurs unless the methylcyclopropane is stabilized by a collision. The butene fraction was found to have the same composition no matter how the excited molecules were formed. Since the initial energy distributions of the excited molecules formed from the two chemical reactions would be very different, this finding can only be explained by the complete redistribution of energy within the excited molecule before isomerization. Furthermore, the rates can be quantitatively described in terms of a single RRKM model although the total rates of isomerization differ in the two cases because of the differing excess energies of excited molecules. Many other similar examples have been studied, and the results obtained strongly support the concept of a rapid intramolecular energy exchange [10].

On the theoretical side, Thiele and Wilson [16] have considered the rate of energy transfer between vibrational motions of an excited linear triatomic molecule and found that the interchange of energy is so rapid that the concept of normal modes is not applicable. Bunker [17] has employed a fast digital computer to simulate unimolecular decompositions. The decomposition of triatomic molecules was considered by calculating the equations of motion and trajectories of the atoms consistent with a given Hamiltonian and initial state of the molecule. The starting states were selected by a Monte Carlo procedure which chooses random molecular configurations consistent with the Hamiltonian. The results are mainly presented as normalized distributions of lifetimes for molecules belonging to an equilibrium distribution at each energy. (The lifetime is the time required for an individual randomly chosen molecular configuration to evolve into products in the absence of collisions.) These distributions can be ultimately transformed into dissociation rate constants as a function of energy. A number of models were considered for nonrotating triatomic molecules. The conventional normal mode description is very poor for all models studied. Harmonic models give rise to a very large number of nonreactive molecules and produce unrealistic rate constants below a pressure of about 100 atm. The computed results are best described by a theory of the RRK type which includes limited molecular anharmonicity. For rotating anharmonic triatomic molecules, the RRKM theory corresponds well with the computed values of high pressure dissociation rates as a function of energy. Intramolecular energy relaxation is complete in about 10^{-11} sec regardless of molecular complexity, and the lifetime distributions of energized molecules are randomly distributed with respect to dissociation as required by the RRKM theory. In terms of the RRKM theory, the assumption of random lifetimes is embodied in the assertion that the activated molecules (A^*) produce activated complexes ($A^\ddagger$) with a single first-order rate constant; that is, once the molecule is "activated," the probability of decomposition is independent of its initial energy state. Any randomly chosen A^* has an equally good chance of becoming $A^\ddagger$. Nonrandom lifetime distributions are likely only for a few triatomic molecules or under exceptional experimental conditions. Thus the computer results are in excellent agreement with the RRKM theory.

An assumption made in all of the theories discussed is that activation and deactivation occur as single steps as opposed to processes in which energy is gained or lost in a series of steps. The assumption that large amounts of energy are transferred in molecular collisions is called the strong collision hypothesis. In essence, it means that consideration of the detailed collision dynamics is not necessary. The energy transfer process can be studied by determining the energizing efficiency of a variety of different gases. The results obtained indicate that the strong collision assumption is valid for most conventional kinetic studies [10].

4-6 ISOMERIZATION OF METHYL ISOCYANIDE

As a particular example of an unimolecular reaction, we consider the isomerization of methyl isocyanide to acetonitrile, which has been extensively studied by Schneider and Rabinovitch [18]. The reaction is

$$CH_3NC \rightarrow CH_3CN$$

and the experimental results at several temperatures are shown in a plot of $\log k_{exp}/k_\infty$ versus $\log P$ in Fig. 4-2, where the lines have been calculated

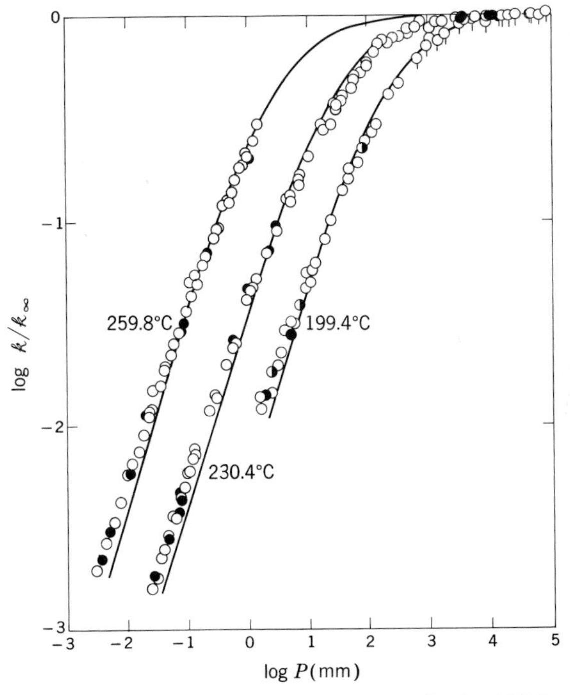

Fig. 4-2 Pressure dependence of unimolecular rate constants for CH_3NC: $\log k_{exp}/k_\infty$ versus $\log P$ at 199.4, 230.4, and 259.8°. For clarity the 260° curve is arbitrarily displaced by one $\log P$ unit to the left in the figure while the 200° curve is displaced the same distance to the right; actually, both these curves would almost coincide with the 230° curve. Vertical marks have been placed under the 200° high-pressure points to assist in distinguishing them from the 230° data. The curves represent the calculated results, adjusted on the pressure axis to coincide with the experimental points at $\log k_{exp}/k_\infty = -1$. [Adapted from F. W. Schneider and B. S. Rabinovitch, *J. Am. Chem. Soc.* **84**, 4215 (1962), Fig. 1. Copyright 1962 by the American Chemical Society. Reproduced by permission of the copyright owner.]

according to Eq. (4-43). In order to make the necessary calculations a model of the transition state must be assumed, since the symmetry numbers, moments of inertia, and vibrational energy levels of the activated complex must be known. In the present case, reasonable estimates of all the necessary parameters were made, and the sums and integrals were evaluated with a computer. As previously indicated, the RRKM theory includes all vibrational energy levels rather than an effective number of oscillators, as in the RRK classical theory. Since vibrational frequencies of the transition-state complex must be assigned, some arbitrariness still exists, but the case being considered is sufficiently tractable so that the agreement between theory and experiment is a convincing indication of the validity of the theory. A detailed consideration of anharmonicity effects indicated that the fit of the theory to the experimental data could be improved. Because of the uncertainty in the collision diameter, this more sophisticated analysis was not made.

The strong collision hypothesis has been carefully studied for this reaction, and a plot of the efficiency of energization relative to that of CH_3NC for a variety of gases is shown in Fig. 4-3 [10, 19]. The efficiency of energization reaches a constant value for energizing molecules with greater than 8–10 atoms, and it is reasonable to assume that this limiting value is the strong collision limit. Thus for pure CH_3NC the strong collision assumption is valid.

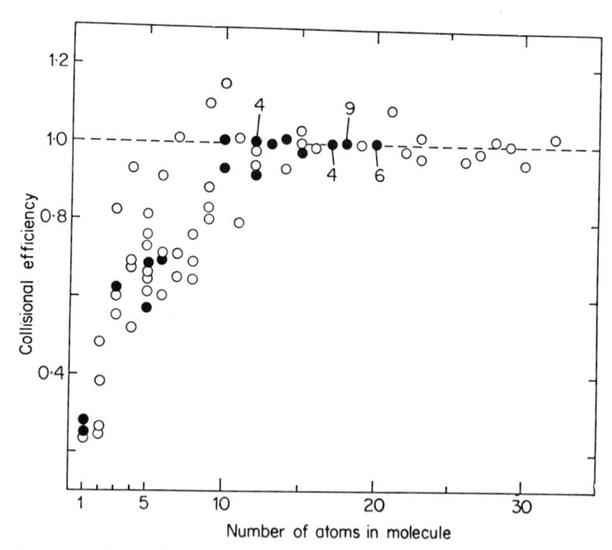

Fig. 4-3 Collisional efficiencies for energization of methyl isocyanide at 280.5°C by various gases relative to CH_3NC. Filled points represent coincident points for the stated number of gases (two if not stated). [Adapted from P. J. Robinson and K. A. Holbrook, "Unimolecular Reactions," Fig. 10-7. Copyright 1972, Wiley (Interscience), New York. Used by permission of John Wiley & Sons, Inc. The data were obtained from Chan *et al.* (19)]

The efficiency also was found to be dependent on the intermolecular potential energy of interaction between the energizer and the reactant.

In conclusion, it appears that the general features of unimolecular reactions are reasonably well described by microscopic treatments involving various molecular parameters. Since there is still room for improvement in the agreement between theory and experiment, it is not unlikely that we shall see further refinements of existing theories or new theoretical approaches to the problem along with additional experimental work. A more extensive discussion of unimolecular decompositions is given by Robinson and Holbrook [10].

Problems

4-1 The water molecule is nonlinear. At 1000°K what is the probability of supplying 40 kcal/mole of activation energy from the vibrational degrees of freedom of the water molecules compared with the probability of supplying it from the component of relative velocity parallel to the line of centers of pairs of water molecules approaching a collision (cf. Problem 2-3)?

4-2 Show that if n is set equal to 2 in the RRK theory [Eq. (4-23)], the expression obtained for k_{exp}/k_{∞} is identical with that found with the Hinshelwood theory [Eq. (4-14)].

4-3 A phenomenological analogy of the RRKM theory for an irreversible unimolecular decomposition in the gas phase can be written as

$A + A \rightarrow A^* + A$	k_1 (activation by collision)
$A^* + A \rightarrow A + A$	k_2 (deactivation by collision)
$A^* \rightarrow X^*$	k_3 (transfer of activation energy within A* to proper bond)
$X^* \rightarrow A^*$	k_4 (return of activation energy to all bonds within A*)
$X^* \rightarrow$ decomposition products	k_5 (slow, rate-determining step)

(a) Find the general expression for the rate of decomposition, $-d(A)/dt$.

(b) What are the limiting expressions for $-d(A)/dt$ for every high concentrations of A and very low concentrations of A?

4-4 The unimolecular decomposition of azomethane, $CH_3—N=N—CH_3$, to ethane and nitrogen has been extensively studied as a function of temperature and pressure. The high-pressure rate constant is

$$k_{\infty} = 3.13 \times 10^{16} e^{-52,440/RT}$$

Some typical experimental values of the rate constants at different pressures and 603°K are [20]:

P (mm):	56.46	33.28	16.21	1.510
k ($sec^{-1} \times 10^5$):	213	176	145	69

(a) Using the classical RRK theory, calculate values of k/k_{∞} at the following pressures: $P = \infty$, $P = 100$ mm, $P = 10$ mm, and $P = 1$ mm. Present your results as a plot of $\log(k/k_{\infty})$ versus P. In performing the necessary graphical integrations, assume $n = 24$ and $d = 13 \times 10^{-8}$ cm. Include the experimental points on your graph.

(b) Carry out the same calculations using the Kassel quantum theory. Assume $m = 25$, $S = 18$, and $d = 4 \times 10^{-8}$ cm. Compare these results with experiment and the results of part (a).

References

1. M. W. Perrin, *Ann. Phys. (Paris)* **11**, 5 (1919).
2. F. Daniels and E. H. Johnston, *J. Am. Chem. Soc.* **43**, 73 (1921).
3. F. A. Lindemann, *Trans. Faraday Soc.* **17**, 598 (1922).
4. G. N. Hinshelwood, *Proc. Roy. Soc. London Ser. A* **113**, 230 (1926).
5. S. W. Benson, "Foundations of Chemical Kinetics," p. 223. McGraw-Hill, New York, 1960.
6. L. S. Kassel, *J. Phys. Chem.* **32**, 225, 1065 (1928); "Kinetics of Homogeneous Gas Reactions," pp. 93–113. Holt, New York, 1932.
7. O. K. Rice and H. C. Ramsperger, *J. Am. Chem. Soc.* **49**, 1617 (1927); **50**, 612 (1928); O. K. Rice, *Proc. Nat. Acad. Sci. USA* **14**, 114, 118 (1928).
8. R. A. Marcus and O. K. Rice, *J. Phys. Chem.* **55**, 894 (1951); R. A. Marcus, *J. Chem. Phys.* **20**, 359 (1952).
9. O. K. Rice, *J. Phys. Chem.* **65**, 1588 (1961).
10. P. J. Robinson and K. A. Holbrook, "Unimolecular Reactions." Wiley (Interscience), New York, 1972.
11. G. M. Wieder and R. A. Marcus, *J. Chem. Phys.* **37**, 1835 (1962).
12. O. K. Rice, *in* "Transfert d'énergie dans les gaz" (R. Stoops, ed.), Inst. Int. Chim. (Douzieme Conseil Chim., 1962), 12th Chemistry Solvay Conference, pp. 17–99. Wiley (Interscience), New York, 1962.
13. N. B. Slater, "Theory of Unimolecular Reactions." Cornell Univ. Press, Ithaca, New York, 1959.
14. E. Thiele and D. J. Wilson, *Can. J. Chem.* **37**, 1035 (1959); E. W. Schlag and B. S. Rabinovitch, *J. Am. Chem. Soc.* **82**, 5996 (1960); B. S. Rabinovitch and K. W. Michel, *ibid.* **81**, 5065 (1959).
15. J. N. Butler and G. B. Kistiakowsky, *J. Am. Chem. Soc.* **82**, 759 (1960).
16. E. Thiele and D. J. Wilson, *J. Chem. Phys.* **34**, 1256 (1961).
17. D. L. Bunker, *J. Chem. Phys.* **37**, 393 (1962); **40**, 1946 (1964).
18. F. W. Schneider and B. S. Rabinovitch, *J. Am. Chem. Soc.* **84**, 4215 (1962).
19. S. C. Chan, B. S. Rabinovitch, J. T. Bryant, L. D. Spicer, T. Fujimoto, Y. L. Lin, and S. P. Pavlow, *J. Phys. Chem.* **74**, 3160 (1970).
20. H. C. Ramsperger, *J. Am. Chem. Soc.* **49**, 912, 1495 (1927).

CHAPTER 5

CHEMICAL REACTIONS IN MOLECULAR BEAMS

5-1 INTRODUCTION

The ideal experimental approach to dynamical problems in chemical or physical systems lies along the molecular, or microscopic, rather than the bulk, or macroscopic, route. For reaction kinetics, advantages which might in principle be obtained from suitable molecular experiments are selection of a specified internal-energy state and translational-energy state or a specified range of such states for the reactants; restriction of reactive encounters to single collisions from which products emerge in directions and in energy states which are not masked or altered by subsequent collisions; identification of the states of the products by application of classical or quantum conservation laws to a reacting system; and direct location of energy thresholds for chemical reactions rather than use of a model to derive energies of activation which are complicated averages involving a large number of widely varying thresholds. If these advantages can, in fact, be realized, the traditional macroscopic rate constant, which depends on temperature, can be resolved into microscopic components which are functions of molecular variables, such as momenta, and quantum numbers characterizing internal states. This resolution has been formulated by Ross and Greene [1] for the special case of a bimolecular reaction in the gas phase; it affords a good illustration of the increase in fundamental information that may be expected from experiments of a microscopic character.

5-2 THEORY

Consider the reaction between A and B to form C and D and let A(i), B(j), C(i'), and D(j') denote isolated molecules of reactants and products in internal quantum states characterized by i, j, i', and j'. Then the rate constant in the forward direction k_f for the reaction

$$A(i) + B(j) \rightleftharpoons C(i') + D(j') \tag{5-1}$$

may be expressed as

$$k_f(i', j', i, j) = \int \cdots \int \frac{p}{m^*} \sigma_R(i', j', i, j, p, \Omega) F_i(p_A) F_j(p_B)\, d\Omega\, dp_A\, dp_B \tag{5-2}$$

where p is the initial relative momentum, m^* the reduced mass of A and B, σ_R the *differential* cross section per unit solid angle for chemical reaction, Ω the solid angle of scattering, and $F_i(p_A)$ the momentum distribution function of A(i). Equation (5-2) may be readily inferred from the general collision relations developed in Chapter 2. If both sides of the equation are multiplied by $c_{A(i)}c_{B(j)}$, we obtain the number of collisions in unit volume and unit time between isolated molecules of A(i) and B(j) which lead to formation of C(i') and D(j'). If Eq. (2-13), with κ set equal to unity, is integrated over the appropriate values of the variables, the resultant relation for Z may also be interpreted as the collision frequency for reaction in unit volume if the variables in Eqs. (2-13) and (5-2) are related in the following manner:

$$Z = k_f(i', j', i, j) c_{A(i)} c_{B(j)}, \qquad 2\pi b\, db = \sigma_R(i', j', i, j, p, \Omega)\, d\Omega$$

$$v_r = |p/m^*|, \qquad dc_1\, dc_2 = c_{A(i)} F_i(p_A)\, dp_A\, c_{B(j)} F_j(p_B)\, dp_B$$

The total cross section $S_R(i', j', i, j, p)$ for the chemical reaction in Eq. (5-1) may be considered as the appropriate inelastic analog of the total elastic scattering cross section given earlier in Eq. (2-15). The relation between the total and differential cross sections for chemical reaction is

$$S_R(i', j', i, j, p) = \int_0^{4\pi} \sigma_R(i', j', i, j, p, \Omega)\, d\Omega \tag{5-3}$$

If classical mechanics is applicable to the reactive collision between A(i) and B(j), Eq. (5-3) may be written as

$$S_R(i', j', i, j, p) = \int_0^{\infty} \int_0^{2\pi} P(i', j', i, j, p, b, \eta) b\, db\, d\eta \tag{5-4}$$

where $P(i', j', i, j, p, b, \eta)$ is the probability of reaction, b the impact parameter previously defined in Chapter 2 in the discussion of elastic scattering, and η the azimuthal angle of scattering (not to be confused with the polar relative angle of deflection χ, mentioned in Chapter 2, which is a function of b [2]). Since the distribution functions in Eq. (5-2) may almost always be taken as the equilibrium or Maxwellian distributions [3], the *temperature-dependent* rate constant for the system of isolated molecules described by Eq. (5-1) becomes a weighted average of the total cross sections for chemical reaction

$$k_f(i', j', i, j, T) = \frac{1}{(\pi m^*)^{1/2}} \left(\frac{2}{kT}\right)^{3/2} \int_0^{\infty} S_R(i', j', i, j, \epsilon) \epsilon e^{-\epsilon/kT}\, d\epsilon \tag{5-5}$$

where the initial kinetic energy of relative motion $\epsilon = p^2/2m^*$, instead of the initial relative momentum, has been used as the variable in the distribution functions. Finally, if we remove from Eq. (5-1) the restriction that A(i), B(j), C(i'), D(j') represent *isolated* molecules, each in a single quantum state, and regard the equation as applicable to a mixture of reactants and products in all possible quantum states, the relation for the rate constant in the forward direction is

$$k_f(T) = \left(\frac{1}{\pi m^*}\right)^{1/2} \left(\frac{2}{kT}\right)^{3/2} \int_0^\infty S_R(\epsilon)\epsilon e^{-\epsilon/kT}\, d\epsilon \tag{5-6}$$

where

$$S_R(\epsilon) = \sum_{i,j} x_{A(i)} x_{B(j)} \left[\sum_{i',j'} S_R(i', j', i, j, \epsilon)\right] \tag{5-7}$$

The quantities $x_{A(i)}$ and $x_{B(j)}$ are the fractions of A and B in given quantum states. If it is assumed that these represent equilibrium fractions (with respect to $x_{C(i')}$ and $x_{D(j')}$ at all times, then k_f in Eq. (5-6) does not vary with time.

It is interesting to note that for the simple model of repelling hard spheres, which can change from reactants to products if the energy along the line of centers on contact exceeds a threshold value ϵ_a so that

$$\begin{aligned} S_R(\epsilon) &= \pi d_{12}^2(1 - \epsilon_a/\epsilon) \qquad \text{for} \quad \epsilon > \epsilon_a \\ S_R(\epsilon) &= 0 \qquad \text{for} \quad \epsilon < \epsilon_a \end{aligned} \tag{5-8}$$

Eq. (5-6) yields

$$k_f(T) = \pi d_{12}^2 \left(\frac{8kT}{\pi m^*}\right)^{1/2} e^{-\epsilon_a/kT} \tag{5-9}$$

Equation (5-9) is identical with the relation obtained from collision theory, Eq. (2-33) for $\kappa = 1$ (unlike reactants), and for a reaction whose steric factor p is unity. It is also worth noting that the synthesis of the macroscopic rate constant $k_f(T)$ in Eq. (5-6) from the microscopic quantities in Eqs. (5-2) to (5-5) strongly parallels portions of the rigorous derivation of the macroscopic transport coefficients, e.g., those of viscosity, thermal conductivity, or diffusion [4, 5].

5-3 GENERAL EXPERIMENTAL BACKGROUND

Before discussing specific cases of reactive scattering of molecular beams, we should understand how such beams are formed and how their intensities may be quantitatively measured. As we describe the rather specialized techniques and procedures, the special suitability of beam studies for elucidating microscopic details of chemical dynamics should become apparent.

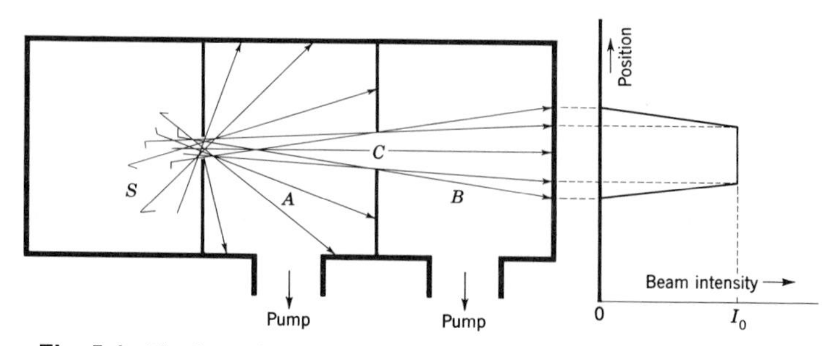

Fig. 5-1 The formation of a beam of molecules. (Adapted from Smith [8].)

One of the fundamental postulates of the kinetic theory of gases is that the motions molecules execute between collisions are very nearly rectilinear. Thus, if the density is reduced to the point where intermolecular collisions are relatively infrequent, the rectilinear paths become quite long, and molecules can be collimated into rays or beams, according to procedures of geometric optics. This was first demonstrated in 1911 by Dunoyer [6], who used the apparatus shown schematically in Fig. 5-1. The beam, whose intensity profile is included in the figure, was produced by filling the source chamber *S* with vapor from heated metallic sodium. Since foreign gas had been previously removed from the apparatus, the temperature in *S* could be controlled to ensure that the mean free path of the sodium vapor was large compared to dimensions of the source aperture between *S* and *A*. Under these conditions molecules in the source chamber which made collisions within a mean free path from the aperture and which had velocity vectors directed toward it passed through the aperture into the intermediate chamber *A*, where they traveled in straight lines until they collided with the walls. By placing a second aperture *C* coaxially with the first, a portion of the effusive rays in *A* was collimated to form a molecular beam in *B*. Since *A* and *B* were highly evacuated, the only intermolecular collisions in *A* and *B* occurred in the relatively infrequent cases when faster molecules overtook slower ones moving along the same path. Thus the beam in *B* was essentially unidirectional and collision-free. Dunoyer found that the deposit of sodium on the cooled end of *B* had the dimensions and form predicted by geometric optics. For example, the deposit had a relatively intense inner core, the umbra, and a less intense outer portion, the penumbra. In high vacuum, objects introduced into the beam path produced well-defined shadows, and introduction of inert gas into *B* diminished the sharpness of the beam profile, a result to be expected if attenuation of the beam was produced by collisions between sodium particles and molecules of the foreign gas.

A schematic representation of a simple beam apparatus which could be used for preliminary experiments on reactive scattering is shown in Fig. 5-2.

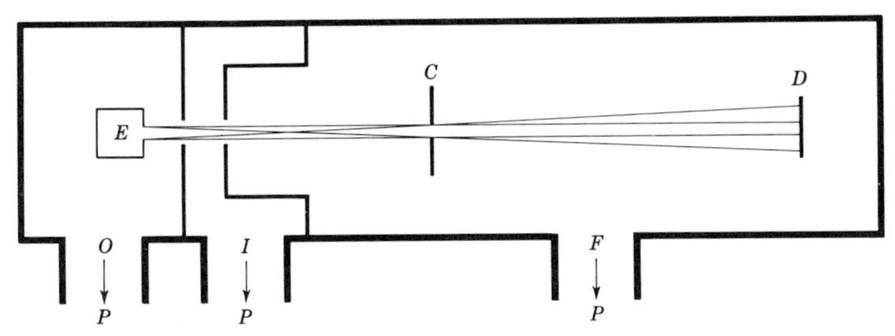

Fig. 5-2 The schematic arrangement of a molecular-beam apparatus. (Adapted from Smith [8].)

The elements of Dunoyer's simple apparatus, which demonstrated that molecular beams could be formed, are readily discernible. The source chamber *O* usually contains an oven *E*, in which condensed beam material may be vaporized or into which low-pressure gas may be introduced. A pump is used to reduce the background pressure in *O*, so that molecules which effuse from the aperture in *E* pass through the source aperture with no diminution in intensity other than that due to the solid angle defined by the dimensions of the apertures in *E* and *O* and the distance between them. An isolation chamber *I* is evacuated with a separate pump for the primary purpose of reducing the background pressure in the reaction chamber *F* to the lowest pressure consistent with the speed of the three pumps which constitute the differential pumping system of the apparatus, the dimensions of the apertures in *E*, *O*, and *I*, and the pressure of gas or vapor in *E*. The collimating aperture *C* in the observation chamber and the source aperture determine the dimensions of the beam, whose flux (particles per unit time) or flux density (flux per unit area) is measured by a suitable detector *D*. In most molecular-beam apparatuses the various apertures are slits, which produce ribbon-shaped beams of narrow width. The intensity of such beams is comparable to that in beams of circular cross section, and the narrow width is helpful in observing small deflections produced by elastic or reactive scattering, or both. Detailed descriptions of a variety of experimental arrangements for the production and detection of molecular beams are given by Fraser [7], Smith [8], and Ramsey [9].

The theoretical flux density or intensity of the molecular beam at the detector plane is readily calculated from simplified-kinetic-theory relations for effusive flow through apertures of finite area but zero length [10]. If we assume that thermal equilibrium at temperature T exists in the source, out of c molecules per unit volume, the number dc moving with speeds between v and $v + dv$ is given by

$$dc = cf(v)\,dv \tag{5-10}$$

where $f(v)\,dv$ is the Maxwellian distribution function for speeds, namely,

$$f(v) = \left(\frac{m}{2\pi kT}\right)^{3/2} 4\pi v^2 \exp\left(-\frac{mv^2}{2kT}\right) dv \tag{5-11}$$

Since the molecules are moving equally in all directions, the number that are moving in directions lying in the solid angle $d\Omega$ is simply $dc\,d\Omega/4\pi$. If $d\Omega$ makes an angle θ with the normal to the source aperture, the number of molecules with speeds in the range dv effusing from unit area of this aperture in unit time is

$$d\Gamma = dc\,d\Omega\,v\,\frac{\cos\theta}{4\pi} = c\left(\frac{m}{2\pi kT}\right)^{3/2} v^3 \exp\left(-\frac{mv^2}{2kT}\right)\cos\theta\,dv\,d\Omega \tag{5-12}$$

Equation (5-12) describes the distribution in direction of molecules that cross unit area of a plane in unit time.

If we are interested in the axial intensity of the beam as measured on a detector opposite the source aperture at a distance r which is large with respect to the aperture dimension, Eq. (5-12) is integrated over all values of v from zero to infinity after setting $\cos\theta = 1$ and $d\Omega = a/r^2$, where a is the area of the source aperture. The result is

$$I = c\bar{v}a/(4\pi r^2) \tag{5-13}$$

where $\bar{v}$, the average speed of molecules in the source, is equal to $(8kT/\pi m)^{1/2}$. At the very low pressures in the source the molecular density c may be set equal to P/kT to obtain a relation for the beam intensity in terms of directly measurable quantities

$$I = Pa/[\pi r^2(2\pi mkT)^{1/2}] \tag{5-14}$$

If cgs units are used in Eq. (5-14), the axial intensity will have units of (particles) $\text{cm}^{-2}\ \text{sec}^{-1}$.

Since unimolecular decompositions or rearrangements cannot occur without collisions (cf. Chapter 4), thermal reactions cannot be observed in a single molecular beam. In fact, the unsuccessful attempts to observe dissociation of iodine [11], internal racemization of pinene [12, 13], and dissociation of nitrogen pentoxide [14] in molecular beams were major experimental refutations of the radiation hypothesis of unimolecular decompositions [15]. The use of two intersecting beams or a beam traversing a cone of molecules effusing from an orifice permits the study of bimolecular reactions under the advantageous conditions associated with beam kinetics. The first studies of this sort were made by Kröger [11], who attempted to measure reactions occurring at the intersection of a beam of cadmium with a beam of iodine or sulfur. Unfortunately, a combination of difficulties (insufficient intensity, poor definition of the beams, low yield associated with appreciable activation energies, and the likelihood of ternary collisions if CdI_2 were to be formed) led to inconclusive results.

Revival of interest in the use of crossed-beam experiments for studying chemical reactions appears to have been catalyzed by the rediscovery of the famous dilute-flame reactions of Polanyi [16]. Many of the systems studied by Polanyi involved bimolecular reaction of alkali metals with halogen compounds, and two factors made these systems particularly suitable for crossed-beam studies. The most important factor was the availability of an extremely sensitive detector almost uniquely suitable for reactions involving atoms of alkali metals or compounds of alkali metals. The detector, known as a surface-ionization detector, depends upon an effect first reported by Langmuir and Kingdon [17]. They found that every cesium atom striking a tungsten wire heated above 1200°K gave up an electron to the wire and returned to the gas phase as a positive ion. If the wire is surrounded by a coaxial cylinder at negative potential, the saturation positive-ion current to the cylinder is a direct measure of the number of atoms striking the wire per second. The effect is observed with potassium and rubidium and can be extended to sodium and lithium by using an oxygenated tungsten wire. In general, surface ionization is observed when the ionization potential of the atom or molecule striking the filament is smaller than the work function of the pure or specially prepared metal comprising the filament. The discovery of Langmuir and Kingdon was applied to the detection of molecular beams by Taylor [18] in Stern's laboratory in Hamburg. The second factor which made many of Polanyi's reactions attractive for beam studies was their extremely large "effective" cross sections, which may be roughly identified with $\pi d_{12}^2 e^{-\epsilon_a/kT}$ of Eq. (5-9). These favorable cross sections are for the most part a consequence of the very low energies of activation associated with the reactions. Although the surface-ionization detector is one of the most sensitive known, the extremely low density of reactants in the volume defined by intersecting beams puts a strain on the task of making accurate intensity measurements of the products of reaction. It is for this reason that reactions with low activation energies are most welcome.

Recent experimental advances have improved both the mode of production of beams and the detection of crossed beam reaction products [19]. For permanent gases beams can be produced by a supersonic expansion of the gas from a source chamber at high pressure, often greater than 100 Torr, through a pinhole nozzel into a vacuum. The collision frequency is high in the nozzel and a short distance beyond, but the collision frequency becomes very low further downstream as a true molecular beam is formed. The collisions in the initial flow bring the molecules to a much more uniform velocity than is found with a thermal effusion source, and a much higher beam intensity is obtained. Also the translational energy of the beam can be controlled easily. The most generally used type of detector is a mass spectrometer: it has great sensitivity and is quite universal in its applicability. Laser induced fluorescence also has proven to be useful for the detection of products because of its great sensitivity. With this method a laser is used to excite the product molecule to a stable excited electronic state, and the product molecule is detected by observing the subsequent emission (fluorescence) from the excited state. The intense research activity in molecular beams indicates that technological improvements will continue to be made rapidly.

5-4 DYNAMICS OF COLLISIONS

As previously discussed in Chapter 2, in a binary collision the conservation laws of energy and linear and angular momentum are often particularly helpful in understanding microscopic features of both reactive and nonreactive scattering [1, 19–22]. If E is the initial kinetic energy of relative motion of the reactants, Z the associated internal energy (rotational, vibrational, or electronic), and $\Delta D_0{}^0$ the difference in dissociation energy of products and reactants (measured from zero-point energy levels), then the final kinetic energy of relative motion of the products E' and the internal energy of the products Z' are given by

$$E' + Z' = E + Z - \Delta D_0{}^0 \tag{5-15}$$

The center-of-mass velocity vector $\mathbf{v}_c$, also called the centroid, is defined in terms of the initial-velocity vectors of the reactants and their respective masses by the relation

$$\mathbf{v}_c = \frac{m_1\mathbf{v}_1 + m_2\mathbf{v}_2}{m_1 + m_2} \tag{5-16}$$

Since it can be shown [23] that the kinetic energy of the total mass moving with the velocity of the center of mass, $\frac{1}{2}(m_1 + m_2)v_c{}^2$, is constant, this term has been omitted from both sides of Eq. (5-15). If the vectors $\mathbf{v}_1$ and $\mathbf{v}_2$ are plotted with a common origin, as shown in Fig. 5-3, the initial relative velocity, $\mathbf{v}_r = \mathbf{v}_1 - \mathbf{v}_2$, intersects $\mathbf{v}_c$ at a point determined by the relations

$$\mathbf{v}_1 - \mathbf{v}_c = m_2\mathbf{v}_r/m \tag{5-17}$$

$$\mathbf{v}_2 - \mathbf{v}_c = -m_1\mathbf{v}_r/m \tag{5-18}$$

where for brevity $m_1 + m_2 = m = m_3 + m_4$, with m_3 and m_4 representing the masses of the products.

From Eqs. (5-17) and (5-18) it is apparent that at large internuclear separations of the particles, an observer moving with the center of mass sees the reactants approach with velocities inversely proportional to the sum of their masses and parallel to the initial-relative-velocity vector. The recoil velocities with which the products recede from the moving center of mass are given by

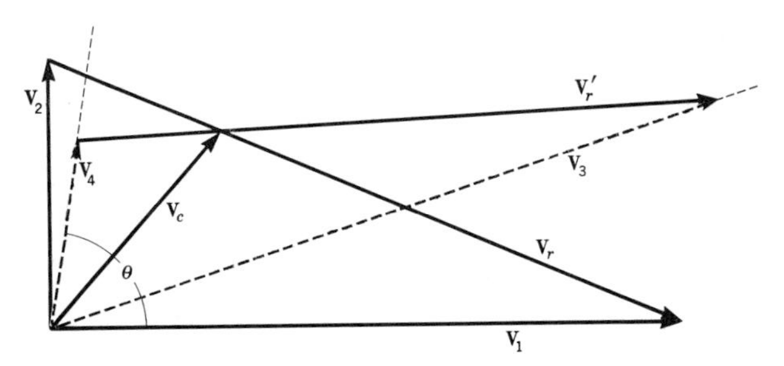

Fig. 5-3 Vector diagram of initial and final velocities of a collision. (Adapted from Ross and Green [1].)

similar relations

$$\mathbf{v}_3 - \mathbf{v}_c = m_4 \mathbf{v}_r'/m \tag{5-19}$$

$$\mathbf{v}_4 - \mathbf{v}_c = -m_3 \mathbf{v}_r'/m \tag{5-20}$$

where the final relative velocity, $\mathbf{v}_r' = \mathbf{v}_3 - \mathbf{v}_4$, may take any direction in space and may therefore be considered to pivot around the end of $\mathbf{v}_c$. Figure 5-3 is a vector diagram of initial and final velocities in a binary collision involving mutually perpendicular initial-velocity vectors. For the situation illustrated in the figure, θ is the laboratory angle, measured from $\mathbf{v}_1$, in the plane of the intersecting reactant beams, into which the product molecule, with velocity $\mathbf{v}_4$, is scattered. In the case of an elastic collision, where $Z' = Z$ and $\Delta D_0{}^0 = 0$, we have the familiar result that $E' = E$ or its equivalent, $|\mathbf{v}_r'| = |\mathbf{v}_r|$. For inelastic collisions, however, the magnitude of $\mathbf{v}_r'$ is restricted to $(2E'/m^{*\prime})^{1/2}$, where $m^{*\prime}$ is the reduced mass of the products, $m_3 m_4/m$, and E' is subject to the condition specified in Eq. (5-15).

From Eq. (5-20) it can be seen that for $m_3 \ll m_4$, so that $m_3/m \ll 1$, and for E' not much larger than E, that is, $|\mathbf{v}_r'| \sim |\mathbf{v}_r|$, the velocity vector $\mathbf{v}_4$ may be confined to a small cone around $\mathbf{v}_c$, so that the product molecule, of mass m_4, will be found near the centroid. Although the intensity of the product is increased by this concentration in a narrow range of angle around the centroid, the problem of improving experimental angular resolution to obtain accurate differential cross sections is correspondingly increased. At the other extreme, if in Eq. (5-20) the magnitude of $m_3 \mathbf{v}_r'/m$ is greater than that of $\mathbf{v}_c$, the product m_4, with velocity $\mathbf{v}_4$, may appear at any laboratory angle. A more detailed analysis of the collision dynamics on which these conclusions are based has been given by Herschbach *et al.* [19–22].

The relations derived thus far have been obtained from conservation of total energy and conservation of linear momentum. In general, however, the total angular momentum is also a collisional invariant, and additional microscopic information can be obtained from this particular conservation law. Before collision, the initial angular momentum of the system is a vector sum of the orbital angular momentum $\mathbf{L} = m^* \mathbf{v}_r b$, where b is the impact parameter, and the rotational angular momenta $\mathbf{J}_1$, $\mathbf{J}_2$ of the reactants. The conservation law for total angular momentum may be written

$$\mathbf{L}' + \mathbf{J}' = \mathbf{L} + \mathbf{J} \tag{5-21}$$

where $\mathbf{J}' = \mathbf{J}_3 + \mathbf{J}_4$ and $\mathbf{J} = \mathbf{J}_1 + \mathbf{J}_2$. For the special case of elastic scattering in a central potential, $\mathbf{J}' = \mathbf{J}$ so that $\mathbf{L}' = \mathbf{L}$. The initial-orbital-angular-momentum vector must be perpendicular to the initial relative velocity, but the initial rotational angular momentum is distributed uniformly in space, as shown in Fig. 5-4a. When, as is frequently the case, $\mathbf{L} \gg \mathbf{J}$ and $\mathbf{L}' \gg \mathbf{J}'$, the final relative-velocity vector $\mathbf{v}_r'$ is again almost perpendicular to $\mathbf{L}$. However, even if it is assumed that the final-relative-velocity vectors are uniformly distributed about $\mathbf{L}$, one may expect to observe an anisotropy in the distribution of scattered product. This can be visualized with the aid of Fig. 5-4b, in which the final-relative-velocity vectors are represented, for a particular $\mathbf{v}_r$ and $\mathbf{L}$, as uniformly spaced radii in the circle perpendicular to $\mathbf{L}$. Now for a given $\mathbf{v}_r$, all values of $\mathbf{L}$ uniformly distributed about $\mathbf{v}_r$ are possible, and the distribution of final-relative-velocity vectors $\mathbf{v}_r'$ is obtained by rotation of Fig. 5-4b about $\mathbf{v}_r$, as shown. In the sphere generated by this rotation, the distribution of $\mathbf{v}_r'$ vectors will be denser along the $\mathbf{v}_r$ axis than near the plane produced by rotation of the orbital momenta [20].

A marked change may occur in the orbital angular momentum during a collision if the ratio of the reduced mass of the reactants to that of the products is greatly different from unity. A corresponding change, as required by Eq. (5-21), must accordingly occur in the rotational angular momentum. For example, if the reduced mass of the products $m^{*\prime}$ is much less than that of the reactants m^* and if $\mathbf{L}$ is large, the products must be in highly excited rotational states ($\mathbf{J}' \gg \mathbf{L}'$) with the orientation of the rotational-momentum vectors approximately parallel to the initial total angular momentum [23].

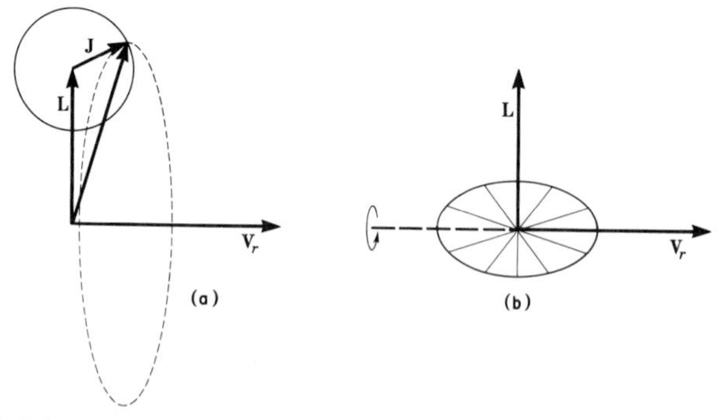

Fig. 5-4 (a) Orientation of initial-angular momentum vectors. (b) Distribution of recoil vectors. (Adapted from Herschbach [20].)

5-5 REACTION OF K WITH HBr

The first significant experiment on reactive scattering of crossed thermal beams was the study of

$$K + HBr \rightarrow H + KBr \tag{5-22}$$

by Taylor and Datz [24]. In terms of beam experiments from which detailed microscopic information can be deduced, theirs may be regarded as a fundamental pioneering study in spite of the fact that their first interpretation of the results was improved in a reinterpretation by Datz *et al.* [21].

Well-collimated beams of K, effusing from ovens between 541 and 837°K, intersected, at 90°, noncollimated cones of HBr effusing from sources between 373 and 460°K. Surface-ionization detectors were used to monitor the fluxes of K and KBr. It was possible to distinguish between the two potassium species, since both K and KBr are surface-ionized by tungsten, whereas platinum surface-ionizes K but is virtually ineffective for KBr [25]. By taking the difference between the positive-ion currents produced by the two surface-ionization detectors, the flux of product was measured as a function of angle with respect to the K beam in the laboratory plane, the plane defined by the direction of the K beam and the normal to the oven from which HBr effused. Various ratios of temperatures in the two ovens were used, so that the results could be analyzed to obtain a threshold energy for the total reaction cross section $S_R(\epsilon)$ of Eq. (5-7).

In this system the mass of the product H is very much less than that of the product KBr, and the final kinetic energy of relative motion of the products E' is not likely to be much different from the initial kinetic energy of the reactants, since $\Delta D_0{}^0$ for the reaction as written in Eq. (5-22) is small, -4.2 kcal/mole. As explained previously, under these special circumstances the KBr tends to stay close to the center of mass, and we may approximate the distribution of products, H + KBr, by the distribution of centroids. The dependence of the total reaction cross section on energy in the region near the threshold energy ϵ^* was assumed to be of the form

$$\begin{aligned} S_R(\epsilon) &= 0 && \text{for} \quad \epsilon < \epsilon^* \\ S_R(\epsilon) &= \pi d_{12}^2 \sum_n c_n(\epsilon^*/\epsilon)^n && \text{for} \quad \epsilon > \epsilon^* \end{aligned} \tag{5-23}$$

where the c_n in the power series are dimensionless adjustable parameters.

To interpret results of experiments like those of Taylor and Datz [24], it is necessary to obtain a theoretical expression for the flux of centroids of products as a function of appropriate direction angles. This expression must include all relevant experimental details, e.g., the intersection angles of the reactant beams; the nature of the velocity distributions in these beams (for beams which are not velocity-selected, as in the present case, these distributions are determined by the source temperatures); and the intersection volume determined by beam dimensions and beam geometry. The expression must also include necessary information concerning the velocity or energy dependence of the reaction cross section, like that represented by Eq. (5-23). The derivation of appropriate expressions for the flux of centroids of products, with or without respect to angular distribution, is involved and cumbersome but not conceptually difficult. The reader is referred to sources where these expressions have been developed for a number of cases of experimental interest [1, 23, 26], and we proceed without further digression to a discussion of the conclusions obtained from the actual experiments.

With respect to Eq. (5-23), two special cases are of interest:

(1) $c_0 = 1$, $c_{n>0} = 0$; and
(2) $c_0 = 1$, $c_1 = -1$; $c_{n>1} = 0$.

The first case represents a cross section of hard-sphere radius d_{12} having characteristics of a step function with respect to ϵ (cf. Chapter 2) and no dependence on orientation and internal excitation. The second case, which has been previously introduced in Eq. (2-29), contains the additional assumption that only the component of relative velocity along the line of centers at impact is effective. In Fig. 5-5 the circles are the fluxes of KBr, $P(\theta)$, measured as a function of angle with respect to the direction of the initial K beam. The solid curves, normalized to unit area, are the corresponding calculated fluxes of centroids in which a number of parametric values of the threshold energy ϵ^* have been used in connection with the second case of Eq. (5-23). The abrupt decrease in the intensity of KBr at small values of θ may result from the experimental difficulty of measuring small fluxes of KBr accurately in the presence of large fluxes of elastically scattered K. An assumed threshold of $N_0\epsilon^*$ $(=E^*)$ in the range 2.5 to 3.0 kcal/mole reproduces the experimental results in the region of the maximum satisfactorily, and this value was suggested by Taylor and Datz [24] as the experimental energy of activation. At larger angles, above about 50°, the same theoretical curves are not in particularly good agreement with experiment. The discrepancy may be a consequence of neglecting the angular spread of the reactant HBr effusive "spray." For example, a spread of about 20° at half height in the HBr and an assumed value of $\epsilon^* \simeq 0$ could produce a reasonable fit. It is concluded that the use of thermal beams and the consequent necessity of averaging over the velocity distributions in the beams may lead to considerable ambiguity in the interpretation of the experimental results for this reaction [1].

The reaction of K with HBr was reinvestigated by Beck *et al.* [23] using a mechanical velocity selector to obtain nearly monoenergetic K beams. (If v_1 was the maximum in a narrow band of selected velocities, at half

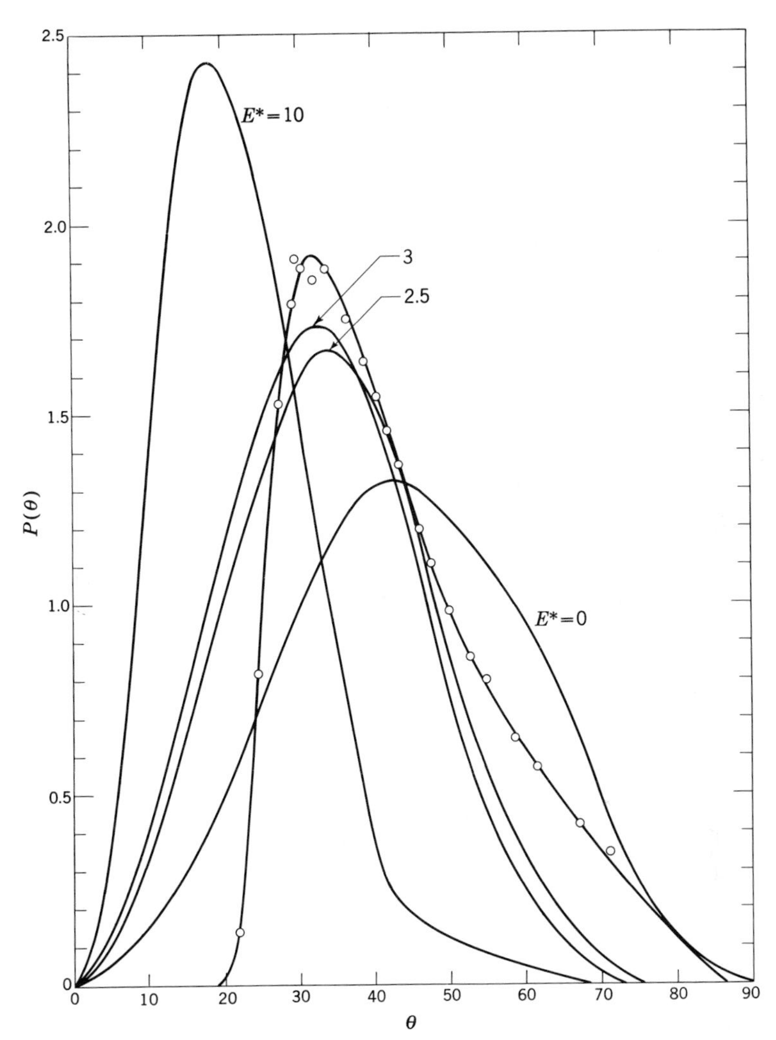

Fig. 5-5 Comparison of observed angular distribution of KBr (○) with calculated distribution of centroids. The energy dependence of the total reaction cross section is taken as that of the second case in Eq. (5-23). The curves are normalized to unit area, and E^* is a parameter in the calculation. (Adapted from Datz *et al.* [21].)

height the spread in velocity was $0.084v_1$). The HBr beam was not velocity-selected but effused from a thermal source at T_2 with a most probable speed of $(2kT_2/m_2)^{1/2}$. The use of velocity selection yielded a great deal of information which was inaccessible when only thermal beams were used. The distribution of angles of intersection of the effusive HBr spray with the well-defined K beam could be inferred, and the threshold for reaction could be more precisely located by changing the velocity of the K beam. If the initial average kinetic energy of the system is taken as $\bar{\epsilon} = \frac{1}{2}m^*(\bar{v}_1{}^2 + \bar{v}_2{}^2)$, where $\bar{v}_2 = (8kT_2/\pi m_2)^{1/2}$, the threshold energy $N_0\epsilon^* = E^*$ was found to be smaller than 0.4 kcal/mole, and for $\bar{E} > 0.4$ kcal/mole, the reaction cross section $S_R(\bar{\epsilon})$ was nearly independent of $\bar{\epsilon}$.

It is interesting to find that a combination of elastic and reactive scattering results in additional information concerning the probability of reaction under certain conditions. In Chapter 2, where an expression was derived for the frequency per unit volume of binary collisions, we found that the *total elastic* cross section for scattering was given as $S(v_r) = \int_0^{b_{\max}} 2\pi b\, db$ for a system with spherically symmetric force fields. It can be shown that the

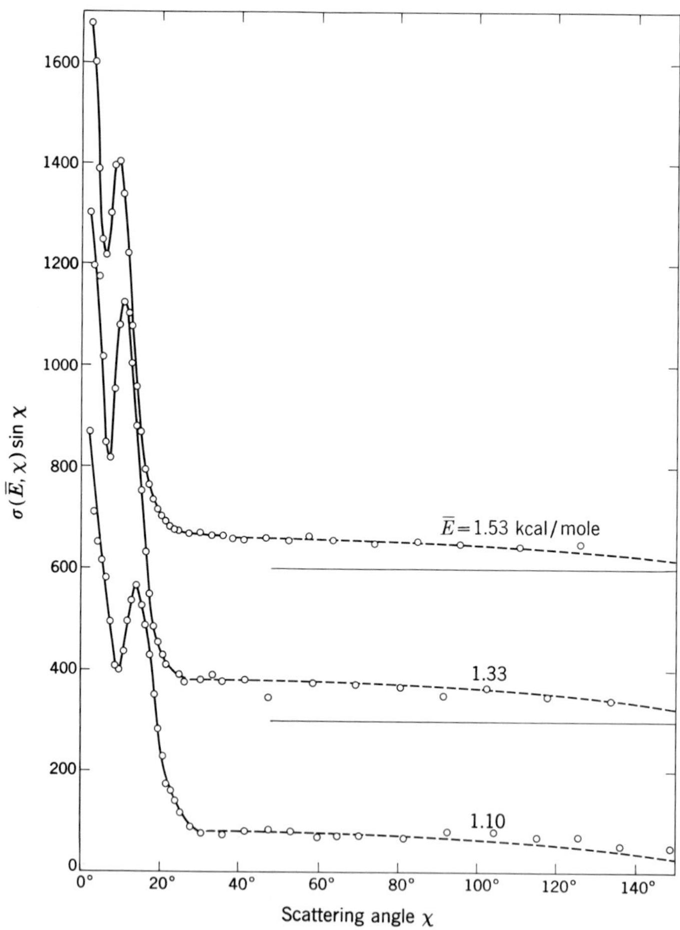

Fig. 5-6 Elastic scattering of the system K–Kr. (Adapted from Beck *et al.* [23].)

differential elastic cross section per unit solid angle for scattering $\sigma(\bar{E}, \chi)$, where χ is the relative angle of scattering, is related to the classical impact parameter b [5, 10] by

$$\sigma(\bar{E}, \chi) \sin \chi \, d\chi = b \, db \tag{5-24}$$

Measurements of $\sigma(\bar{E}, \chi)$ as a function of χ were made for the nonreactive analog of K + HBr, namely, K + Kr [27], as well as for K + HBr [23, 27]. The results for K + Kr at three different values of $\bar{E}$ are shown in Fig. 5-6 and that for K + HBr in Fig. 5-7. In both figures the ordinates are in arbitrary units, but a large difference exists in the scattering pattern of the two systems at large values of χ. This difference was ascribed to chemical reaction. In analyzing the results shown in Figs. 5-6 and 5-7, it was assumed that the interactions were a composite of central finite repulsive and attractive forces (cf. Fig. 2-1d) and that, in general, the two-dimensional trajectory would be that shown in Fig. 2-4c. The potential was recast in the form

$$U\left(\frac{r}{r_m}\right) = \frac{\epsilon}{1 - 6/\alpha}\left\{\frac{6}{\alpha}\exp\left[\alpha\left(1 - \frac{r}{r_m}\right)\right] - \left(\frac{r_m}{r}\right)^6\right\} \tag{5-25}$$

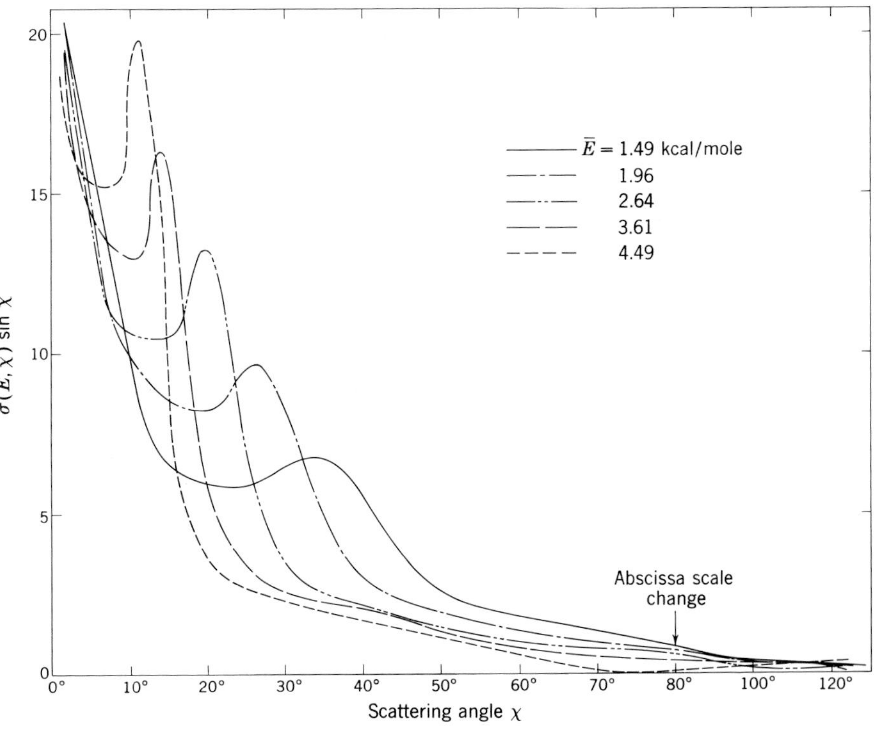

Fig. 5-7 Elastic scattering of the system K–HBr. (Adapted from Ross and Green [1].)

where r_m is the position of the potential minimum and ϵ its depth. For the systems under consideration, α was set equal to 12.

For a double-valued function, like that in Eq. (5-25), there are certain values of $\bar{E}$ for which the relative angle of deflection has a minimum value, called the rainbow angle χ_r, where the classically calculated value of $\sigma(\bar{E}, \chi) \sin \chi$ becomes infinite. A semiclassical analysis [28] removes this infinity and predicts instead the maxima (and accompanying minima) observed experimentally. It is important to observe that rainbow scattering in general is not observed unless velocity selection is used, since each initial relative velocity is associated with a different rainbow angle. Thus, the fine structure of the scattering pattern, as shown in Figs. 5-6 and 5-7, will be masked if the spectrum of initial relative velocities is too broad. In the present examples the spread in initial relative velocities was kept small by using velocity selection on the K beams and by minimizing the contribution of the Maxwellian distributions of the Kr and HBr beams. The minimization resulted for the most part from having the heavier Kr and HBr particles issue from sources whose temperatures were lower than that of the K source. Thus, the average velocity of Kr or HBr, $\bar{\mathbf{v}}_2$, was significantly smaller than the selected velocity of K, $\mathbf{v}_1$. In addition, an intersection angle of 90°, as opposed to a smaller angle, reduced somewhat further the effect on $\mathbf{v}_r$ of a spread in $\bar{\mathbf{v}}_2$. For an assumed form of the potential such as Eq. (5-25), the parameters ϵ and r_m can be determined from the position of the rainbow angle and from the negative slope of $\sigma(\bar{E}, \chi) \sin \chi$ versus χ [27, 29, 30].

The experimental determination of differential cross sections at a given value of $\bar{E}$ is relatively simple, since it merely involves measurement of the initial axial intensity of velocity-selected beam atoms (in the present case, K) and the scattered intensities of these same atoms at selected values of χ. In most cases, scattering near the relatively small angles near the rainbow region is unaffected by the chemical reaction. The potential parameters, obtained as already indicated, can therefore be used to calculate a hypothetical elastic differential cross section for the same system (assumed to be nonreactive) at larger angles, where the effects of reactive scattering have a large effect on the intensity of scattered reactant. For example, if $\sigma_c(\bar{E}, \chi)$ is the elastic differential cross section calculated on the assumption that no reaction occurred, then for larger angles the experimental differential cross section $\sigma(\bar{E}, \chi)$ is smaller than $\sigma_c(\bar{E}, \chi)$, since some beam atoms are lost due to reaction. Thus, the reaction probability $P(\bar{E}, \chi)$ for specified values of $\bar{E}$ and χ is simply

$$P(\bar{E}, \chi) = \frac{\sigma_c(\bar{E}, \chi) - \sigma(\bar{E}, \chi)}{\sigma_c(\bar{E}, \chi)} \tag{5-26}$$

Since the deflection angle χ is related to the impact parameter b through the intermolecular potential, $P(\bar{E}, \chi)$ may be converted into $P(\bar{E}, b)$. As described by Ross and Greene [1], it is possible to extrapolate the probability of reaction as a function of a fixed value of the potential at the distance of closest approach; that is, $\bar{E}$ varies but r_0 does not. This extrapolation leads to a threshold ($P = 0$) at $\bar{E} = 0.15$ kcal/mole compared with an upper limit of 0.4 kcal/mole from direct measurements on KBr [23]. With the threshold value of the impact parameter (b_{th}) known in terms of $\bar{E}$, the total reaction cross section at a specified $\bar{E}$ is simply

$$S_R(\bar{E}) = 2\pi \int_0^{b_{\text{th}}} P(\bar{E}, b) b \, db \tag{5-27}$$

The use of elastic scattering in reactive systems as already described involves a number of approximations and assumptions. However, the method may represent a virtually independent procedure for evaluating $S_R(\bar{E})$ with practically no recourse to direct measurements of the yield of reaction product.

A summary of the most important microscopic information obtained from this study of elastic and reactive scattering is given in Table 5-1. Since the entries in the table involve a number of assumptions peculiar to the reaction of K and HBr, these assumptions are repeated for clarity. The

TABLE 5-1

Distribution of Energy in the Products of the Reaction of K + HBr[a]

$\bar{E}$ (kcal/mole)		1.49			4.49	
Angular momentum of HBr, $j = 2$ (g cm^2/mole sec)		0.00155			0.00155	
Rotational energy of HBr, $j = 2$ (kcal/mole)		0.15			0.15	
Potential at distance of closest approach (kcal/mole)	0.15	0.6	1.2	0.15	0.6	1.2
Impact parameter b (Å)	3.72	2.92	1.67	3.84	3.53	3.21
$m^* v_r b$, nearly equal to angular momentum of KBr (g cm^2/mole sec)	0.068	0.053	0.030	0.121	0.111	0.101
Rotational energy of KBr (kcal/mole)	2.52	1.55	0.51	8.07	6.83	5.65
Rotational quantum number	104			186		
Sum of final relative kinetic energy and vibrational energy of KBr (kcal/mole)	3.17	4.14	5.18	0.59	1.83	2.42
Maximum vibrational state of KBr	5	6	8	1	3	4

[a] Adapted from Ross and Green [1].

rotational angular momentum $\mathbf{J}_2$ of HBr was much less than the initial orbital angular momentum $\mathbf{L}$ and could be neglected. The final orbital angular momentum $m^{*\prime}v_r'b$ was much less than the initial value because of a large reduction in m^* and a likely reduction (not large) in b. Although the reaction is exothermic, $\Delta D_0{}^0 = -4.2$ kcal/mole, this energy is so small that no large change in $\bar{E}$ could be expected to occur. To a first approximation, therefore, the rotational angular momentum of KBr, $\mathbf{J}_2'$, could be taken as nearly equal to the initial orbital angular momentum $\mathbf{L}$. The entries in the table were calculated according to these assumptions for a rigid-rotor harmonic-oscillator model for HBr and KBr.

5-6 REACTION OF K WITH CH_3I

The reaction of K + HBr is rather unique in the sense that it has a very small value of $m^*_{\text{products}}/m^*_{\text{reactants}}$ and of $\Delta D_0{}^0$ (so that $E' \simeq E$). One of the reactions for which the reduced mass of the products is more nearly comparable to that of the reactants is

$$K + CH_3I \rightarrow CH_3 + KI \tag{5-28}$$

for which $\Delta D_0{}^0 = -22$ kcal/mole. The reaction was studied by Herschbach *et al.* [31], who used crossed thermal beams and two different surface-ionization detectors, pure tungsten and platinum alloy, to distinguish between the fluxes of K and KI. The authors actually studied a series of reactions of the type in Eq. (5-28) by substituting Rb and Cs for K. In all their experiments the temperatures of the sources of the reactants as well as the angle of intersection of the beams were varied.

The results for K + CH_3I are illustrated in Fig. 5-8. Figure 5-8a shows the actual detector signals as a function of laboratory angle with reference to the incident K beam, while Fig. 5-8b shows the intensity of KI product (for two separate experiments) which is obtained from the difference between the ordinates of the two curves in Fig. 5-8a. The velocity-vector diagram in Fig. 5-9 is an alternative and in some ways a more informative representation of the experimental results for the perpendicular intersection of the beams specified in Figs. 5-8a and 5-8b. The concentric circles with their center at the end of the center-of-mass vector represent the loci of points which KI molecules may reach in the time required for the center of mass to traverse the distance shown in the figure. Each circle corresponds to a different value E' of the final kinetic energy of relative motion of the products. Thus Fig. 5-9 is actually a comparison of the observed distribution of products with that permitted by the conservation laws, where the lengths of the vectors for CH_3I and K correspond to the average speeds at the

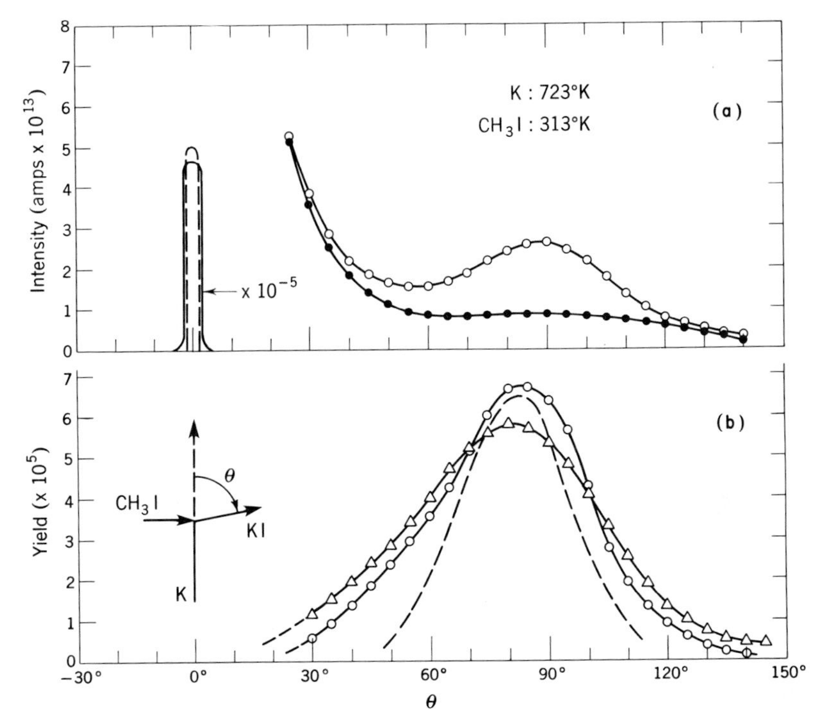

Fig. 5-8 Scattering of K + CH_3I. (a) Intensity of scattered signal as measured by Pt (●) and W (○) surface-ionization detectors. Pt measurements are normalized to W measurement in main K beam. Main K beam, shown at laboratory angle = 0°, is attenuated 7% by CH_3I crossed beam. (b) Detected KI distribution for two experiments. (Adapted from Herschbach [22].)

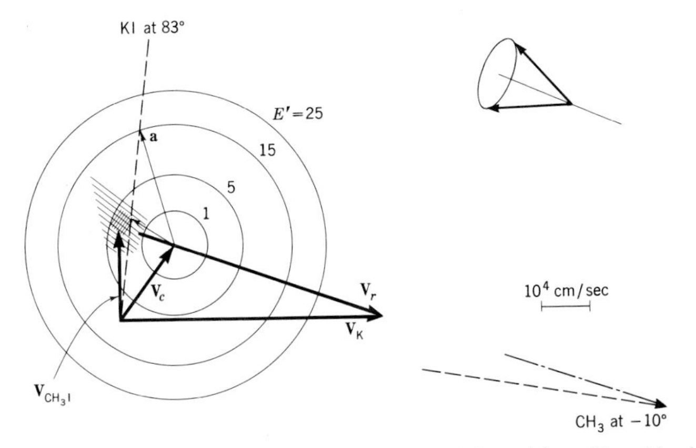

Fig. 5-9 Vector diagram for the reaction K + CH_3I. (Adapted from Herschbach [22].)

temperatures in Fig. 5-8a. The intersection of $\mathbf{v}_c$ with the relative-velocity vector $\mathbf{v}_r$ has been obtained from Eqs. (5-17) and (5-18). Finally, the total energy available, i.e., the magnitude of the right-hand side of Eq. (5-15), is about 24 kcal/mole, consisting of $\Delta D_0{}^0 = -22$ kcal/mole, $E \cong 1.3$ kcal/mole, and $Z \cong 0.6$ kcal/mole (made up, for the most part, of rotational energy of CH_3I, since there is very little vibrational excitation of this molecule at 313°K).

The observed broad peak of KI near 83° in the laboratory system corresponds to reactive scattering, in which an observer stationed at the center of mass would see KI recoil backward (and CH_3 scatter forward) with respect to the initial K beam. The displacement of the KI peak by about 30° from $\mathbf{v}_c$ shows that most of the contributions to the reactive scattering must be associated with values of the final kinetic energy of relative motion E' greater than about 1 kcal/mole, since at this lower energy the KI product would be restricted to the range $30° \lesssim \theta \lesssim 75°$. These limits on θ are the extreme values of the angle between the axis of the K beam and the points of intersection of lines from the origin of the centroid $\mathbf{v}_c$ to tangents to the circle for $E' = 1$ kcal/mole. As larger values of E' are assumed, the range of θ in which KI may appear increases, until for $E' \gtrsim 7$ kcal/mole KI may appear at any laboratory angle. This latter situation (full range of θ) occurs as soon as E' attains a value such that the radius of its circle is equal to the length of the centroid. Reactions of $K + CH_3I$ associated with large values of E', such as the vector **a** in Fig. 5-9, can contribute to the observed peak at $\theta = 83°$ only if the direction of the recoil vector deviates markedly from the negative direction of the initial relative velocity $\mathbf{v}_r$. Since these recoil vectors must have cylindrical symmetry about $\mathbf{v}_r$, contributions from scattering associated with large values of E' should be directly observed by measurements of out-of-plane scattering. This prediction is based on the fact that if recoil vectors, such as **a**, associated with large E', are rotated about $\mathbf{v}_r$, cones of large volume are generated. Since the large volume is associated with a large area at the base of the cone, KI products can appear out of the plane defined by the intersection of the beams of K and CH_3I at angles larger than actually observed for the in-plane scattering. Since the KI is found to be strongly peaked near the plane of the incident beams, it is apparent that for $K + CH_3I$, scattering close to the negative direction of $\mathbf{v}_r$, with small values of E', predominates. Variation of the angle of intersection of the reactant beams confirmed this conclusion in terms of agreement between predictions of the new vector diagrams (similar to Fig. 5-9) and the in-plane and out-of-plane measurement of the angular distribution of KI [22]. To account for all the experimental observations it was found that about 50% of the KI recoil vectors must lie within the doubly shaded region of Fig. 5-9 and about 90% within the singly shaded region. Once the

KI distribution of vectors was determined, the most probable recoil velocity of CH_3I could be estimated from Eqs. (5-19) and (5-20).

Since it has been established that the products contain a small amount of kinetic energy of relative motion, it follows that a large fraction of the energy of reaction (about 90%) appears in the products as internal excitation. Without velocity selection it is not possible to allocate this energy to specific internal degrees of freedom. It is also interesting to observe that velocity selection improves the experimental resolution, as shown by the dashed curve in Fig. 5-8b. This curve was obtained by assuming that all the KI returned at 180° from the original K direction and that E' for the products was 1.6 kcal/mole.

An alternative representation of the scattering data is the KI flux velocity-angle distribution shown in Fig. 5-10. In this figure the contours represent the relative amounts of product appearing at a given angle and recoil velocity in the center-of-mass coordinate system. Thus ten times more product is found at the contour labeled 10 than at the contour labeled 1. The zero of θ is taken in the direction of the initial relative velocity component of K. All KI molecules with the same recoil velocity will lie in a circle centered at the center of mass and the dashed circle corresponds to the largest recoil

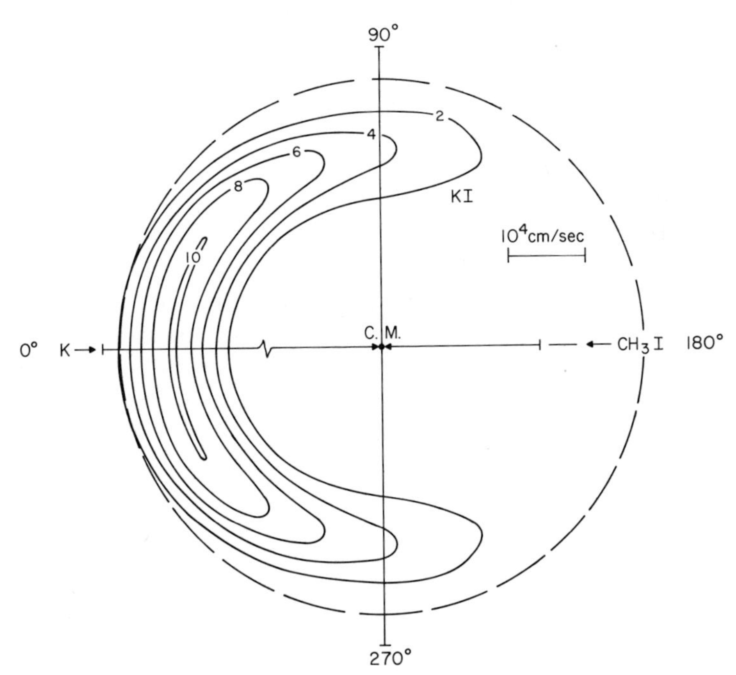

Fig. 5-10 Flux (velocity-angle) contour map for the KI product from the reaction $K + CH_3I \rightarrow KI + CH_3$. (Adapted from Rulis and Bernstein [33] and Levine and Bernstein [34].)

velocity consistent with the conservation of energy. The closer a point is to the center of mass, the higher the internal energy of the KI product. From Fig. 5-10, it can be readily assessed that the KI is scattered predominantly backward with a considerable amount of internal energy.

It is possible to estimate the reaction cross section from the information shown in Fig. 5-8. In reactive scattering of crossed beams, elastic scattering frequently predominates, since extremely weak interactions are sufficient to deflect a molecule from a thin beam, so that it misses a detector of small aperture. Such a situation is shown in the traces at $\theta = 0$ in Fig. 5-8a, where, in an intersection volume of about 0.04 cc, collisions between about 10^{10} molecules/cc of K ($\sim 10^{-6}$ mm) with about 10^{12} molecules/cc of CH_3I ($\sim 10^{-4}$ mm) result in a 7% attenuation of the K beam. The beam is attenuated according to the relationship

$$I = I_0 e^{-clS} \tag{5-29}$$

where I is the observed beam intensity, I_0 the beam intensity before scattering, c the concentration of scattering gas, l the path length for scattering, 0.5 cm in the present case, and S the scattering cross section. The observed scattering corresponds to an extremely large *total* cross section, about 1400 Å^2, in agreement with the findings of Rothe and Bernstein [32]. In the case of Figs. 5-8a and 5-8b, integrated intensity of the KI shows that about 0.1% of the potassium appears as KI. Thus the estimated reaction cross section from Eq. (5-29) is about 20 Å^2. Since variation of the source temperatures has little effect on the magnitude of this cross section and since the absolute value of the cross section for reaction is large, of the order of kinetic theory cross sections, it is apparent that there is little or no activation energy for the reaction of K + CH_3I.

The detailed dependence of the total reaction cross section on the relative translational energy of the reactants has been determined [32] and is found to rise from zero at about 1 kcal/mole to a maximum of about 40 Å^2 at 4 kcal/mole. At higher energies the cross section slowly decreases because the increased amount of available energy makes it increasingly difficult to form a stable KI molecule. The average fraction of the total available energy released into product translation was found to be about 0.58, slightly higher than in the original work. The reaction cross section for this reaction also has been estimated using elastic scattering results as described for the K + HBr reaction, and the results obtained are in reasonable accord with the measured values [35].

A further refinement of the molecular beam studies of K + CH_3I was made by orienting the molecules in a six-pole electric field so that the alkali atom could be made to approach the CH_3I principally either from the I end or from the CH_3 end [36, 37]. As expected, the cross section is considerably greater when the K approaches from the I end (greater than or equal to 1.5 times the cross section for approach from the other end). This provides a direct measurement of the steric factor p invoked in simple kinetic theory to take into account orientation effects for chemical reactions.

Extensive computer simulation experiments also have been carried out for the reaction of K + CH_3I. A major difficulty in carrying out the simulation is calculation of an appropriate potential energy surface. Blais and Bunker [38] considered the methyl group as a single mass so that the three masses K, CH_3, and I define a plane. A potential energy function was constructed with the following properties: a Morse potential for the K–I interaction; an attractive potential between K and CH_3 which is gradually eliminated as I and CH_3 come together; and a hard-sphere-like potential between K and CH_3. The parameters in the potential were adjusted to give a reasonable potential energy contour map and agreement of the calculation with molecular beam experiments. As initial conditions the angular orientation of CH_3I was selected by a Monte Carlo procedure, along with the relative velocity vector, the impact parameter, the rotational velocity and the remaining coordinates. The effect of the rotational and vibrational energy state of CH_3I and the mass of the alkali metal atom also was considered. The trajectories are generally uncomplicated, and the interaction time between reactions sufficiently short that no intermediate can be said to be formed. The product scattering angle is peaked at an angle in good agreement with that observed [31, 38]; a small peak not observed experimentally, which is probably an artifact of the calculation, also was found. The distributions of total internal energy of alkali halide are broadly peaked at the energy of reaction, and the distribution of rotational energies of KI corresponds closely to complete conversion of the total initial angular momentum to product rotation.

More recently, Karplus and Raff [39] have carried out extensive computer experiments for the reaction between K and CH_3I and have found that the dimensionality of the calculation and the details of the potential are of critical importance for some considerations. They used a potential energy function similar to that employed by Blais and Bunker and calculated both two- and three-dimensional classical trajectories. As usual, the initial conditions were selected by use of a Monte Carlo procedure, and the trajectories were examined to see if reaction occurs. In both dimensions, most of the reaction energy appears as internal excitation of products. The calculated differential cross sections are markedly different for two and three dimensions and both disagree with the molecular beam experiments. The total reaction cross section obtained from the computer calculations is about 400 $Å^2$ in both two and three dimensions, as compared to the experimental result of about 10 $Å^2$. Karplus and Raff found the situation could be greatly improved by modifying the potential energy function. If an additional term is added which decreases the K–I attraction for an iodine atom that is part of a CH_3I, the total reaction cross section is reduced to 20 Å^2, and the calculated differential cross section is closer to the experimental results. Also, only a single peak is obtained in the angular product distribution. The inclusion of a repulsive interaction between K and CH_3 [40] predicts the experimental steric factor observed with oriented CH_3I beams. All of the calculations predict too large a conversion of translational energy into internal product energy since the calculations were not based on the more recent results of Rulis and Bernstein [33]. The only potential energy surface predicting the observed energy partitioning also predicted a long-lived collision complex of CH_3I and K, which is contrary to experimental results. Nevertheless, the computer simulation studies are able to reproduce the main features of the experimental results, *providing* the potential energy surfaces are adjusted appropriately. Thus the combination of molecular beam experiments and computer simulation provides information both about the potential energy surfaces and about the collision dynamics of chemical reactions.

5-7 LIFETIMES OF COLLISION COMPLEXES

In the two systems discussed thus far, the product scattering is very anisotropic. This indicates the lifetime of a molecular collision is very short relative to rotation of the "complex" formed between reactants (i.e., $< \sim 10^{-13}$ sec). If this were not the case, the scattering of product would be symmetric about the center of mass since all information about initial geometry of the collision would be lost after a few rotations. In fact, long-lived complexes have been observed in many cases. For example, the flux (velocity-angle) contour map for CsF formed in the reaction

$$SF_6 + Cs \rightarrow CsF + SF_5 \tag{5-30}$$

is shown in Fig. 5-11 [34, 41]. Considerable symmetry in the backward and forward scattering patterns around the center of mass can be seen. We briefly consider the conditions necessary for formation of a long lived complex (cf. Levine and Bernstein [34] and Wolfgang [42] for more complete discussions of this subject).

In order for a collision complex to be stable, the plot of potential energy versus reaction coordinate must resemble that shown in Fig. 5-12. The most important feature of this energy profile is the existence of a minimum between reactants and products. The quantity ϵ_m is the minimum energy necessary for decomposition of this intermediate. The barriers α and β are generally small or nonexistent, depending on the details of the trajectory, and need not be considered further. In order to have a long-lived collision complex, the

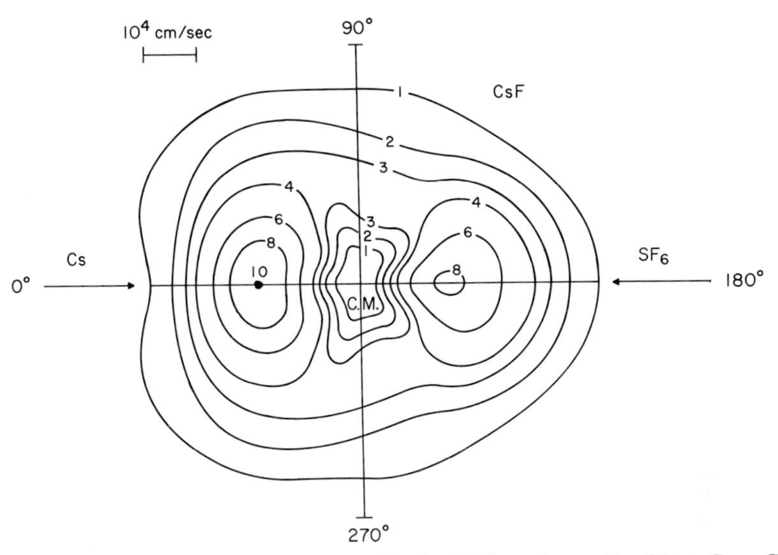

Fig. 5-11 Flux (velocity-angle) contour map for the CsF from the reaction $SF_6 + Cs \rightarrow Cs + SF_5$. (Adapted from Levine and Bernstein [34] and Riley and Herschbach [41].)

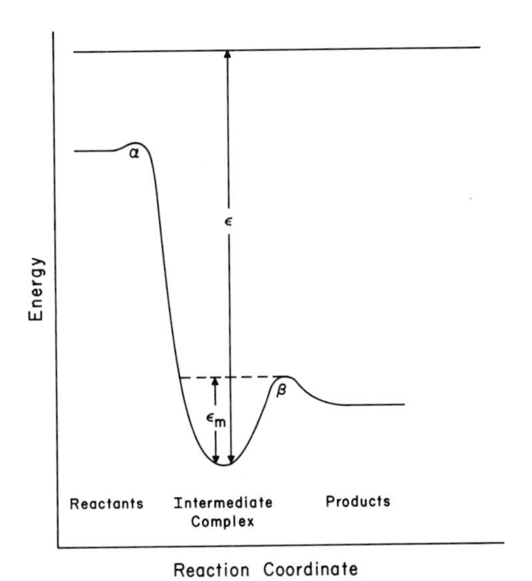

Fig. 5-12 Schematic cross section of a potential energy surface involving an intermediate complex.

complex must have a lifetime toward unimolecular dissociation which is long compared to times of molecular vibrations and rotations, i.e. $> \sim 10^{-12}$ sec, and the relative translational energy of the reactants must be convertible into internal excitation of the product. The lifetime for unimolecular decomposition can be estimated from the RRK theory of unimolecular decomposition as [Eq. (4-16)]

$$\frac{1}{k(\epsilon)} \approx 10^{-13}\left(\frac{\epsilon - \epsilon_m}{\epsilon}\right)^{1-n/2} \quad \text{sec} \tag{5-31}$$

Therefore, the lifetime is maximized when the minimum energy for decomposition ϵ_m and the number of vibrational degrees of freedom, $n/2$, are large, and the total energy ϵ is small. The efficient conversion of translational energy into internal energy depends on many factors; the most important factors are necessary internal rearrangements and energy transfer must be rapid compared to collision times, and a sufficient number of internal excited levels of the complex must exist so as to match the total energy available. Also, a centrifugal potential must be overcome for grazing collisions. Both criteria indicate that complex formation is favored when the reactants are polyatomic since then $n/2$ is large and a high density of internal energy levels exist. Thus, for the reaction of K + HBr, $n/2$ is small and ϵ_m is small so that a stable collsion complex would not be expected. In the case of K + CH_3I, the very large value of ϵ would require an unusually large value of ϵ_m in order to form a stable collision complex. On the other hand, for the reaction of SF_6 + Cs, $n/2$ is larger (≤ 18), energy partitioning is rapid, and $\epsilon/(\epsilon - \epsilon_m)$ is significantly greater than unity (it was deduced to be 1.3–1.6 [42]). Collision complexes tend to occur at low relative energies (ϵ small) so that a reaction may occur via complex formation at low energies and via a direct collision at high energies. This behavior is observed for the reaction [43]

$$C_2H_4^+ + C_2H_4 \rightarrow C_3H_5^+ + CH_3 \tag{5-32}$$

Molecular beams provide one of the few methods for assessing the importance of stable collision complexes in chemical reactions.

This chapter has considered only a few examples of molecular beam studies. More comprehensive discussions are available [19, 34, 43–45].

Problems

5-1 A thermal molecular-beam source at 200°C contains liquid potassium in equilibrium with potassium vapor at 1×10^{-2} Torr. The source aperture is a rectangular slit 0.5 cm long and 1×10^{-2} cm wide. Other apertures in the collimating system are designed to preserve the ribbonlike character of the beam. Calculate the following:

(a) The total flux (in molecules per second) of potassium issuing from the source into the evacuated isolation chamber *I* in Fig. 5-2.

(b) The total axial intensity of the beam at a plane perpendicular to the source and 10 cm distant from it.

(c) The corresponding intensity when a velocity selector is used which transmits only atoms with relative speeds in the range $1.90 < v/v_{mp} < 2.10$, where v_{mp} is the most probable speed of the atoms in the source.

(d) Calculate the positive-ion current which would be read for case (c) on a surface-ionization detector using an incandescent tungsten filament with an active length of 0.5 cm and a diameter of 2.5×10^{-3} cm. The length of the detector is parallel to the 0.5-cm dimension of the source aperture.

It may be assumed that (a) the intensity along the length of the beam is constant; (b) the efficiency of the surface-ionization detector is unity.

5-2 A thermal beam of Cs atoms produced, without velocity selection, in a source at 500°K intersects at 90° a similar beam of CH_3I from a source at 300°K. The concentration of CH_3I in the volume defined by the intersection of the two beams is 1.2×10^{12} molecules/cc. In traversing the path length for scattering, 0.5 cm, within this volume, the axial intensity of the Cs beam is reduced by 10% as the result of elastic scattering and chemical reaction.

(a) Calculate the total scattering cross section, in Å^2, for Cs + CH_3I.

(b) The flux of CsI was obtained as a function of angle θ with respect to the Cs beam from the difference in currents of two surface-ionization detectors, one sensitive to Cs + CsI, the other sensitive only to Cs [22]. The normalized function for the in-plane distribution of CsI may be represented by

$$f(\theta) = 5.4 \times 10^{-8}\theta^2 \exp(-4 \times 10^{-4}\theta^2) \quad \text{deg}^{-1}$$

where $f(\theta)\,d\theta$ is the fraction of total CsI flux produced by the intersection of the two beams which is scattered into an angular region between θ and $\theta + d\theta$ for values of θ between 0 and 180°. Calculate the reaction cross section for

$$\text{Cs} + \text{CH}_3\text{I} \rightarrow \text{CsI} + \text{CH}_3$$

assuming that the out-of-plane scattering is negligible.

Hint: The difference between the value of $f(\theta)$ for $\theta = 180°$ and $\theta = \infty$ may be neglected.

5-3 Consider the following reaction in a system of thermal beams, not velocity-selected, which intersect at 90°:

$$\text{Cs} + \text{HI} \rightarrow \text{CsI} + \text{H}$$

The source temperature of the Cs beam is 1600°K, and that of the HI beam is 400°K.

For this particular system, the following may be assumed: the impact parameters for the initial and final systems b and b' are equal; the rotational and vibrational energies of HI are negligible compared to the sum of $\Delta D_0{}^0$, the heat of reaction at 0°K, and E, the kinetic energy of relative motion of the reactants; **J**, the rotational angular momentum of HI, is negligible compared to the orbital angular momentum of the reactant system; the vibrational energy of CsI is negligible compared with the total energy of the products; the relative speed of the reactants v_r is very much smaller than that of the products.

(a) Use the conservation laws for angular momentum and energy to show *analytically* that as b (and b') increase relative to r_{CsI}, the bond length of CsI, a large fraction of the total available energy $-\Delta D_0{}^0 + \frac{1}{2}m^* v_r{}^2$ tends to be converted into rotational energy of CsI, $J'^2/(2m^*_{\mathrm{Cs+I}} r^2_{\mathrm{CsI}})$.

(b) On the basis of the assumptions given above and the following additional numerical data, calculate the rotational energy in kcal/mole of CsI: $b = b' = 4.60$ Å; $r_{\mathrm{CsI}} = 3.41$ Å, $\Delta D_0{}^0 = -7.1$ kcal/mole.

(c) Find the rotational quantum number which characterizes the CsI, assuming that all such molecules are in the same rotational level.

5-4 When a beam of Rb atoms from a source at 600°K intersects at 90° a beam of CH_3I from a source at 300°K (with no velocity selection), reaction occurs according to

$$\mathrm{Rb} + \mathrm{CH_3I} \rightarrow \mathrm{CH_3} + \mathrm{RbI}$$

and the final kinetic energy of relative motion of the products E' peaks at about 2 kcal/mole. It may be assumed that the angle between the final relative velocity $\mathbf{v}_r'$ and the velocity of the center of mass $\mathbf{v}_c$ is 90°.

(a) Construct a vector diagram ("Newton diagram") similar to Fig. 5-9 and indicate the in-plane laboratory angle where the peak intensity of RbI would be expected to appear.

(b) Account for any major differences between the figure obtained in part (a) and Fig. 5-9.

(c) Assume that the in-plane peak intensity predicted in part (a) has been experimentally observed. What additional experimental observations would be expected if a substantial fraction of the products had values of E' equal to about 5 kcal/mole?

References

1. J. Ross and E. F. Greene, *in* "Transfert d'énergie dans les gaz" (R. Stoops, ed.), p. 363, 12th Chemistry Solvay Conference, Inst. Int. Chim. (Douzième Conseil Chim., 1962). Wiley (Interscience) New York, 1962.
2. M. A. Eliason and J. O. Hirschfelder, *J. Chem. Phys.* **35**, 19 (1961).
3. I. Prigogine and E. Xhrouet, *Physica* **15**, 913 (1949).
4. S. Chapman and T. G. Cowling, "The Mathematical Theory of Non-Uniform Gases," 2d ed. Cambridge Univ. Press, London, and New York 1952.
5. J. O. Hirschfelder, C. F. Curtiss, and R. B. Bird, "The Molecular Theory of Gases and Liquids." Wiley, New York, 1954.
6. L. Dunoyer, *Compt. Rend.* **152**, 594 (1911).
7. R. G. J. Fraser, "Molecular Rays." Cambridge Univ. Press, London and New York, 1931.
8. K. F. Smith, "Molecular Beams." Methuen, London, 1955.
9. N. F. Ramsey, "Molecular Beams." Oxford Univ. Press, London and New York, 1956.
10. E. H. Kennard, "Kinetic Theory of Gases." McGraw-Hill, New York, 1938.
11. M. Kröger, *Z. Phys. Chem.* (*Leipzig*) **117**, 387 (1925).
12. J. E. Mayer, *J. Am. Chem. Soc.* **49**, 3033 (1927).
13. G. N. Lewis and J. E. Mayer, *Proc. Nat. Acad. Sci. USA* **13**, 623 (1927).
14. F. O. Rice, H. C. Urey, and R. N. Washburne, *J. Am. Chem. Soc.* **50**, 2402 (1928).
15. M. W. Perrin, *Ann. Phys.* (*Paris*) **11**, 5 (1919).
16. M. Polanyi, "Atomic Reactions." Ernest Benn–Benn Bros., London, 1932.

17. I. Langmuir and K. H. Kingdon, *Proc. Roy. Soc. London Ser. A*. **21**, 380 (1923); see also Ref. 7, p. 43.
18. J. B. Taylor, *Z. Phys.* (*Leipzig*) **57**, 242 (1929).
19. D. R. Herschbach, *Pure Appl. Chem.* **47**, 61 (1976).
20. D. R. Herschbach, *J. Chem. Phys.* **33**, 1870 (1960).
21. S. Datz, D. R. Herschbach, and E. H. Taylor, *J. Chem. Phys.* **35**, 1549 (1961).
22. D. R. Herschbach, *Discuss. Faraday Soc.* **33**, 149 (1962).
23. D. Beck, E. F. Greene, and J. Ross, *J. Chem. Phys.* **37**, 2895 (1962).
24. E. H. Taylor and S. Datz, *J. Chem. Phys.* **23**, 1711 (1955).
25. S. Datz and E. H. Taylor, *J. Chem. Phys.* **25**, 389, 395 (1956).
26. M. Ackerman, E. F. Greene, A. L. Moursund, and J. Ross, *J. Chem. Phys.* **41**, 1183 (1964).
27. D. Beck, *J. Chem. Phys.* **37**, 2884 (1962).
28. E. A. Mason, *J. Chem. Phys.* **26**, 667 (1957).
29. K. W. Ford and J. A. Wheeler, *Ann. Phys.* (*NY*) **7**, 259 (1959).
30. F. A. Morse and R. B. Bernstein, *J. Chem. Phys.* **37**, 2019 (1962).
31. D. R. Herschbach, G. H. Kwei, and J. A. Norris, *J. Chem. Phys.* **34**, 1842 (1961).
32. E. W. Rothe and R. B. Bernstein, *J. Chem. Phys.* **31**, 1619 (1959).
33. A. M. Rulis and R. B. Bernstein, *J. Chem. Phys.* **57**, 5497 (1972).
34. R. D. Levine and R. B. Bernstein, "Molecular Reaction Dynamics." Oxford Univ. Press, London and New York, 1974.
35. E. F. Greene and J. Ross, *Science* **159**, 587 (1968).
36. P. R. Brooks and E. M. Jones, *J. Chem. Phys.* **45**, 3449 (1966).
37. R. J. Beuhler, R. B. Bernstein, and K. H. Kramer, *J. Am. Chem. Soc.* **88**, 5331 (1966).
38. N. C. Blais and D. L. Bunker, *J. Chem. Phys.* **37**, 2713 (1962).
39. M. Karplus and L. M. Raff, *J. Chem. Phys.* **41**, 1267 (1967); L. M. Raff and M. Karplus, *J. Chem. Phys.* **44**, 1212 (1966).
40. M. Karplus and M. Godfrey, *J. Am. Chem. Soc.* **88**, 5332 (1966).
41. S. J. Riley and D. R. Herschbach, *J. Chem. Phys.* **58**, 27 (1973).
42. R. Wolfgang, *Acc. Chem. Res.* **3**, 48 (1970).
43. D. R. Herschbach, *Adv. Chem. Phys.* **10**, 319 (1966).
44. J. L. Kinsey, *Int. Rev. Sci. Phys. Chem.* **9**, 173 (1972).
45. M. A. D. Fluendy and K. P. Lawley, "Molecular Beams in Chemistry." Chapman and Hill, London, 1974.

CHAPTER 6

ENERGY TRANSFER AND ENERGY PARTITIONING IN CHEMICAL REACTIONS

6-1 INTRODUCTION

The important role of both intramolecular and intermolecular energy transfer in gas-phase chemical reactions has been stressed in the discussions of unimolecular decompositions and molecular beam studies. In this chapter, intermolecular energy transfer and energy partitioning in chemical reactions is considered more explicitly. The dissociation of homonuclear molecules is discussed since the details of the energy distribution appear to play a major role in the reaction mechanism. The reaction

$$F + H_2 \rightarrow HF + H \tag{6-1}$$

is also discussed as an example of a reaction where several different experimental and theoretical methods have led to a detailed picture of the reaction energetics.

6-2 ENERGY RELAXATION

In thermal molecular collisions, vibrational, rotational, and translational energy can be interchanged in an inelastic collision. For the moment, energy transfer between different electronic states is not considered. The general convention is to use the designations V, R, and T for the three types of energy and to write the source of energy first and where it is transferred to second. For example, a case of V–T energy transfer is

$$HCl\,(v = 1) + Ar \rightarrow HCl\,(v = 0) + Ar \tag{6-2}$$

since HCl in its first vibrational state is converted to its ground vibrational state, and the vibrational energy must appear as translational energy of the products. The reverse of Eq. (6-2) is, of course, a case of T–V energy transfer. In general, all possible permutations of energy transfer are possible, although the efficiency of energy transfer between different degrees of freedom varies considerably.

The efficiency of energy transfer is generally characterized by the relaxation time for energy transfer or by the average number of collisions required for a transfer event to take place. The relaxation time is defined by writing the rate equations for the energy transfer processes required to establish equilibrium. Of course, if the system is at equilibrium, no net energy transfer can be observed. If the population of HCl in the first vibrational state exceeds the equilibrium population [Eq. (6-2)], equilibration can be achieved by three processes: emission of radiation, characterized by the first-order rate constant k_r; collisions with the wall characterized by the rate constant k_w; and collisions with Ar characterized by the rate constant k_{10}. [The deactivation of HCl (v = 1) by HCl itself is considerably more efficient than deactivation by Ar, but is neglected for the sake of simplicity and because the concentration of HCl is much smaller than that of Ar.] If the reverse reaction is considered negligible, the rate equation describing the approach to equilibrium is

$$\frac{d(\mathrm{HCl}\ (\mathrm{v}=1))}{dt} = [k_r + k_w + k_{10}(\mathrm{Ar})](\mathrm{HCl}\ (\mathrm{v}=1))$$

$$= \frac{1}{\tau}(\mathrm{HCl}\ (\mathrm{v}=1)) \qquad (6\text{-}3)$$

Therefore, the approach to equilibrium is a first-order process characterized by a relaxation time

$$\tau = [k_r + k_w + k_{10}(\mathrm{Ar})]^{-1} \qquad (6\text{-}4)$$

A more complete analysis (cf. Levine and Bernstein [1]) which includes the reverse reaction shows the time dependence of the approach to equilibrium is still exponential but the characteristic relaxation time is

$$\tau = [k_{10}(\mathrm{Ar}) + k_{01}(\mathrm{Ar}) + \textstyle\sum k_i]^{-1} \qquad (6\text{-}5)$$

where $\sum k_i$ indicates the sum of rate constants for all processes that are not gas-phase collision processes for both the forward and reverse reactions, and k_{01} is the rate constant characterizing the reverse of Eq. (6-2). The quantity $\sum k_i$ is often negligible, and in any event can be obtained by extrapolation of the relaxation time to zero pressure. Since the concentration of buffer gas is proportional to the pressure, the quantity $P\tau$ is often reported; a second-order rate constant in the units of M^{-1} sec^{-1} also is a useful parameter and can be obtained as $RT/P\tau$ (P in atm, $R = 0.0821$ liter atm deg^{-1} mol^{-1}). Some order-of-magnitude values of $RT/P\tau$ for various types of energy transfer processes are given in Table 6-1. Also included are the average number

TABLE 6-1

Typical Rate Constants and Collision Numbers for Energy Transfer

Transfer process	$RT/P\tau$ (M^{-1} sec^{-1})[a]	Z_{ij}[b]
V–V	10^6–10^7	10^2–10^3
V–R	10^4–10^5	10^4–10^5
R–R	10^8–10^9	1–10
V–T	10^3–10^4	10^5–10^6
R–T	10^6–10^7	10^2–10^3

[a] 298°K.

[b] Average number of collisions for energy transfer between the i and j energy modes.

of collisions required for a transfer to take place. These numbers should not be taken too literally since many variations from them occur. Obviously a great range in efficiencies exists. For R–R (rotational) energy transfer, transfer occurs every 1–10 collisions, while at the opposite extreme for V–T and T–V, transfer occurs only once in 10^5 to 10^6 collisions. In general, $\tau_{V-T} > \tau_{V-V} > \tau_{R-R}$ for a given buffer gas.

6-3 THE LASER AND ENERGY TRANSFER

The laser is the single most powerful tool in studying energy transfer processes. The intense monochromatic radiation produced by the laser permits excitation of molecules to specific vibrational and rotational energy levels and also allows the determination of very high resolution spectra. The basic principle of laser operation is to obtain a population inversion in energy levels such that for a nondegenerate system more molecules are in the higher energy state than in the lower energy state [2]. For a degenerate system, the condition for lasing is that $g_jN_i/g_iN_j > 1$ where N_i and g_i are the population and degeneracy of the upper energy level and N_j and g_j are the corresponding quantities for the lower energy level. The emission of a photon as a molecule goes from the higher state to the lower state stimulates a cascade of similar emission. The laser radiation then becomes intense and either depletes the population inversion for pulsed lasers or reaches a steady state for continuous wave lasers. As an example, consider the three-level scheme in Fig. 6-1 which approximates the situation found with a ruby laser. Excitation of Cr^{3+} ions in the crystal by the absorption of light from an intense flashlamp raises the system to level three. Most of the excitation is rapidly transferred by a radiationless transition to level two. This continues until the population in level two is greater than in level one. Stimulated emission then can occur, and in

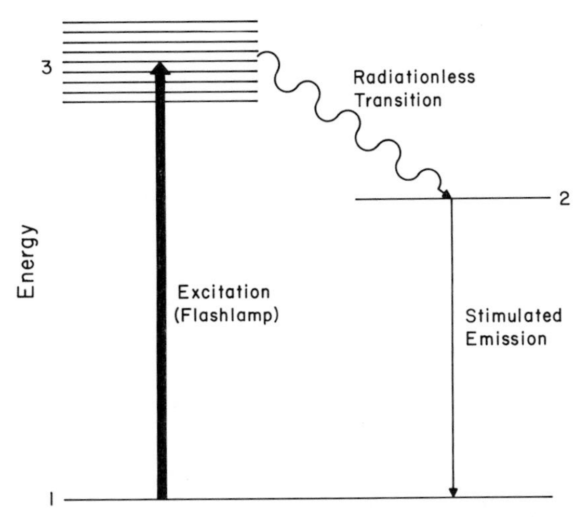

Fig. 6-1 Schematic energy level diagram for the ruby laser.

a properly adjusted optical resonator a very intense pulse of monochromatic radiation at 6943 Å is generated. In gas lasers (for example, He–Ne, Co_2–N_2), a continuous electric discharge is generally used to pump the molecules into the higher energy state. Some wavelength tuning is possible because population inversions can be created for several different transitions. The actual tuning is done by the optical resonator. Probably the most useful type of lasers are organic dye lasers because they can be continuously tuned over the visible, near ultraviolet, and near infrared. The principle is exactly the same: the pumping source is generally another laser or a flashlamp which excites molecules into a higher electronic state. The large manifold of different energy states available to organic dyes permits tuning over a large wavelength region. Finally, mention should be made of chemical lasers where the population inversion is produced by chemical reactions.

As an example of a gas laser and of the information which can be obtained about energy transfer, we consider some of the energy transfer processes which occur in gaseous CO_2 [1–3]. A simplified vibrational energy diagram for CO_2 is shown in Fig. 6-2 together with an indication of the normal mode motions of the molecule. The numbers k, l, m indicate the quantum numbers of the symmetric stretching mode, the degenerate bending mode, and the antisymmetric stretching mode. The CO_2 laser operates on the transition between the asymmetric (001) and symmetric (100) stretching fundamental or (020) bending overtone levels. In the laser, nitrogen molecules are vibrationally excited by collisions with electrons produced by electrical discharge of a CO_2–N_2–He mixture:

$$N_2\,(v = 0) + e^- \rightarrow N_2\,(v = 1) + e^- \tag{6-6}$$

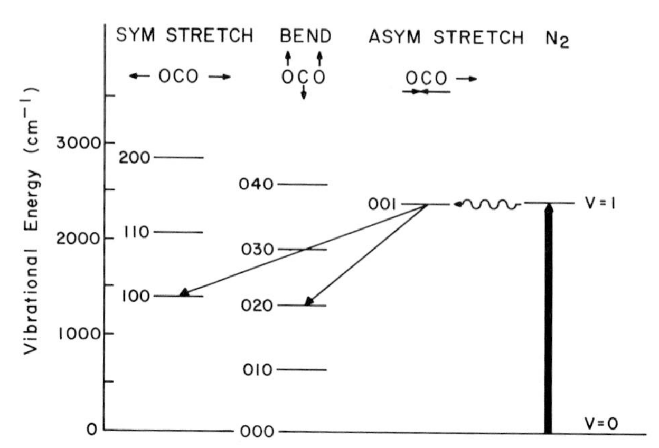

Fig.6-2 Simplified vibrational energy diagram for CO_2. The normal modes of vibration and the transitions involved in the CO_2 laser also are indicated. The wavy arrow shows the near resonant energy transfer from N_2 (v = 1) to CO_2 (001).

The excited nitrogen then undergoes an efficient *V–V* transfer with CO_2:

$$N_2\,(v = 1) + CO_2\,(000) \rightleftharpoons CO_2\,(001) + N_2\,(v = 0) - 51 \quad \text{cal} \tag{6-7}$$

The lasing operation is caused by induced emission from the 001 level through the two processes

$$CO_2\,(001) \rightarrow CO_2\,(100) + h\nu\,(10.6\ \mu)$$
$$CO_2\,(001) \rightarrow CO_2\,(020) + h\nu\,(9.6\ \mu)$$

In order for this laser to be efficient, the collision-induced loss of molecules from the 001 state must be much slower than the collision-induced loss from the 100 and 020 states so as to induce a population inversion. In fact, the relaxation time for depletion of the 001 state is unusually long, indicating that intramolecular *V–V* energy transfer from this state is unusually inefficient; about 20,000 collisions are required for an energy transferring event. On the other hand, the 100 and 020 states are rapidly depleted by the processes

$$CO_2\,(020) + M \rightarrow CO_2\,(100) + M - 294 \quad \text{cal}$$
$$CO_2\,(100) + CO_2\,(000) \rightarrow 2CO_2\,(010) + 154 \quad \text{cal}$$
$$CO_2\,(020) + CO_2\,(000) \rightarrow 2CO_2\,(010) - 140 \quad \text{cal}$$

The first reaction requires about 10 collisions for equilibration and the second and third about 50 collisions. With these processes the whole manifold of symmetric stretch and bending vibrational energy levels are rapidly brought into equilibrium, while the asymmetric stretch manifold is essentially uncoupled primarily because states close to the 001 level do not exist. The 010 state is finally depleted by the rather inefficient *V–T* transfer

$$CO_2\,(010) + M \rightarrow CO_2\,(000) + M + 1.91 \quad \text{kcal} \tag{6-8}$$

Thus energy transfer considerations are required to explain the operation of the CO_2 laser; conversely, the CO_2 laser can be used to study energy transfer processes in CO_2. It should be mentioned that the depletion of the 001 state can be monitored by spontaneous emission (fluorescence). The predominant relaxation mechanism is through collisions at pressures usually employed for energy transfer studies; however, a small fraction of the molecules relax through radiative decay providing an ideal monitoring mechanism. A complete analysis of the CO_2 laser is more complex than given because the details of CO_2–N_2 energy transfer processes, which have been studied, also must be considered [3].

6-4 THEORETICAL CONSIDERATIONS OF TRANSLATIONAL ENERGY TRANSFER

Detailed theories of vibrational–translational energy transfer have been developed and are rather complex [1, 4, 5]. Some insight into the problem can be obtained by considering the simple model of a diatomic harmonic oscillator molecule AB colliding with an atom M [1, 6]. If only colinear collisions are considered, two limiting possibilities exist. If the oscillator is very "stiff," the AB molecule acts as a rigid body so that the collision between M and AB is elastic with no change in the total kinetic energy; this is the *adiabatic* limit. The opposite extreme is a very loose coupling between A and B. In this case, when M collides with A, it will have essentially no effect on B. Therefore, the kinetic energy acquired by A becomes motion relative to B and can be considered to be vibrational energy. If the AB molecule is initially at rest, and M has an initial velocity $\mathbf{v}_M$, the conservation of momentum and energy can be written as

$$m_M \mathbf{v}_M = m_M \mathbf{v}_M' + m_A \mathbf{v}_A', \qquad \tfrac{1}{2} m_M v_M^2 = \tfrac{1}{2} m_M v_M'^2 + \tfrac{1}{2} m_A v_A'^2$$

where the primed velocities are the final velocities and the m_i are the masses. These equations can be easily solved to give the final velocity of A, $\mathbf{v}_A' = 2m_M \mathbf{v}_M/(m_M + m_A)$. The vibrational energy transferred, ΔE_V, is the relative translational energy of A with respect to B:

$$\Delta E_V = \frac{1}{2} \frac{m_A m_B}{m_A + m_B} v_A'^2 = \frac{2 m_M^2 m_A m_B}{(m_A + m_B)(m_M + m_A)^2} v_M^2$$

The initial relative translational energy E_T is due to the relative motion of M with respect to AB:

$$E_T = \frac{1}{2} \frac{m_M (m_A + m_B)}{m_M + m_A + m_B} v_M^2$$

Therefore, the fraction of initial translational energy converted to vibrational energy is

$$\frac{\Delta E_V}{E_T} = \frac{4m_M m_A m_B(m_M + m_A + m_B)}{(m_A + m_B)^2(m_M + m_A)^2} \quad (\leq 1) \tag{6-9}$$

Thus the maximum possible T–V energy transfer is determined by the relative masses of the three atoms. For example, if all three masses are equal, the maximum energy transfer is 0.75. This theory is obviously oversimplified, but leads to qualitatively correct predictions for energy transfer from atoms to diatomic molecules at high collision energies. Similar considerations are applicable to more complex situations. A quantum mechanical theory is needed for a quantitative theory of vibrational energy transfer. However, generally the probability for energy transfer to occur decreases as more vibrational quantum number changes are required and as more energy must be transferred into translation [5].

The spacing of rotational energy levels is very much closer than vibration, and rotational energy transfer is generally very efficient at ordinary temperatures. Qualitatively this can be understood by noting that the duration of the collision is much shorter than the rotational period of most molecules at ordinary temperatures. Therefore, the rotational motion sees a collision as a sudden impulse which efficiently transfers energy. In terms of the model presented for vibrational energy transfer, this corresponds to loose coupling between A and B in the diatomic molecule; this loose coupling means the collision time is short compared to the vibrational period, and the impulse perturbation then leads to efficient energy transfer. Since rotational periods of molecules are generally much longer than vibrational periods, rotational energy transfer might be expected to be much more efficient than vibrational energy transfer and such is the case.

6-5 FURTHER CONSIDERATION OF *V–V*, *R–R*, AND *V–R* ENERGY TRANSFER

The multitude of available lasers makes it possible to excite almost any molecule to any desired specific vibrational energy level. Many methods also are available for monitoring the kinetics of vibrational relaxation in laser excited gases, for example, infrared fluorescence, absorption of light from a probe laser, and Raman scattering. Probably the most popular type of experiment utilizes double resonance [7]. With this method, a large amplitude laser pulse is used to excite a molecule to a specific vibrational energy state, and a second weaker intensity laser is used to monitor changes in energy level populations. The second laser beam must be sufficiently weak so as not to perturb the population of energy levels.

As a specific example, we return again to CO_2. The kinetics of energy transfer among the levels of the symmetric stretch and bend vibrations was studied by using one laser to create population changes and a second to monitor level populations through transient absorption [8]. A 9.6-μ pulsed CO_2 laser pumped molecules from the 020 level to the 001 level, approximately equalizing the populations of these two levels. A second CO_2 laser operating at 10.6 μm (the 100 to 001 transition) was used to probe the absorption at a power level too low to affect the level populations. The time course of the absorption was as follows: during the pump laser pulse, the transmission increased as the population of upper laser state (001) increased; a further increase in transmission occurred following the pulse with a rise time of about 10^{-6} sec; the transmission probe then decreased with a relaxation rate of about 3×10^{8} sec^{-1} atm^{-1} between 1.3 to 21×10^{-4} atm of gas; finally, on a much longer time scale, the probe transmission decreased further as the 001 state was depopulated. The very rapid process, which is complete in about five collisions, is attributed to the reaction

$$CO_2\,(100) + CO_2\,(000) \rightarrow CO_2\,(020) + CO_2\,(000)$$

while the process with a rate of 3×10^{8} sec^{-1} atm^{-1} is probably due to the reaction

$$CO_2\,(020) + CO_2\,(000) \rightarrow CO_2\,(100) + CO_2\,(000)$$

Although the study of *V–V* energy transfer is difficult because of the rapidity of the transfer, many examples have been studied, a few of which are listed in Table 6-2 [9].

The intermolecular forces causing *V–V* transfer vary widely. However, a very important rule is that the efficiency of energy transfer is generally greatest when the amount of vibrational energy available for transfer is the same as the vibrational energy level spacing of the acceptor. Such a *resonant*

TABLE 6-2

Selected Examples of *V–V* Energy Transfer[a]

Reaction	ΔE (cal/mole)
CH_4 (symmetric bend) + M $\rightarrow$ CH_4 (asymmetric bend) + M	649
HCl (v = 1) + HBr (v = 0) $\rightarrow$ HCl (v = 0) + HBr (v = 1)	935
HCl (v = 2) + HCl (v = 0) $\rightarrow$ HCl (v = 1) + HCl (v = 1)	−295
HCl (v = 1) + O_2 (v = 0) $\rightarrow$ HCl (v = 0) + O_2 (v = 2)	−69
CO_2 (001) + SF_6 (v_i = 0) $\rightarrow$ CO_2 (100) + SF_6 (v_3 = 1)	37

[a] Adapted from Moore [9].

energy transfer reaction is most efficient when the energy levels are exactly matched, and the efficiency drops off rapidly as mismatching occurs. The adiabatic character of vibrational motions in molecular collisions suggests that short-range repulsive forces are most effective in producing vibrational transitions. However, Sharma and Brau [10] have shown that relatively long-range forces can be a dominant source of vibrational energy transfer. In particular, they have been able to explain the unusual negative temperature dependence of the *V*–*V* transfer of Eq. (6-7) with a theory based on long-range interactions.

Since the energy involved in rotational energy transitions is very small, the rate of *R*–*R* and *R*–*T* energy transfer is usually very rapid. The most common method of studying energy transfer between rotational energy levels again is double resonance experiments, either microwave–microwave or microwave–infrared or infrared–infrared. The principle is very simple and is illustrated in Fig. 6-3 for a four-level double resonance experiment [11]. For microwave–microwave double resonance, the gas is irradiated with a high power of microwaves at a frequency ν_P which pumps a substantial fraction of the molecules from state 2 to state 1 (Fig. 6-3a). The population of energy levels is now not the equilibrium Boltzmann distribution, and the adjustment of the populations among the energy levels by collisions (usually with a buffer gas) can be monitored by a second source of microwave radiation with frequency ν_S. If the pumping power is turned off, intermolecular collisions will restore the system to equilibrium. The relaxation to equilibrium comes about by altering the populations of rotational states that are in some way coupled to the two states whose population distribution was

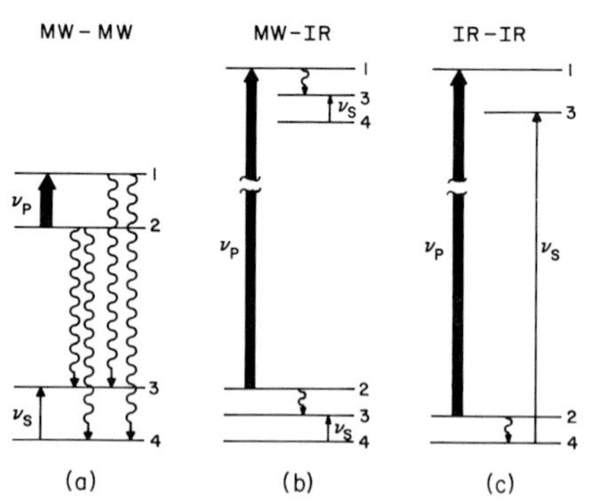

Fig. 6-3 Energy level schemes for a four-level double resonance experiment. The molecules are pumped from level 2 to level 1 by intense radiation with frequency ν_P. The non-Boltzmann distribution is then transferred to other levels by collision, and the transfers are monitored by a weak intensity signal with frequency ν_S. MW designates microwave and IR infrared radiation. (Adapted from Oka [11].)

altered by the original pumping. The time dependence of this process also can be monitored by the second source of microwave radiation. Infrared lasers also have been used for pumping, causing a non-Boltzmann population in a single rotational level (for the ground vibrational level); the equilibration of the system following pumping can be monitored either with microwave or infrared radiation (Figs. 6-3b and 6-3c). These methods permit the study of state-to-state rotation–rotation and rotation–translation energy transfer. In particular, steady-state populations of the rotational states can be studied, thus avoiding the difficulties of working in the real time domain.

The theory of *R–R* transfer requires complex quantum mechanical calculations, and well-defined selection rules can be obtained. In a sense, the problem is easier than *V–V* transfer since to a good approximation the vibrational motion of the molecule can be ignored, whereas *V–V* transfer is almost always accompanied by extensive rotational transfer so that *V–R* and *R–R* processes must be considered.

The rotational energy levels of ammonia have been extensively studied with microwave–microwave double resonance [11]. The rotational levels of ammonia are split due to the molecular inversion process. An allowed transition exists between the upper and lower inversion doublet state. These doublets have been systematically saturated with an intense microwave pumping field and intensity changes in other doublets have been examined. The results obtained give information about the preferred pathways for rotational energy transfer and the relative rates of inelastic processes that are involved. (The absolute rates have been measured using laser infrared pumping.) Selection rules similar to those associated with dipole transitions appear to be obeyed in most types of collisions. In particular, processes with $\Delta j = 0$ or 1 (j is the rotational quantum number) are favored and parity changes sign. The $\Delta j = 0$ rate is much faster than the rate for $\Delta j = 1$, and the rates of processes with $\Delta j > 1$ are much smaller still. In the collisions of ammonia with rare gases, these selection rules are not as pronounced, but strong symmetry restrictions still exist.

The results obtained with ammonia suggest long-range interactions, such as dipole–dipole, are probably responsible for the large rotational inelastic cross sections. The potential energy of interaction need not be isotropic. In fact, the more anisotropic a molecule is, the more easily it gains or loses rotational energy. Actually the same statement is applicable to vibrational energy transfer.

The occurrence of *V–R* energy transfer is quite common since as mentioned previously it often accompanies *V–V* energy transfer. For example, in CO_2 the process

$$CO_2\,(001, j = 19) + CO_2\,(000, j_i) \rightarrow CO_2\,(000, j = 19 \pm 1) + CO_2\,(001, j_i \pm 1) \qquad (6\text{-}10)$$

has been postulated to occur [12]. In many cases, *V–R* energy transfer actually may be rate determining in vibrational energy relaxation, and a simple theory has been developed for *V–R* energy transfer [13]. The model for this theory is a stationary vibrating diatomic molecule and a rigid rotator. Translational energy transfer is assumed to be negligible. As might be anticipated, *V–R* energy transfer is most efficient when the frequency of vibration is low, and the moment of inertia of the rotator is small.

To summarize, two important factors controlling the rates of energy transfer involving vibration and rotation are *resonance* and *long-range interactions*. The combination of lasers and microwave techniques can be used to study a wide array of vibrational and rotational energy transfer processes.

6-6 ELECTRONIC ENERGY TRANSFER

Electronic excitation energy is so large that collisions are often adiabatic with respect to electronic energy transfer; consequently, *V*, *R*, and *T* energy transfer can be thought of as occurring within a given electronic energy state, usually the ground state. However, in many cases electronic energy can be efficiently converted to vibrational energy to yield highly vibrationally excited molecules. For example, electronically excited mercury atoms (112 kcal/mole is required for excitation) can vibrationally excite CO:

$$\mathrm{Hg}^* + \mathrm{CO}\,(\mathrm{v}=0) \rightarrow \mathrm{Hg} + \mathrm{CO}\,(\mathrm{v} \leq 9) \tag{6-11}$$

In many cases, bond dissociation occurs because of the large amount of energy available; an example is the reaction

$$\mathrm{Hg}^* + \mathrm{H}_2 \rightarrow \mathrm{Hg} + \mathrm{H} + \mathrm{H} \tag{6-12}$$

An excited electronic state can be formed through absorption of visible and ultraviolet radiation. The excited state generally decays quite rapidly, in about 10^{-8} sec, through emission; however, for transitions that are quantum mechanically forbidden, the excited state can have very long lifetimes, even hours. Generally when studying electronic energy transfer, the radiative lifetime makes an important contribution to the relaxation of the electronically excited species so that the relaxation time must be studied as a function of pressure to precisely determine the collisional quenching contribution to the relaxation time. The possibility of intramolecular (radiationless) energy transfer where the electronic energy is transferred to vibrational energy within the excited molecule also can be of importance.

An important new factor in electronic energy transfer is the possibility of potential curve and energy surface crossing. For a diatomic molecule, the potential energy can be represented in a curve such as those shown in Fig. 6-4 for the covalent and ionic forms of KI. At large separations, the ionized

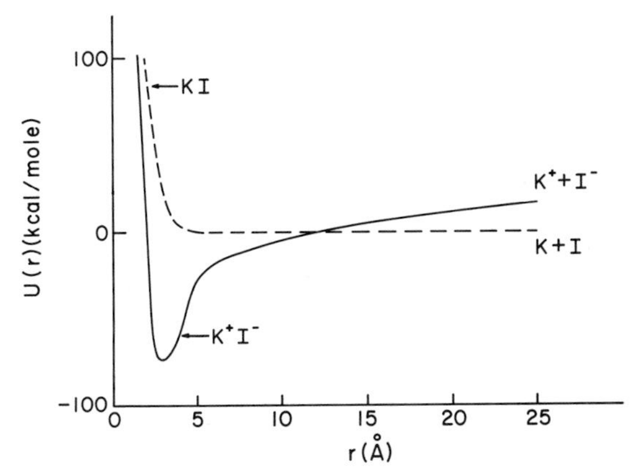

Fig. 6-4 Potential energy curves for the covalent and ionic forms of potassium iodide.

atoms, K^+ and I^-, have a greater energy than the neutral atoms. However, an internuclear distance exists at which the energies of the two electronic states are equal. The fact that this curve crossing exists makes the processes

$$K^+ + I^- \rightarrow K + I$$

very efficient since the electronic states can be switched with essentially no change in the kinetic energy of the nuclei. Since the energy difference between ions and atoms is 29.5 kcal at large distances, this represents a large transfer of energy. For polyatomic molecules, a multidimensional surface is needed to represent an electronic energy state. In this case, the crossing of two energy surfaces makes the energy transfer process efficient. In general, electronic energy transfer is efficient when the energy gap is small.

The He–Ne laser depends on electronic energy transfer for its operation. The He atoms are excited by electron impact from the singlet S ground state to the triplet excited state:

$$\mathrm{He}(1\,{}^1S_0) + e^- \rightarrow \mathrm{He}(2\,{}^3S) + e^- \tag{6-13}$$

The excited He then transfers its energy to Ne, producing an electronically excited Ne:

$$\mathrm{He}(2\,{}^3S) + \mathrm{Ne}(1\,{}^1S_0) \rightarrow \mathrm{He}(1\,{}^1S_0) + \mathrm{Ne}(2s) \tag{6-14}$$

The energy gap for this transfer is ≤ 0.15 eV. The Ne(2s) state, which contains four sublevels, decays radiatively to the 2p configuration, which is 0.7 eV lower in energy and contains ten sublevels, with a lifetime of about 10^{-7} sec. An inverted electronic energy level distribution is maintained because the decay of the 2p state to the 1s configuration is much more rapid, with a lifetime of about 10^{-8} sec. Thus the 2p level becomes depleted relative to the

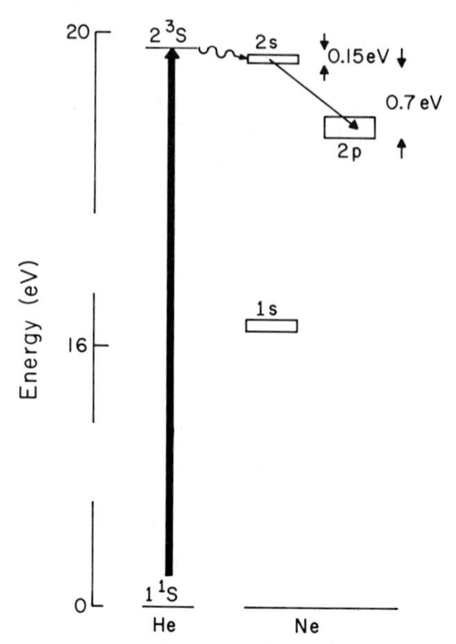

Fig. 6-5 Schematic energy levels for the He–Ne laser. The He is excited to the $2\,^3S$ state and excites Ne to the $2s$ state by electronic energy transfer. The laser light comes from conversion of Ne in the $2s$ state to Ne in the $2p$ state. Since both the $2s$ and $2p$ states are multileveled, a variety of light frequencies can be obtained by appropriate tuning.

$2s$ level, and a lasing action can be established. An energy diagram for this laser is presented in Fig. 6-5.

We now proceed to consider the mechanism of some chemical reactions in which energy transfer and energy partitioning play an important role. More detailed consideration of molecular energy transfer processes can be found in Refs. 1, 3, 5, and 14.

6-7 DISSOCIATION OF DIATOMIC MOLECULES

The dissociation of diatomic molecules and the reverse reaction, recombination of atoms, is a deceptively simple reaction which has been extensively studied for many years [15, 16]. As mentioned in Chapter 2, atomic recombinations to form stable diatomic molecules generally require termolecular collisions so that the reaction mechanism in its simplest form can be written as

$$A_2 + X \rightleftharpoons A + A + X \tag{6-15}$$

where A is the atomic species and X is a third body such as A, A_2, or an inert gas. The recombination of atoms has been studied primarily by using photolysis or electrical discharge to produce atoms. However, the most extensive studies of Eq. (6-15) have been carried out in shock tubes where the temperature is raised rapidly (generally to 1000–10,000°K), and the rate of dissociation is then studied at the higher temperature. Since the equilibrium constant can be accurately calculated from thermodynamic tables, both the association and dissociation rate constants can be readily determined. We now briefly digress to explain the principles of the shock tube.

Theory of Shock Tubes

When a sound wave is propagated through a fluid medium, the properties of reflection, refraction, and superposition may be represented by linear differential equations if the amplitude of the wave is sufficiently small. For very violent disturbances, i.e., when the amplitudes are very large, the wave motion becomes nonlinear, and a phenomenon known as a shock front appears. Shock fronts are regions where extremely rapid changes occur in the thermodynamic state of the fluid medium. To a good approximation, these rapid changes can be represented as mathematical discontinuities. In a system in which the flow was initially continuous at $t = 0$, such discontinuities can occur at $t = t$, after which the shock fronts can propagate without further change through the medium. This propagation is referred to as the shock wave.

For many physical and chemical investigations, shock waves are produced in a long straight tube of uniform cross section, which is divided by a thin diaphragm into two sections, each filled with a suitable gas at different initial pressures. The arrangement is known as a shock tube. If the diaphragm is suddenly ruptured by increase of pressure on the high side or by mechanical piercing, a shock wave travels into the low-pressure chamber of the shock tube, and a rarefaction wave continuous in the properties of the fluid medium travels in the opposite direction, as shown in Fig. 6-6. In such an arrangement adiabatic compression by the shock wave of the gas ahead of it can increase the pressure in the expansion section, so that temperatures as high as 20,000°K are reached, and the velocity of the shock front can attain values as high as 10 times sonic velocity.

An introduction to the general theory of shock tubes is given in Appendix C and in many general references [17–19]. For our purposes, we note that measurement of the velocity of the shock front U permits calculation of the properties of the shocked gas. The Mach number of the unshocked gas, which is the ratio of the shock velocity to the velocity of sound in the unshocked gas, can be written as

$$\text{Mach}_{\text{u}} = \frac{|U|}{(\gamma R T_{\text{u}}/M)^{1/2}} = \left[\frac{P_{\text{s}}/P_{\text{u}} + \beta}{\gamma(1 - \beta)}\right]^{1/2} \tag{6-16}$$

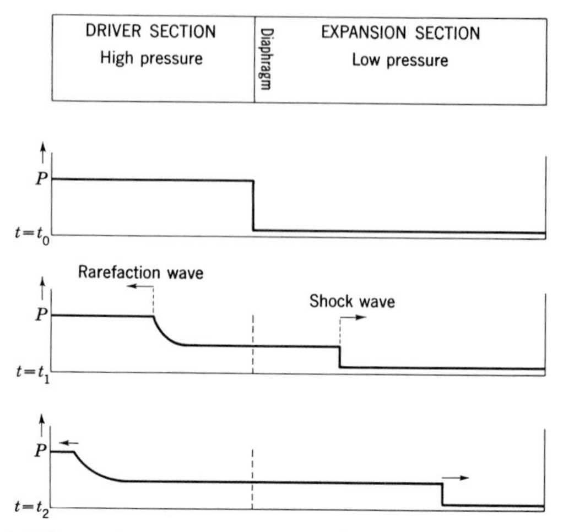

Fig. 6-6 A simple shock tube and pressure–distance profiles.

where γ is the ratio of the constant pressure specific heat to the constant volume specific heat, C_p/C_v, M the molecular weight, R the gas constant, T the temperature, P the pressure, $\beta = (\gamma - 1)/(\gamma + 1)$, and the subscripts s and u refer to the shocked and unshocked gas, respectively. Once P_s/P_u is known, the temperature and density of the shocked gas, T_s and P_s, can be readily calculated since

$$\frac{T_s}{T_u} = \frac{(1 + \beta P_s/P_u)\, P_s}{(P_s/P_u + \beta)\, P_u} \tag{6-17}$$

Three advantages of shock tubes for studying chemical reactions merit mention. First, shock heating is extremely rapid and homogeneous; translational temperatures of the order of 1000 to 3000°K can be reached in times as short as 10^{-9} sec, which is about the time required for a molecule to experience about 10 collisions at STP. (The actual thickness of the shock front is of the order of magnitude of a few mean free paths of the gas ahead of the shock front.) It is thus possible, for example, to dissociate diatomic molecules completely into atoms. In other cases, it is possible to study rates of very rapid reactions occurring after inhibition periods which under ordinary conditions are relatively long. By using shock tubes, the rapid rise in temperature serves to initiate the rapid reaction. Second, the extremely high temperatures obtainable in shock tubes permit investigation of reactions whose rates at conventional temperatures are prohibitively slow. Finally, heterogeneous effects in shock-tube kinetics can be virtually eliminated, since in the maximum time during which most experiments take place, about 1 msec, only a very small fraction of the gas molecules in the tube is able to diffuse to the wall and participate in surface reactions [17]. The specialized and difficult experimental techniques required in order to obtain reliable and reproducible rate information from shock-tube investigations represent the most serious disadvantages of the method. However, optical or spectroscopic methods having rapid response times can usually be used for observing kinetic processes through windows incorporated in the shock tube.

Results and Mechanistic Interpretation

A surprising result consistently found is that the dissociation rate constant has an activation energy which is less than the dissociation energy. This implies the association rate constant has a *negative* activation energy, and in fact direct measurements of the association rate constant confirm that it decreases with increasing temperature. Table 6-3 lists average values of the activation energies for the dissociation, in the presence of Ar, of some homonuclear diatomic molecules. In calculating E_a, it has been assumed that $k = A \exp(-E_a/RT)$. Values of the thermodynamic dissociation energies $\Delta E_T{}^0$ and the temperature range of the investigations are also listed. The entries in the table indicate that activation energies for the association rate constants range from about -3 to -13 kcal/mole. Many third bodies other than argon have been investigated, and their efficiencies and temperature coefficients vary. As might be expected, the efficiency of a third body tends to increase with molecular complexity, i.e., with increasing internal degrees of freedom. For simplicity, we confine this discussion to cases in which Ar is the third body.

A possible explanation of the temperature dependence of the rate constant in terms of classical kinetic theory is to assume that the activation energy for dissociation is exactly equal to the dissociation energy but that the internal

TABLE 6-3

Activation Energies and Values of n for $Ar + A_2 \rightarrow Ar + 2A$

A_2	E_a (kcal/mole)	$\Delta E_T{}^0$ (kcal/mole)[a]	n[a]	Temperature range (°K)	Reference
I_2	30	34.6	4.7	850–1650	20–22
Br_2	41	44.4	2.7	1300–2700	23, 24
Cl_2	48	56.4	5.0	1700–2600	25, 26
F_2	29	36.8	6.4	1300–1600	27, 28
H_2	97	102.8	2.5	2300–5300	29–32
O_2	108	112.4	1.8	3400–7500	33, 34
N_2	211	224	2.6	6000–10,000	35, 36

[a] An average value of T has been assumed.

degrees of freedom of the diatomic molecule (and X) contribute to the activation process. In this case, according to Eq. (2-35), the dissociation constant is of the form

$$k_d = \text{const}\, T^{1/2}(\Delta E_T{}^0/RT)^{n/2} \exp(-\Delta E_T{}^0/RT) \qquad (6\text{-}18)$$

where n is the number of square terms in the classical expressions for the energies of rotation and vibration of A_2 and X. If the third body is Ar, n has a maximum value of 4. The temperature dependence of the equilibrium constant according to elementary statistical mechanics is

$$\frac{k_d}{k_a} = \text{const}\, T^{1/2}[1 - \exp(-h\nu/kT)] \exp(-\Delta E_0{}^0/RT) \qquad (6\text{-}19)$$

where ν is the characteristic vibrational frequency of the diatomic molecule and $\Delta E_0{}^0$ is the change in internal energy for the hypothetical gas-phase reaction at 0°K. The temperature dependence of the association rate constant can therefore be written as

$$k_a = \frac{\text{const}\exp[-(\Delta E_T{}^0 - \Delta E_0{}^0)/RT]}{T^{n/2}[1 - \exp(-h\nu/kT)]}$$

The difference between $\Delta E_T{}^0$ and $\Delta E_0{}^0$ is small, so that at very high temperatures ($h\nu/kT \ll 1$)

$$k_a \simeq \text{const}/T^{n/2-1}$$

and at sufficiently low temperatures ($h\nu/kT \gg 1$)

$$k_a \simeq \text{const}/T^{n/2}$$

Although the actual situation in most of the shock-tube experiments lies between these two extremes, the temperature dependence of the association constant for a given reaction is frequently represented by a single index over the entire experimental range of temperature. Physically this negative temperature dependence simply means that the probability of a given ternary collision resulting in recombination decreases as the energy of the collision is increased. Since the function of the third body is to take away excess vibrational energy, the negative temperature coefficient is consistent with an unfavorable balance, as T increases, between the increase in vibrational energy and the increase in the frequency of ternary collisions.

Equation (6-20) is the most convenient relation for obtaining values of n, since the formal temperature dependence of k_d, unlike that of k_a, does not change with T. Thus, if we define $RT^2\, d\ln k_d/dT$ as E_a, Eq. (6-18) leads to

$$E_a = \Delta E_T{}^0 - (n - 1)(RT/2) \qquad (6\text{-}20)$$

Table 6-3 lists values of n calculated by substituting values of E_a, $\Delta E_T{}^0$, and the average temperature into Eq. (6-20). The fact that in some cases n is greater than the maximum theoretical value of 4 would seem to rule out the generality of the mechanism associated with Eq. (6-18), but the experimental error in the measurements is rather large, and values of n differing by ± 1 from the entries in Table 6-3 would also represent the experimental results satisfactorily. Thus while some discrepancies exist, the mechanism represented by Eq. (6-18) cannot be ruled out. In any event, this mechanism is classical in nature and does not include the quantum mechanical nature of energy levels.

An alternative mechanism can be proposed which is based on the derivation, in Chapter 2, of the rate of termolecular collisions. This derivation postulated formation of an unstable complex between two of the atoms and subsequent collision of the complex with the third body. This mechanism can be described by

$$\begin{aligned} &\mathrm{A + A \rightleftharpoons A\cdots A}, \qquad K = (\mathrm{A\cdots A})/(\mathrm{A})^2 \\ &\mathrm{A\cdots A + X} \underset{k_1}{\overset{k_{-1}}{\rightleftharpoons}} \mathrm{A_2 + X} \end{aligned} \tag{6-21}$$

(Initial formation of an $A \cdots X$ complex followed by reaction with an A atom is an equivalent alternative mechanism.) Equation (6-21) leads to $k_a = Kk_{-1}$ and $k_d = k_1$. If the initial equilibrium reaction is exothermic ($\Delta E_T^{0\prime} < 0$) and the rate constant k_{-1} has no activation energy, the temperature dependence of the rate constants according to simple kinetic theory is

$$k_a = \text{const}\, T^{1/2} \exp(-\Delta E_T^{0\prime}/RT) \tag{6-22}$$

$$k_d = \text{const}\, T^{1/2} \exp(-E_a/RT) \tag{6-23}$$

where $E_a - \Delta E_T^{0\prime} = \Delta E_T{}^0$. This mechanism formally fits the data, but it does not seem likely that the exothermicity required of the initial complex formation is a general phenomenon. Again, detailed consideration of energy transfer processes is not included. However, this *chaperone* mechanism almost certainly does occur to some extent.

A more complete mechanism can be formulated by considering the vibrational energy level ladder of a diatomic molecule as shown schematically in Fig. 6-7. The concentration of molecules in the ith state changes with time due to transitions within the energy level ladder and due to dissociation to atoms and association of atoms to form a molecule in the ith state. Thus the rate equation can be written as

$$\begin{aligned} -\frac{d(\mathrm{A_2}(i))}{dt} &= \sum_j k_{ij}(\mathrm{A_2}(i))(\mathrm{Ar}) - \sum_j k_{ji}(\mathrm{A_2}(j))(\mathrm{Ar}) \\ &\quad + k_{id}(\mathrm{A_2}(i))(\mathrm{Ar}) - k_{di}(\mathrm{A})^2(\mathrm{Ar}) \end{aligned} \tag{6-24}$$

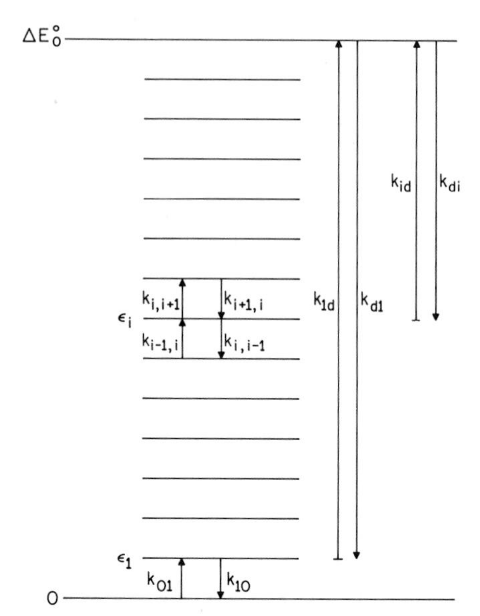

Fig. 6-7 Schematic drawing of energy levels for a diatomic molecule indicating the origin of the master equation.

where the rate constant k_{ij} represents the rate of transfer from the ith to the jth energy level, k_{id} is the rate constant for dissociation to atoms from the ith energy level, and k_{ij} and k_{di} are the rate constants for the reverse processes. (More correctly, a distinction should be made between rate constants and flux coefficients as discussed in Chapter 1.) This equation is an example of a *master equation*, and one exists for each vibrational state. The overall rate equation for the association–dissociation process is

$$-\frac{d(A_2)}{dt} = \sum_{i=0}^{n} k_{id}(A_2(i))(Ar) - \sum_{i=0}^{n} k_{di}(A)^2(Ar) \tag{6-25}$$

A complete solution to this problem would require solving all of the coupled differential equations simultaneously and specifying all of the rate constants. Obviously this is not possible in general, but the question can be asked what properties of the system are necessary to explain the experimental observations. This question has been extensively examined with methods ranging from simplified models to large-scale computer solutions of the differential equations [16, 37, 38]. A basic conclusion reached is that during the course of the reaction a *nonequilibrium* distribution of vibrational states exists. This is because the higher vibrational states are more likely to dissociate and therefore become depleted (relative to a Boltzmann distribution) as the

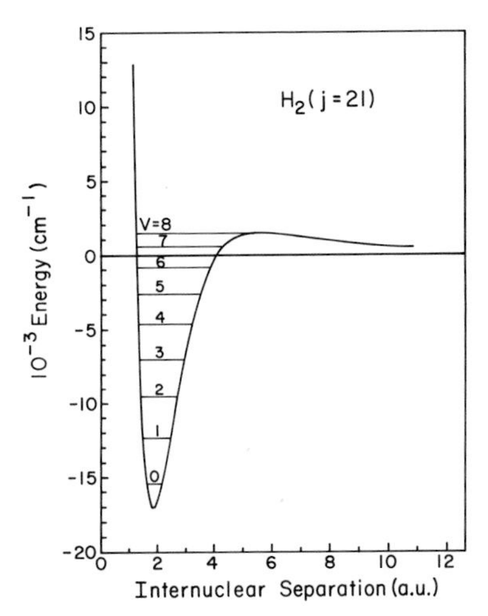

Fig. 6-8 Vibrational energy-level diagram for the $j = 21$ state of H_2. [Adapted from H. O. Pritchard, *Acc. Chem. Res.* **9**, 99 (1976), Fig. 3. Copyright 1976 by the American Chemical Society. Reproduced by permission of the copyright owner.]

reaction proceeds, thereby reducing the dissociation rate. This depletion becomes greater with increasing temperature. The probability of dissociation apparently varies with the vibrational state, but the quantitative nature of the variation is still a matter of controversy. However, even these properties are not sufficient to fully account for the temperature dependence of the rate constants. An important factor we have not yet considered is the role of rotation. Angular momentum restrictions result in a coupling of rotation and vibration. For example, the vibrational energy level diagram for the $j = 21$ state of H_2 is shown schematically in Fig. 6-8, and it is apparent some quasi-bound states are produced (v = 7 and 8) by the centrifugal barrier so that the overall rate will depend on j. If the rate equations are modified to include rotational effects, the negative temperature coefficient observed in the recombination process can be partially accounted for. In a recent computer study of the differential equations governing the dissociation of H_2, it was suggested that in addition to the above factors the individual association rate constants k_{di} must themselves have a negative temperature coefficient to explain the experimental results [8]. In this work, the distinction between the total flux coefficient and the rate constant for dissociation also was demonstrated.

Surprisingly, the simple dissociation of diatomic molecules and the reverse recombination process is far from understood. Apparently three different effects are important: complex formation (the chaperone mechanism), the nonequilibrium distribution of vibrational energy states, and rotation–vibration energy coupling. The extent to which each of these factors contributes to the mechanism depends on the particular diatomic molecule and third body (X) under consideration. A better understanding of the energy distribution and coupling in this reaction remains a goal for future kineticists.

6-8 DYNAMICS OF $F + H_2 \rightarrow HF + H$

As a final example, the reaction

$$F + H_2 \rightarrow HF + H \qquad (\text{and/or } F + D_2 \rightarrow DF + D) \tag{6-26}$$

is discussed. This reaction has been the subject of a great deal of experimental and theoretical investigation because it forms the basis for the HF chemical laser. The HF is produced with inverted vibrational energy level populations, and thus a lasing action is possible. The detailed energetics of this reaction has been studied by three techniques: crossed molecular beams [39, 40], the chemical laser [41, 42], and infrared chemiluminescence [43].

The molecular beam technique already has been discussed in great detail, as have the principles of laser operation. With the chemical laser, the light emitted from the excited product molecules, HF or DF in this case, is monitored as a function of time. The F atoms are usually produced by flash photolysis of a fluorine containing compound such as F_2 or CF_3I. By varying the tuning, the details of the time dependence of the vibrational energy level populations can be studied. The rotational relaxation time is usually too short to be studied by this technique. With the chemiluminescence method,

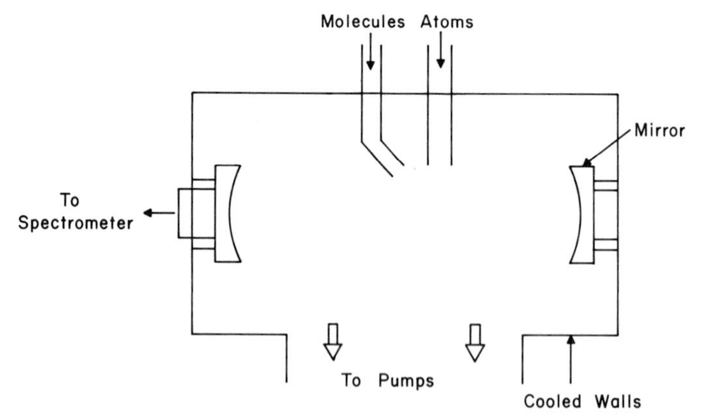

Fig. 6-9 Schematic drawing of chemiluminescence apparatus. (Adapted from Polanyi and Woodall [43].)

the light emitted by the excited *nascent* (initial) products of a reaction is subjected to spectroscopic analysis. The objective is the measurement of the *relative* populations of excited thermal states of the product molecules. Visible, ultraviolet, and infrared emission all have been measured. Typically, uncollimated beams of the reactants are rapidly mixed under very low pressures to avoid collisional relaxation. The bulk of the molecules are rapidly pumped away, but the product molecules can also be removed rapidly by condensation at the reaction vessel walls with liquid nitrogen so as to minimize relaxation processes. A schematic drawing of a typical apparatus is shown in Fig. 6-9. The radiation emitted from the reaction zone (or condensed product molecules) is focused into a spectrometer, and from the relative intensities of the lines, relative populations of excited states can be inferred. The chemiluminescence method gives only relative rates for the formation of specific vibrational and rotational energy states. Absolute rates must be obtained from other types of experiments. With this method, however, it has been possible to deduce *nascent* internal state distributions.

In the molecular beam experiment, a beam of F atoms obtained by thermal dissociation of F_2 was crossed with a beam of D_2, and the angular distribution of DF was measured. The results obtained for an initial relative translational energy of 1.68 kcal/mole are summarized in Fig. 6-10 as a center-of-mass flux velocity-angle contour map. The circles are the highest possible relative translational energy of products consistent with the specified final vibrational state v′. The rotational energy was assumed to be zero, and in fact only low-energy rotational levels are highly populated. This is due to conservation of angular momentum restrictions. The DF is mostly scattered backward, and the intensity peaks in Fig. 6-10 obviously coincide with the vibrational energy spacings. As the initial relative translational energy is decreased, the scattering of DF is shifted backward with respect to the F velocity. Suitable integration of the intensities over all angles and velocities allows the relative populations of the vibrational energy levels to be calculated. These relative populations are summarized in Table 6-4, along with similar data obtained from chemical laser and infrared chemiluminescence experiments. Unfortunately the initial translational energies were not the same, but the results are in reasonable accord. The infrared chemiluminescence experiments were able to provide relative populations for both rotational and vibrational energy states of HF and DF. The relative populations of the rotational energy states for a given vibrational energy are plotted in Fig. 6-11 for the reaction involving deuterium. The height of each line is the relative probability of the final vibrational–rotational energy state. This distribution is plotted versus the fraction of final energy in translation. Obviously, for any final state the energy conservation relationship

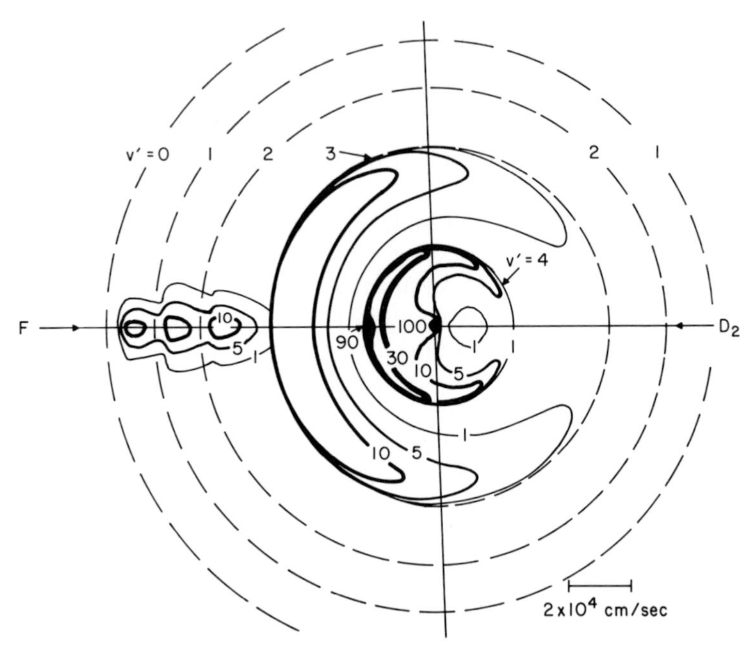

Fig. 6-10 Flux velocity–angle contour map for DF obtained from the reaction $F + D_2 \rightarrow DF + D$. The initial relative translational energy was 1.68 kcal/mole. The circles represent the largest possible value of the final velocity of DF consistent with the vibrational quantum number v′. [Adapted from Y. T. Lee, *in* "Physics of Electronic and Atomic Collisions, VII ICPEAC 1971," Fig. 4a. Reproduced by permission of copyright owner, North–Holland Publishing Co., Amsterdam.]

TABLE 6-4

Relative Populations in the Vibrational States of Nascent DF for the Reaction $F + D_2 \rightarrow DF + D$[a]

v′	Molecular beam[b]	Chemical laser[c]	Infrared chemiluminescence[d]
0	0.04	0.10	—
1	0.07	0.24	0.29
2	0.18	0.56	0.67
3	1.0	1.0	1.0
4	3.5 (0.75)	0.4	0.66

[a] Normalized to 1 in the v′ = 3 state.

[b] Schafer *et al.* [39] and Lee [40]; initial relative translational energy is 2.6 kcal/mole except for the value in parentheses where it is 0.8 kcal/mole.

[c] Berry [42].

[d] Berry [42] and Polyani and Woodall [43].

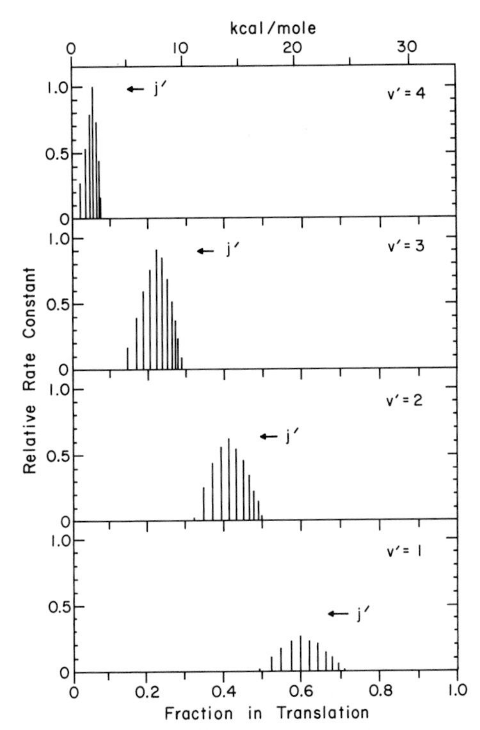

Fig. 6-11 Distribution of the rotational energy level (j') populations for nascent DF formed in the reaction F + $D_2 \rightarrow$ DF + D arranged according to vibrational energy levels (v′) versus the fraction of energy in translation. (Adapted from Polanyi and Woodall [43].)

$E = E_T' + E_R' + E_V'$ must be satisfied. Thus, in this case, a quantitative description of the energy partitioning in translation, vibration, and rotation can be obtained.

Many attempts have been made to construct potential energy surfaces for the reaction(s) of Eq. (6-26). Both semiempirical [44] and ab initio [45] calculations have been made, and some significant differences still exist in the potential energy surfaces that have been constructed. A LEPS potential energy surface for this reaction is shown in Fig. 6-12. The energy barrier with this surface is only 0.9 kcal/mole. The reaction is highly exothermic with $\Delta E_0 = -29$ kcal/mole. The potential energy surface has a saddle point quite far to the right on the H_bF coordinate. Surfaces with an early saddle point are called attractive potential energy surfaces; those with a "late" saddle point (such as the reverse reaction!) are called repulsive potential energy surfaces. On an attractive surface, the heat of reaction is liberated as the reactants are approaching. In the case under consideration, the HF bond length is considerably greater than normal when the heat of reaction is

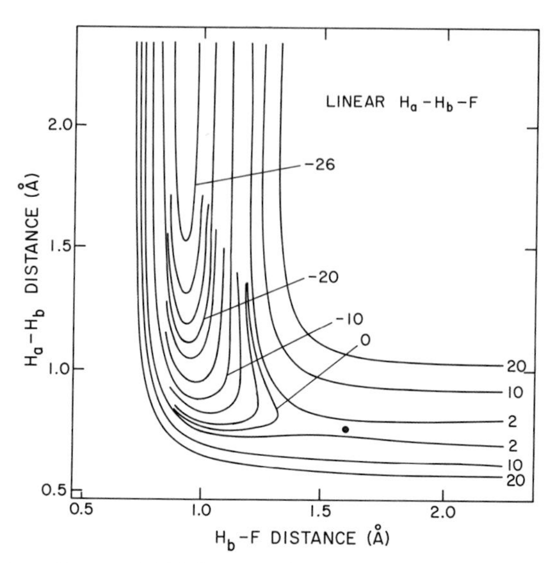

Fig. 6-12 Contour diagram of LEPS potential energy surface for $F + H_2$ system. (Adapted from Muckerman [44].)

released; consequently, much of the heat of reaction is converted into HF vibrational energy. On a repulsive surface, the heat of reaction is released as the products are separating, and the bond length of the product is nearly normal. Therefore, the energy of reaction usually appears as translational energy, rather than vibrational energy. A detailed interpretation of potential energy surfaces in terms of attractive and repulsive character depends on the heat of reaction and on the relative masses of the reactants so that broad generalizations are not possible.

In spite of the uncertainties in the potential energy surface, trajectory calculations have been carried out for the reaction of $F + H_2$ (or D_2) [44, 46, 47]. The results obtained are found to depend on the details of the potential energy surface, but as might be expected with a suitably adjusted surface, the experimental results can be reasonably reproduced in terms of vibrational energy population distributions and other reaction characteristics. In this case, the experimental results and trajectory calculations provide information about the potential energy surface; however, the agreement between experiment and calculations is a necessary but not sufficient condition for establishing the validity of a potential energy surface.

The combination of theoretical calculations and the available experimental methods should provide an increasingly intimate understanding of the dynamics of simple reaction mechanisms; this in turn will provide a basis for understanding complex chemical processes.

Problems

6-1 A mixture of HCl and DCl in an excess of Ar is irradiated by a brief pulse of light from an HCl chemical laser. The time dependence of the fluorescence from the v = 1 levels of HCl and DCl following laser excitation looks as shown in Fig. P6-1a. The pertinent energy level diagrams for HCl and DCl look schematically as shown in Fig. P6-1b. Explain the observed time dependence of the HCl and DCl fluorescence in terms of the energy level diagram and *specific* energy transfer processes.

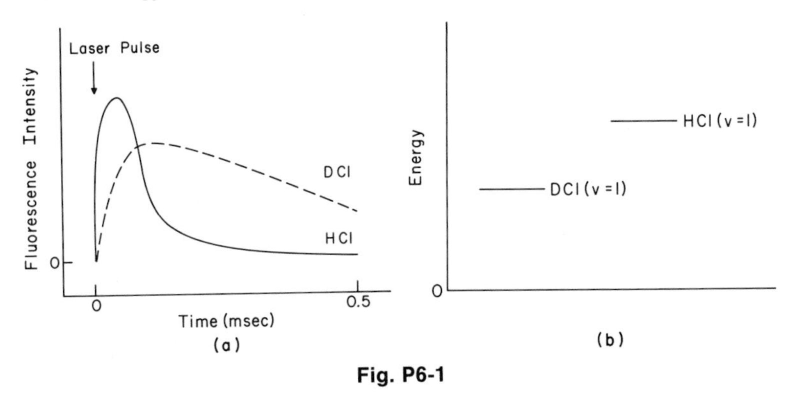

Fig. P6-1

6-2 An electronically excited state of Hg can decay either through a unimolecular fluorescence

$$Hg^* \xrightarrow{k_r} Hg + h\nu$$

or through an energy transfer reaction with N_2

$$Hg^* + N_2(v = 0) \xrightarrow{k_{EV}} Hg + N_2(v')$$

(a) Derive the kinetic equation and the relaxation time for the decay of Hg*.

(b) Using the data given in Table P6-2, derive the rate constants for the two above processes. (The data were obtained at a constant Hg density; adapted from Maitland [48].)

TABLE P6-2

No N_2 present		Plus 9.74×10^{-4} atm N_2	
Relative fluorescence intensity	Time (μsec)	Relative fluorescence intensity	Time (μsec)
1	0	1	0
0.606	5	0.585	3
0.360	10	0.342	6
0.223	15	0.200	9
0.135	20	0.117	12

6-3 A shock is generated in N_2 gas initially at 298°K such that the measured velocity of the shock front is 2.5×10^5 cm/sec. Calculate the temperature of the gas in the shock tube assuming (a) that the vibrational degrees of freedom do not contribute to the heat capacity and (b) that the vibrational degrees of freedom are fully excited.

6-4 For the reaction

$$F + HCl(v) \rightarrow Cl + HF(v')$$

the relative rate constants given in Table P6-4 have been measured [49]. Calculate

TABLE P6-4

v	v′	k (relative)	v	v′	k (relative)
0	1	0.4	1	1	~0
0	2	1.0	1	2	2
0	3	0.1	1	3	4.8
			1	4	0.3

the total relative rate constants for the production of HF when the vibrational temperature of HCl is 298°K and when it is 3000°K. Assume a Boltzmann distribution of vibrational states of HCl. The vibrational frequency of HCl is 2989 cm^{-1}. [Recall that $E = (v + \frac{1}{2})hv$.]

References

1. R. D. Levine and R. B. Bernstein, "Molecular Reaction Dynamics." Oxford Univ. (Clarendon) Press, London and New York, 1974.
2. C. B. Moore, *Acc. Chem. Res.* **2**, 103 (1969).
3. G. W. Flynn, *in* "Chemical and Biochemical Applications of Lasers" (C. B. Moore, ed.), p. 163. Academic Press, New York, 1974.
4. I. W. M. Smith, *Acc. Chem. Res.* **9**, 161 (1976).
5. T. L. Cottrell and J. C. McCoubrey, "Molecular Energy Transfer in Gases." Butterworth, London, 1961.
6. B. H. Mahan, *J. Chem. Phys.* **52**, 5221 (1970).
7. J. I. Steinfeld and P. L. Houston in "Treatise on Spectroscopy" (J. I. Steinfeld, ed.), Vol. I. Plenum, New York, 1977.
8. C. K. Rhodes, M. J. Kelly, and A. Javan, *J. Chem. Phys.* **48**, 5730 (1968).
9. C. B. Moore, *Adv. Chem. Phys.* **23**, 41 (1973).
10. R. D. Sharma and C. A. Brau, *J. Chem. Phys.* **50**, 924 (1969).
11. T. Oka, *Adv. At. Mol. Phys.* **9**, 127 (1973).
12. P. K. Cheo and R. L. Abrams, *Appl. Phys. Lett.* **14**, 47 (1969).
13. C. B. Moore, *J. Chem. Phys.* **43**, 2979 (1965).
14. R. D. Levine and J. Jortner, eds., "Molecular Energy Transfer." Wiley, New York, 1976.
15. R. D. Kern, *in* "Comprehensive Chemical Kinetics" (C. H. Bamford and C. F. H. Tipper, eds.), Vol. 18. Elsevier, New York, 1976.
16. H. Johnston and J. Birks, *Acc. Chem. Res.* **5**, 327 (1972).
17. J. N. Bradley, "Shock Waves in Chemistry and Physics." Wiley, New York, 1962.

18. E. F. Greene and J. P. Toennies, "Chemical Reactions in Shock Waves." Arnold, London, 1964.
19. S. H. Bauer, *Science* **141**, 3584 (1963).
20. D. Britton, N. Davidson, W. Gehman, and G. Schott, *J. Chem. Phys.* **25**, 804 (1956).
21. D. Britton, N. Davidson, and G. Schott, *Discuss. Faraday Soc.* **17**, 58 (1954).
22. J. Troe and H. G. Wagner, *Z. Phys. Chem.* (*Frankfurt am Main*) **55**, 326 (1967).
23. D. Britton, *J. Phys. Chem.* **64**, 742 (1960).
24. D. Britton and N. Davidson, *J. Chem. Phys.* **25**, 810 (1965).
25. T. A. Jacobs and R. R. Giedt, *J. Chem. Phys.* **39**, 749 (1963).
26. R. A. Carabetta and H. B. Palmer, *J. Chem. Phys.* **46**, 1333 (1967).
27. C. D. Johnson and D. Britton, *J. Phys. Chem.* **68**, 3032 (1964).
28. D. J. Seery and D. Britton, *J. Phys. Chem.* **70**, 4074 (1966).
29. E. A. Sutton, *J. Chem. Phys.* **36**, 2923 (1962).
30. R. W. Patch, *J. Chem. Phys.* **36**, 1919 (1962).
31. J. P. Rink, *J. Chem. Phys.* **36**, 262 (1962).
32. A. L. Myerson and W. S. Watt, *J. Chem. Phys.* **49**, 425 (1968).
33. O. L. Anderson, United Aircraft Corp. Res. Lab. Rep. R-1828-1 (1961).
34. M. Camas and A. Vaughn, *J. Chem. Phys.* **34**, 460 (1961).
35. S. R. Byron, *J. Chem. Phys.* **44**, 1378 (1966).
36. B. Cary, *Phys. Fluids* **8**, 26 (1965).
37. H. O. Pritchard, *Acc. Chem. Res.* **9**, 99 (1976).
38. S. H. Bauer, D. Hilden and P. Jeffers, *J. Phys. Chem.* **80**, 922 (1976).
39. T. P. Schafer, P. E. Siska, J. M. Parson, F. P. Tully, Y. C. Wong, and Y. T. Lee, *J. Chem. Phys.* **53**, 3385 (1970).
40. Y. T. Lee, *in* "Physics of Electronic and Atomic Collisions, VII ICPEAC 1971," pp. 357–372. North-Holland Publ., Amsterdam, 1972.
41. J. H. Parker and G. C. Pimentel, *J. Chem. Phys.* **51**, 91 (1969).
42. M. J. Berry, *J. Chem. Phys.* **59**, 6229 (1973).
43. J. C. Polanyi and K. B. Woodall, *J. Chem. Phys.* **57**, 1574 (1972).
44. J. T. Muckerman, *J. Chem. Phys.* **54**, 1155 (1971).
45. C. F. Bender, P. K. Pearson, S. V. O'Neil, and H. F. Schaefer, III, *J. Chem. Phys.* **56**, 4626 (1972).
46. R. L. Jaffe, J. M. Henry, and J. B. Anderson, *J. Chem. Phys.* **59**, 1128 (1973).
47. P. Whitlock and J. T. Muckerman, *J. Chem. Phys.* **61**, 4618 (1974); and J. T. Muckerman, private communications cited in Refs. 1 and 42.
48. C. G. Maitland, *Phys. Rev.* **92**, 637 (1953).
49. J. L. Kirsch and J. C. Polanyi, *J. Chem. Phys.* **57**, 4498 (1972).

CHAPTER 7

KINETICS IN LIQUID SOLUTIONS

7-1 INTRODUCTION

Reactions in liquids differ markedly from reactions in the gas phase because of the presence of solvent molecules, which are always in intimate contact with the reactants and, in fact, often interact strongly with them. The most important consequence of this interaction is that ions are often stable species in liquid systems. This is because the energy required to dissociate molecules into ions is usually more than compensated by the energy released from the process of ion solvation. According to the results obtained earlier, Eqs. (2-69) and (2-95), the specific rate constant in condensed phases for a nonideal system can be expressed in terms of transition-state theory as

$$k = \frac{kT}{h} K^{\ddagger} \frac{\gamma_A{}^a \gamma_B{}^b \cdots}{\gamma_{M^{\ddagger}}} \tag{7-1}$$

or, in terms of a macroscopic empirical theory for a bimolecular reaction, as

$$k = \nu_0 \frac{\kappa 4\pi N_0 a^3}{3000} e^{-U(a)/kT} e^{-\epsilon_a/kT} \tag{7-2}$$

where the symbols have been previously defined. In liquid solutions, elementary steps will rarely be more than bimolecular with respect to solute, since the probability of a three-body solute collision is small and, unlike the gas phase, solvent molecules are readily available for energy transfers. This does not, of course, imply anything about the reaction order or the composition of the activated complex, which may contain any number of solute molecules.

We shall now examine Eqs. (7-1) and (7-2) in some detail to see how they can be used to correlate reaction rates in solution. For the sake of simplicity, we shall usually consider only second-order reactions, although the extension to more complex situations is straightforward. In the case of transition-state theory, solute–solute and solute–solvent interactions can be regarded as causing deviations from some arbitrary reference state. The free-energy change required to go from the reference state to the actual state ΔG_i^n may be related to the activity coefficients as in ordinary thermodynamics, that is,

$$\Delta G_i^n = RT \ln \gamma_i \tag{7-3}$$

Thus the ratio $\gamma_A{}^a \gamma_B{}^b \cdots / \gamma_{M\ddagger}$ is equal to $\exp(-\Delta G^n/RT)$, where $\sum_i \nu_i \Delta G_i^n = \Delta G^n$ (ν_i is equal to the stoichiometric coefficient and has a positive sign for the transition state and a negative sign for reactants) is the difference in the free energy between the activated complex and reactants when going from the reference state to the actual state. Thus the effect of nonideality is simply to add ΔG^n to the free energy of activation $\Delta G^{0\ddagger}$. Equation (7-2) can be used when considering solute–solute interactions if the potential energy of interaction for the reactants is known, but solvent–solute interactions cannot be handled with this equation. Of course, both Eqs. (7-1) and (7-2) will give identical results for a given situation, as will be illustrated.

7-2 INTERMOLECULAR POTENTIALS IN LIQUIDS

Before proceeding further, a brief discussion of intermolecular potentials in liquids will be useful. The interaction of solute with solvent and nonionic solute–solute interactions can sometimes be described with potential functions similar to those used in the gas phase. However, writing down exact equations for even a relatively simple system is quite difficult because of the fact that each solute interacts with a large number of solvent molecules in a relatively unordered structure. We shall find that in liquid solutions, long-range electrostatic forces often predominate over the short-range van der Waals and repulsive forces, which are of great importance in gas-phase interactions. In the absence of long-range interactions, a square-well potential is often used because of its mathematical convenience; if long-range interactions are present, the potential usually used is a hard-sphere cutoff plus the appropriate long-range terms.

First we discuss solutions in which the ionic strength is negligible, i.e., infinitely dilute solutions. The electrical potential of a single ion ψ_i is

$$\psi_i = z_i e/(\varepsilon r) \tag{7-4}$$

and the potential energy of interaction (or the work required to take the charges from r to infinity) between two point charges is simply

$$U = \tfrac{1}{2}(z_1 e \psi_2 + z_2 e \psi_1) = z_1 z_2 e^2/(\varepsilon r) \tag{7-5}$$

where z_1 and z_2 are the ionic valences, e the electronic charge, ε the macroscopic dielectric constant, and r the distance of separation between the

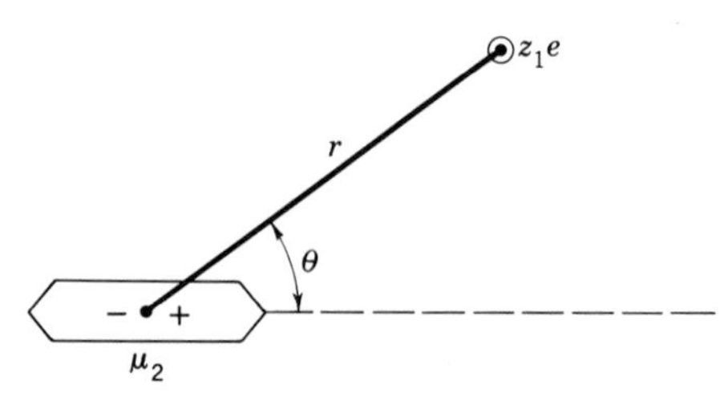

Fig. 7-1 Ion–dipole interaction.

charges [or the distance between the ion and a test charge in the case of Eq. (7-4)]. The potential energy, or work required to charge an ion and bring it from r to infinity (with respect to a unit charge at $r = 0$), for a single ion is

$$U_i = \tfrac{1}{2} z_i e \psi_i = z_i^2 e^2/(2\varepsilon r)$$

This interaction energy can often be of the order of magnitude of chemical bond energies: for example, if $\varepsilon = z_1 = z_2 = 1$ and $r = 4$ Å, U is equal to 80 kcal/mole. Of course, in liquids ε is greater than 1, so that the interaction energy is correspondingly lowered.

The potential energy of interaction of a point charge with a dipole is

$$U = \frac{z_1 e \mu_2}{\varepsilon r^2} \cos\theta \tag{7-6}$$

where μ_2 is the dipole moment and θ is the angle between the axis of the dipole and the point charge (cf. Fig. 7-1). This interaction can be important for both solvent–solute and solute–solute interactions. Assuming $\mu = 2$ debyes, $\varepsilon = z_1 = 1, \theta = 0°$, and $r = 4$ Å, $U = 8.5$ kcal/mole. While this energy is somewhat smaller than most bond energies, heats of reaction are of this order of magnitude. The angle-averaged potential is more useful, since in most situations it will be the effective intermolecular potential. This average is obtained by averaging the ion–dipole potential over all angles with respect to a Boltzmann energy distribution

$$\bar{U} = \frac{\int_0^\pi e^{-(U'/kT)\cos\theta} U' \cos\theta \sin\theta \, d\theta}{\int_0^\pi e^{-(U'/kT)\cos\theta} \sin\theta \, d\theta}$$

$$\bar{U} = -U'\left(\coth\frac{U'}{kT} - \frac{kT}{U'}\right) \tag{7-7}$$

and

$$\bar{U} \cong \frac{z_1^2 e^2 \mu_2^2}{3\varepsilon^2 kT r^4} \qquad \text{for} \quad \frac{U}{kT} < 1 \tag{7-8}$$

where $U' = z_1 e\mu_2/(\varepsilon r^2)$. Note that this average interaction is always attractive. Using the same parameters as above at 298°K, we obtain $\bar{U} = -7.9$ kcal/mole. Since this potential decreases as the inverse square of the dielectric constant, it is considerably lower in liquids.

Finally we shall consider the effect of ionic strength s on the intermolecular potential. The potential of an ion as a function of ionic strength can be calculated from the Debye–Hückel theory of electrolytes for low ionic strengths ($<10^{-2}$ M for 1 : 1 electrolytes and even smaller concentrations for higher valences) and is [1]

$$\psi_i = \frac{z_i e}{\varepsilon r} \frac{e^{\beta a}}{1 + \beta a} e^{-\beta r}, \qquad \beta^2 = \frac{8\pi N_0 e^2}{1000 \varepsilon k T} s, \qquad s = \tfrac{1}{2} \sum_i c_i z_i^2 \tag{7-9}$$

Here a is the distance of closest approach between ions. It is usually assumed to be the same for all ions, for although such an assumption is obviously not strictly valid, the convenience it embodies usually compensates for the slight error it entails. The potential energy of interaction, therefore, is

$$U = \frac{z_1 z_2 e^2}{\varepsilon r} \frac{e^{\beta a}}{1 + \beta a} e^{-\beta r} \tag{7-10}$$

and the potential energy of a single ion is

$$U_i = \frac{z_i^2 e^2}{2\varepsilon r} \frac{e^{\beta a}}{1 + \beta a} e^{-\beta r} \tag{7-11}$$

We shall often be interested in the potential at $r = a$. Equations (7-10) and (7-11) are then

$$U = \frac{z_1 z_2 e^2}{\varepsilon a} - \frac{z_1 z_2 e^2 \beta}{\varepsilon(1 + \beta a)} \tag{7-12}$$

$$U_i = \frac{z_i^2 e^2}{2\varepsilon a} - \frac{z_i^2 e^2 \beta}{2\varepsilon(1 + \beta a)} \tag{7-13}$$

Both Eqs. (7-12) and (7-13) contain a simple coulombic term plus an additional term embodying the effect of the ion atmosphere. The presence of an ion atmosphere decreases the interaction between two charges, as might be expected. No reliable potential function exists for high ionic strengths.

The effect of ionic concentrations on other intermolecular potentials is difficult to assess quantitatively, although qualitatively, increasing ionic strength decreases the strength of the interactions between charges. Thus increasing the ionic strength decreases the potential energy for interactions involving similar charges, and increases it in the case of interactions between unlike charges.

A point that should be stressed is that a macroscopic dielectric constant has been assumed in all these equations. In actuality, the dielectric constant changes rapidly in the immediate vicinity of a charged particle, so that the effective dielectric constant may be quite different (smaller) from the macroscopic dielectric constant. Also, in all cases involving dipoles, the distance between the ends of the dipole has been assumed to be negligible compared to r. In general, the effect of solute–solute interactions on the rate constant can be predicted by inserting the appropriate intermolecular potential into Eq. (7-2).

7-3 REACTIONS IN IDEAL SOLUTIONS

The calculation of $\Delta G''$ depends on the reference state chosen, and we shall choose various reference states, depending upon the problem of interest. In order to assess the effect of a liquid solvent, let us take as a reference state an ideal gas at a given temperature and pressure and calculate $\Delta G''$ for going from an ideal gas to a dilute ideal solution at the same temperature and pressure. This $\Delta G''$ is simply equal to

$$RT \ln \frac{K^{\ddagger}{}_{l}}{K^{\ddagger}{}_{g}} = RT \ln\left(\frac{C^{\ddagger}{}_{l} C_{Ag} C_{Bg} \cdots}{C^{\ddagger}{}_{g} C_{Al} C_{Bl} \cdots}\right) \tag{7-14}$$

where the subscript l designates the liquid phase and g the gas phase. For ideal gases, the pressure $P_i = C_{ig}RT$; for ideal liquid solutions, $P_i = X_i P_i{}^0$, where $P_i{}^0$ is the vapor pressure of pure i and X_i is the mole fraction of i in solution. Finally, for dilute solutions,

$$X_i = C_{il}/C_s = C_{il} V_s{}^0$$

where C_s is the concentration of solvent and $V_s{}^0$ is the solvent molar volume. Inserting these results into Eqs. (7-1) and (7-14), we obtain

$$\frac{k_l}{k_g} = \frac{K^{\ddagger}{}_{l}}{K^{\ddagger}{}_{g}} = \frac{P_A{}^0 P_B{}^0 \cdots}{P^0_{M\ddagger}} \left(\frac{V_s{}^0}{RT}\right)^{n-1} \tag{7-15}$$

where n is the number of reactants in the activated complex. From thermodynamics

$$P_A{}^0 = \exp(-\Delta H^0_{Av}/RT)\exp(\Delta S^0_{Av}/R) = \exp(-\Delta G^0_{Av}/RT)$$

where ΔH^0_{Av} and ΔS^0_{Av} are the enthalpy and entropy of vaporization of A, respectively. Using the free-volume theory of liquids [2], the term $\exp(\Delta S_{Av}/R)$ can be written as RT/V_{fA}, where V_{fA} is the free volume per mole of A, i.e., the molar volume minus the actual volume of the molecules. Empirically V_f turns out to be equal to about 0.5 cc/mole for most liquids (within an order of magnitude). Therefore,

$$k_l = k_g \frac{(V_s^0)^{n-1} V^{\ddagger}{}_f}{V_{fA} V_{fB} \cdots} \exp\left(\frac{\Delta H_v^0}{RT}\right) \tag{7-16}$$

where n is the number of reactant molecules in the activated complex, and ΔH_v^0 is the difference in heats of vaporization of $M^{\ddagger}$ and the reactants.

Thus, for a first-order reaction, the preexponential factor should be the same in both phases, while for second-order reactions, the preexponential factor should be greater in the liquid by the factor $V_s^0 V^{\ddagger}{}_f / V_{fA} V_{fB}$. In general, $V_s^0 / V_{fi} \approx 100$ and $V^{\ddagger}{}_f / V_s^0 \approx n/100$, where n is the number of reactants, so that this effect should become quite large for higher-order reactions. On the other hand, ΔH_v^0 is about zero unless the activated complex is much more or much less polar than the reactants. A highly polar activated complex would have a high heat of vaporization, and hence the reaction would have a smaller enthalpy of activation in liquid solution. Actually, very few experiments allowing a test of these results have been carried out; however, in the few cases studied, both the activation energy and preexponential factor are practically unchanged in going from the gas phase to inert solvents [3].

7-4 REACTIONS BETWEEN POLAR MOLECULES

If a simple electrostatic model (neglecting ionic-strength effects) is considered, the effect of solvent on reaction rates of polar molecules can be assessed by calculating the free energy of solvation of a spherical molecule of radius r containing a point dipole of magnitude μ_i at its center. The value obtained by Kirkwood [4] with a reference state of $\varepsilon = 1$ (all other state variables being held constant) is

$$\Delta G_i^n = -\frac{\mu_i^2}{r^3} \frac{\varepsilon - 1}{2\varepsilon + 1} \tag{7-17}$$

Therefore the rate constant for the reaction

$$A + B \rightleftharpoons X^{\ddagger} \rightarrow \text{products}$$

can be written as

$$\ln k = \ln k_0 + \frac{1}{kT} \frac{\varepsilon - 1}{2\varepsilon + 1} \left[\frac{\mu_{\ddagger}^2}{r_{\ddagger}^3} - \frac{\mu_A^2}{r_A^3} - \frac{\mu_B^2}{r_B^3} \right] \tag{7-18}$$

where k_0 is the rate constant in a medium of unit dielectric constant. As might be expected, this equation predicts that the rate constant increases with increasing dielectric constant if the activated complex is more polar than the reactants. Although Eq. (7-18) has been successfully used to correlate some kinetic results, its success has been limited. The main reason for this is that the energy of such interactions is quite small, so that van der Waals interactions and repulsive forces are not negligible; also, the macroscopic dielectric constant is not adequate for describing the molecular interactions. Obviously dipole–dipole and dipole–induced dipole interactions between solutes can be taken into account using Eqs. (7-1) and (7-2) together with the appropriate intermolecular potentials, but the results would really not be an adequate description of any real system for reasons similar to those already given.

7-5 REACTIONS BETWEEN IONS AND DIPOLES

Since the magnitude of ion–dipole interactions is not very large, deviations from nonideality should not be large. Therefore we expect "normal" preexponential factors and activation energies. The few representative examples cited in Table 7-1 all have approximately the same preexponential factors and activation energies. Therefore reactions having activation energies of approximately 20 kcal/mole and preexponential factors of approximately 10^{11} to 10^{12} M^{-1} sec^{-1} at room temperature can be considered as "normal" solution reactions.

A quantitative evaluation of the effect of ion–dipole interactions on the rate constant in an infinitely dilute solution of ions can be obtained by inserting the ion–dipole potential energy of interaction from Eqs. (7-7) and (7-8) into Eq. (7-2) or by calculating the free energy $\Delta G''$ necessary to bring the ion and dipole together and then using transition-state theory. Since

$$\Delta G'' = -U'\left(\coth \frac{U'}{kT} - \frac{kT}{U'}\right) \cong -\frac{z_1{}^2 e^2 \mu_2{}^2}{3\varepsilon^2 kTr^4} \tag{7-19}$$

both Eqs. (7-1) and (7-2) can then be written as

$$\ln k = \ln k_0 + \frac{1}{3}\left(\frac{z_1 e \mu_2}{\varepsilon kTr^2}\right)^2, \qquad U < kT \tag{7-20}$$

TABLE 7-1

Kinetic Constants for Ion–Dipole Reactions

Reaction	E_a (kcal/mole)	A (M^{-1} sec^{-1})	Reference
$CH_3I + C_2H_5O^-$	19.5	2.42×10^{11}	5
$C_2H_5Br + OH^-$	21.4	4.30×10^{11}	6
$C_2H_5I + C_2H_5O^-$	20.7	1.99×10^{11}	7
$C_3H_7I + C_6H_5O^-$	22.5	3.53×10^{11}	8

Here k_0 can be regarded as the rate constant in a medium of infinite dielectric constant. This model predicts that increasing the dielectric constant slows down the reaction rate. Actually, quantitative verification of Eq. (7-20) has not been found and, in fact, is not to be expected, since the free energies involved are of the same order of magnitude as free energies of solvation.

An alternative approach is to attribute the effect of the solvent–solute interaction to the change in free energy caused by the fact that the reactant ion and the transition-state complex have different radii, i.e., neglecting the effects of the dipoles ($\gamma_2 = 1$). The calculation of ΔG^n for this case parallels previous examples and gives

$$kT \ln \gamma_i = \Delta G_i^n = \int_0^{z_i e} \int_r^\infty \frac{q}{\varepsilon r^2}\, dq\, dr = \frac{1}{2}\frac{z_i^2 e^2}{\varepsilon r} \tag{7-21}$$

Therefore the rate constant can be written

$$\ln k = \ln k_0 + \frac{z_1^2 e^2}{2\varepsilon kT}\left(\frac{1}{r} - \frac{1}{r^\ddagger}\right) \tag{7-22}$$

Since $r^\ddagger > r$, both Eqs. (7-21) and (7-22) predict qualitatively similar results. Here again, quantitative correlation with experiment has not been established. Several more complex theories have been developed, but it is doubtful whether they will be of much use in correlating experimental results because of the inherent difficulties in treating such weak interactions.

7-6 REACTIONS BETWEEN IONS

Reactions between ions are of great importance in liquids. The effect of solvent, ionic charge, and ionic strength on the rate constant can be taken into account by making use of the Debye–Hückel potential and calculating the appropriate activity coefficients. Usually, activity coefficients in ionic solutions are calculated neglecting the potential of the ion itself; i.e., only the effect of other ions on the potential is taken into account. However, we use the complete potential to obtain a general expression for the rate constant, and we then consider some special cases. Since the potential energy of an ion is simply the work of charging and taking the ion from r to infinity, ΔG_i^n is given directly by Eq. (7-13) and is

$$kT \ln \gamma_i = \Delta G_i^n = \frac{z_i^2 e^2}{2\varepsilon a} - \frac{z_i^2 e^2 \beta}{2\varepsilon(1 + \beta a)} \tag{7-23}$$

Note that the reference state for the activity coefficient depends on the variable under consideration. The reference state with respect to the dielectric constant is a hypothetical solvent with infinite dielectric constant at a given temperature and pressure. On the other hand, with respect to ionic concentration, the reference state is a solution of zero ionic concentration (at a given T and P). The equation for the rate constant follows directly from the transition-state formalism:

$$\ln k = \ln k_0 - \frac{z_1 z_2 e^2}{\varepsilon a k T} + \frac{z_1 z_2 e^2 \beta}{\varepsilon k T(1 + \beta a)} \tag{7-24}$$

This equation also follows from Eq. (7-2) if the intermolecular potential energy given in Eq. (7-12) is used.

First consider the case where $\beta = 0$ (zero ionic strength), which is equivalent to a simple coulombic interaction. Equation (7-24) predicts that $\ln k$ versus $1/\varepsilon$ should be a straight line, the slope of the line being negative for reactions between ions of the same sign, and positive for charges of opposite sign. This prediction has been tested for several different reactions, and the results are in semiquantitative agreement with theory [9]. Actually, quantitative agreement would be surprising, since the macroscopic dielectric constant for mixed solvents is probably not the correct parameter to use for molecular interactions.

The effect of electrostatic interactions on the experimentally observed rate parameters, namely, the preexponential factor and activation energy, can be predicted quite easily for infinitely dilute solutions. The nonideal free energy can be divided into two terms, the nonideal entropy

$$\Delta S^n = -\left(\frac{\partial\, \Delta G^n}{\partial T}\right)_P = \frac{z_1 z_2 e^2}{\varepsilon a}\left(\frac{\partial \ln \varepsilon}{\partial T}\right)_P \tag{7-25}$$

and the nonideal enthalpy

$$\Delta H^n = \Delta G^n + T\,\Delta S^n = \frac{z_1 z_2 e^2}{\varepsilon a}\left[1 + T\left(\frac{\partial \ln \varepsilon}{\partial T}\right)_P\right] \tag{7-26}$$

The nonideal entropy contributes to the preexponential factor and, since $(\partial \ln \varepsilon/\partial T)_P$ is usually negative, predicts that the preexponential factor will be larger than normal for reactions between oppositely charged ions and smaller than normal for reactions between like-charged ions. Actually, this is just what we should predict on a purely physical interpretation of the entropy of activation. For ions of the same sign, the transition state is a more highly charged ion, which is more strongly solvated than the reactants, so that a relative decrease in entropy in forming the transition state is to be expected. For a reaction between ions of opposite charge, the transition state is desolvated, and hence the entropy is increased. This is generally the experimental behavior observed, and three illustrations are cited in Table 7-2. Also included is an exception to this rule, indicating that the situation is

TABLE 7-2

Kinetic Constants of Ion–Ion Reactions

Reaction	A (M^{-1} sec^{-1})	$\Delta S^{0\ddagger}$ (eu)	E_a (kcal/mole)	Reference
$Co(NH_3)_5Br^{2+} + OH^-$	4.2×10^{17}	20.1	23.5	10
$S_2O_3^{2-} + SO_3^{2-}$	2.3×10^{6}	−31.4	14.5	11
$Cr(H_2O)_6^{3+} + SCN^-$	2.3×10^{13}	0.7	25.7	12

not so simple as the above analysis indicates. The effect on the enthalpy or energy of activation depends on the particular solvent. In general, the activation energy is relatively insensitive to the dielectric properties of the solvent and long-range interactions because the dominant contribution to the activation energy for most reactions is from short-range repulsive forces which are found when the reactants are brought close together. This is largely a property of the reactants themselves.

A brief qualitative consideration of the inverse reaction, namely, the production of ions, or ionization, is illuminating. Since the products are ions and the reactant is usually a neutral molecule, the transition state has an intermediate structure. This means that the transition state is more polar than the reactants and therefore more solvated. But such a situation causes a large entropy decrease, and ionization processes should have relatively large negative entropies of activation or abnormally small preexponential factors. Raising the dielectric constant should, of course, increase the rate of reaction. These predictions have been well verified by experiment [13].

Including the effect of ionic strength would not alter the conclusions significantly, but a quantitative evaluation of ion–ion interactions is of considerable interest. Let us consider the effect of ionic strength on the rate constant in a medium of fixed dielectric constant at a given temperature. We can then write Eq. (7-24) as

$$\ln k = \ln k_0' + 2z_1z_2Bs^{1/2} \tag{7-27}$$

where B is defined by Eqs. (7-24) and (7-27) for the limiting case of very low ionic strength and $(1 + \beta a)$ is replaced by unity. (At 25°C in H_2O, $\log k/k_0' = 1.02z_1z_2s^{1/2}$). This equation predicts that the rate constant should increase as the ionic strength increases for reactions between ions of like sign and decrease with increasing ionic strength for oppositely charged ion reactants. This effect is called the primary salt effect, and Eq. (7-27) is quantitatively obeyed at low ionic strengths ($s < 0.01$ for 1:1 electrolytes) (cf. [14]). Significant deviations occur, however, at high ionic strengths. This is simply a reflection of the inadequacy of the Debye–Hückel potential at high ionic strengths. More exact equations can be obtained by using an empirical formula for the activity coefficient, such as

$$\ln \gamma_i = -\frac{z_i^2 B_i s^{1/2}}{1 + s^{1/2}} + \sum_j D_{ij} c_j \tag{7-28}$$

where the empirical constants D_{ij} are summed only over ions of sign opposite to i [15].

7-7 SECONDARY SALT EFFECT

In many cases, particularly acid–base catalysis, one of the ionic species whose concentration appears in the rate law is in rapid equilibrium with other species. This ionic concentration, then, depends on the ionic strength. For example, assume that the concentration of H^+ appears in the rate law and that the following type of equilibrium exists:

$$HA^z \rightleftharpoons H^+ + A^{z-1}$$

$$K = \frac{(H^+)(A^{z-1})}{(HA^z)} \frac{\gamma_{H^+}\gamma_{A^{z-1}}}{\gamma_{HA^z}} \tag{7-29}$$

or

$$\ln(H^+) = \ln K + \ln \frac{(HA^z)}{(A^{z-1})} + \ln \frac{\gamma_{HA^z}}{\gamma_{A^{z-1}}\gamma_{H^+}} \tag{7-30}$$

Thus the concentration of H^+ is dependent on the ionic strength through the activity-coefficient ratio. For example, assuming the Debye–Hückel limiting law to be applicable, we find

$$\ln(H^+) = \ln K + \ln \frac{(HA^z)}{(A^{z-1})} - 2(z - 1)Bs^{1/2} \tag{7-31}$$

This secondary salt effect can cause either an increase or decrease in the rate constant and can be either larger or smaller than a primary salt effect. (If the rate expression is written in terms of the true hydrogen-ion concentration, a secondary salt effect obviously does not exist.) In some instances, both the primary and secondary salt effects occur, in which case cancellation of effects may cause the reaction rate to appear to be independent of ionic strength.

7-8 OTHER SALT EFFECTS

Even the activity coefficients of neutral molecules often depend on the ionic strength (especially at high ionic strengths); therefore, reactions involving neutral molecules can be influenced by varying the ionic strength. Qualitatively, the general effect of ionic strength on reaction rates can be predicted by using the transition-state model. For example, if the transition state is more polar than the reactant molecules, adding ions increases the rate, since each end of the dipole in the transition state surrounds itself with ions of the opposite charge and hence is stabilized.

This discussion points out the fact that the ionic strength should be carefully controlled in any kinetic experiment. Furthermore, if salt effects are observed, they must be interpreted with care in terms of mechanism because of the many possible sources of these effects.

7-9 CONVERSION OF AMMONIUM CYANATE TO UREA

Let us consider a reaction which has been thoroughly studied by kinetic methods, the synthesis of urea in aqueous and alcoholic solutions according to the reaction

$$NH_4^+ + CNO^- \rightarrow (NH_2)_2CO \tag{7-32}$$

The equilibrium constant for this reaction (at room temperature) is about $10^4\ M^{-1}$, so that the rate of the back reaction can usually be neglected. The first kinetic study of this reaction was published in 1895 by Walker and Hambly [16], who found the rate law

$$R = k(NH_4^+)(CNO^-) \tag{7-33}$$

The rate was increased if potassium cyanate or ammonium sulfate was added to the solution, but the addition of ammonia had very little effect on the rate, leading to the conclusion that the reaction occurred between the ions and not between ammonia and cyanic acid. Shortly later, in 1912, F. D. Chattaway [17] pointed out that because the equilibrium between the possible pairs of reactants is rapidly established,

$$NH_4^+ + CNO^- \rightleftharpoons NH_3 + HCNO \tag{7-34}$$

the two mechanisms are kinetically indistinguishable, and the rate can be written either as in Eq. (7-33) or as

$$R = k'(NH_3)(HCNO) = \frac{k'K_w}{K_A K_B}(NH_4^+)(CNO^-) \tag{7-35}$$

where $K_w/K_A K_B$ is the equilibrium constant for the reaction in Eq. (7-34). Addition of ammonia to the reaction mixture would not increase the rate, since the cyanic acid concentration is reduced a corresponding amount. It is worth noting that at 25°, K_A is $2 \times 10^{-4}\ M$, $K_w = 10^{-14}\ M^2$, and K_B is $2 \times 10^{-5}\ M$, so that $K_w/K_A K_B$ is 2.5×10^{-6} and the concentrations of ammonia and cyanic acid are ordinarily very small (unless one or the other is independently added to the reaction mixture).

In an effort to distinguish between these two mechanisms, extensive studies were made to test the predictions of electrostatic theories for reactions between ions. Warner and Stitt [18] found a large negative salt effect; at high dilutions the slope of a plot of $\ln k$ versus $s^{1/2}$ was a straight line with a slope correctly predicted by Eq. (7-27). Studies in mixed solvents [19] (water–alcohols, water–dioxane) revealed that a plot of $\ln k$ versus $1/\varepsilon$ was linear, as predicted by Eq. (7-24) ($\beta = 0$). Moreover, reasonable values of the distance of closest approach were obtained. Also, the change in activation energy or enthalpy upon changing the dielectric constant as given by Eq. (7-26) was found to correlate the data quite well.

The discussion thus far might lead one to say that the experimental evidence indicates that the initial collision process occurs between the ions. Actually, because of the rapidly established equilibrium [Eq. (7-34)], the ionic mechanism is no more probable on a kinetic basis than the collision process involving neutral molecules. To see why this is so we need only write down the rate equation for both mechanisms, including activity coefficients. For the ionic reactants,

$$R = k(NH_4^+)(CNO^-)\frac{\gamma_+\gamma_-}{\gamma_{M^\ddagger}} \tag{7-36}$$

and for the neutral reactants

$$R = k'(NH_3)(HCNO)\frac{\gamma_{NH_3}\gamma_{HCNO}}{\gamma_{M^\ddagger}} \tag{7-37}$$

However, since

$$(NH_3)(HCNO) = \frac{K_w}{K_A K_B}(NH_4^+)(CNO^-)\frac{\gamma_+\gamma_-}{\gamma_{NH_3}\gamma_{HCNO}}$$

Eq. (7-37) can be written as

$$R = \frac{k' K_w}{K_A K_B}(NH_4^+)(CNO^-)\frac{\gamma_+\gamma_-}{\gamma_{M^\ddagger}} \tag{7-38}$$

Thus both mechanisms are indistinguishable as far as electrostatic theories are concerned.

It is clear that kinetic studies cannot distinguish between the two mechanisms. One possibility is that a consideration of the preexponential factors and activation energies of the rate constants k and k' might lead to a decision. From experiments in aqueous solution at an ionic strength of 0.0376 M (Svirbely and Warner [19]),

$$k = 4 \times 10^{12} e^{-23{,}200/RT} \quad M^{-1}\,\text{sec}^{-1}, \qquad k' \approx 10^{8} e^{-11{,}000/RT} \quad M^{-1}\,\text{sec}^{-1}$$

Both these rate constants have reasonable preexponential factors and activation energies for the types of reaction being considered. However, if we look at the chemistry of the process, the reaction between neutral molecules seems more probable. This is because the unpaired electrons of the nitrogen can easily combine with the positively polarized carbon of cyanic acid, followed by the sterically easy synthesis of urea (I).

```
H—N═C═O → H—N═C—O⁻ → HN—C═O
                |         |   |
HNH           HNH⁺        H  NH
 H              H             H
```

I

On the other hand, the synthesis of urea through reaction of ammonium and cyanate ions is difficult to imagine (though certainly not impossible), since the carbon–nitrogen bond which must be formed is blocked by the presence of the fourth hydrogen atom in NH_4^+.

The reaction just discussed points out several features which are worth noting. *The rate law, in general, gives information only about the composition of the activated complex. It does not indicate the detailed pathway to the complex because of the assumption of rapid establishment of equilibrium between the complex and the reactants.* Whenever reactants are in rapid equilibrium with other species, i.e., rapid compared to the rate-controlling step, kinetic and thermodynamic quantities become intertwined in an inseparable manner. All the equations derived for the dependence of the rate constant on ionic strength, dielectric constant, etc., are also applicable to thermodynamic equilibria; in general, it is not possible to distinguish between these two situations. The two proposed mechanisms are, in principle, kinetically indistinguishable. However, we must always remember that a reaction mechanism must be chemically reasonable, and this criterion often causes one mechanism to be preferred over another.

7-10 ELECTRON-TRANSFER REACTIONS BETWEEN METAL IONS

As a second example of how one interprets kinetic data in terms of mechanism, some simple electron-transfer (oxidation–reduction) reactions in solution are discussed. A complete review of this subject is not intended; instead, we consider mainly the reduction of $Co(NH_3)_5X^{3-n}$ by $Cr^{2+}(aq)$ and $Cr(bipy)_3^{2+}$. Here X^{-n} can be any of a number of ligands, such as H_2O, Cl^-, Br^-, OH^-, etc., and bipy is α,α'-bipyridine. At a constant pH, the rate law for the oxidation–reduction reaction is first order in each of the reactants. The second-order rate constant is often dependent on pH because the reactants can exist in different ionized forms which react at different rates. We

TABLE 7-3

Second-Order Rate Constants for the Reduction of Some Co(III) Complexes[a]

Oxidant	k for Cr^{2+}	Reference	k for $Cr(bipy)_3^{2+}$	Reference
RNH_3^{3+}	8.9×10^{-5} [b]	23	6.9×10^2	21
ROH_2^{3+}	0.5[c]	24	5×10^4	25
RF^{2+}	$>2 \times 10^6$	25	1.8×10^3	25
RCl^{2+}	$>2 \times 10^6$	25	8×10^5	25
RBr^{2+}	$>2 \times 10^6$	25	5×10^6	25
RI^{2+}	$>2 \times 10^6$	25		
RN_3^{2+}	$\sim 3 \times 10^5$	25	4.1×10^4	25
RSO_4^{+}	18	25	4.5×10^4	25
$ROAc^{2+}$	0.18	26	1.2×10^3	25

[a] k is in M^{-1} sec^{-1} at 25°, $s_{NaClO_4-HClO_4} = 1.0$ M unless otherwise stated, $R = (NH_3)_5Co$, $OAc = CH_3COO^-$.
[b] $s = 0.4$ M.
[c] $s = 1.2$ M.

ignore this complication, however. A summary of some of the second-order rate constants obtained is given in Table 7-3.

At first glance, one might assume that all these reactions proceed by the same mechanism, one involving a simple bimolecular process, but this is not true. A clue to this problem is the observation that the rate constants show considerably more variation with Cr^{2+}(aq) as the reducing agent compared to the bipyridine complex as reductant. Let us consider as examples the reaction of $(NH_3)_5CoCl^{2+}$ with Cr^{2+}(aq) and with $Cr(bipy)_3^{2+}$. When the cobalt complex is prepared with radioactive chlorine, all the radioactivity ends up with the product Cr^{3+} (as $CrCl^{2+}$) if aquated chromous ion is the reactant [20]; if the bipyridine complex is the reductant, no radioactivity is associated with the product ion. The fact that $Cr(bipy)_3^{3+}$ is formed in the latter case, rather than $CrCl^{2+}$, was shown through spectral analysis of the reaction product [21]. Moreover, it can be shown experimentally by putting free radioactive chloride ion in the reaction mixture that the radioactivity cannot be due to a reaction whereby the chloride ion is first free in solution and then combines with Cr^{3+}. This suggests that the Cl penetrates the coordination sphere of Cr^{2+}(aq), forming the intermediate complex

$$(NH_3)_5Co—Cl—Cr(H_2O)_5^{2+}$$

and that electron transfer is accompanied by atom transfer of Cl^-. In the reaction involving $Cr(bipy)_3^{2+}$, this penetration does not occur, so that electron transfer must take place with both inner coordination spheres intact. Thus, even though the observed rate laws are identical, the mechanisms of these two reactions are quite different.

As a result of these and related studies, electron-transfer reactions of metal ions are generally classified as proceeding via "outer-sphere" or "inner-sphere" mechanisms [22]. In the former case, the reactants are usually substitution-inert; the electron transfer is much more rapid than substitution in the first coordination sphere. In the latter case, an atom or a group of atoms is part of the coordination sphere of both reactants, so that the coordination sphere of at least one of the partners has been entered; in this case, the oxidizing agent and oxidized product must undergo substitution slowly compared to electron transfer, but the reducing agent must be labile with respect to substitution. Thus all the reactions cited in Table 7-3 involving $Cr^{2+}(aq)$ occur by an inner-sphere mechanism, whereas those involving $Cr(bipy)_3^{2+}$ occur via an outer-sphere mechanism. As might be expected, reaction rates involving inner-sphere mechanisms are much more sensitive to the nature of X^{-n} than those for outer-sphere mechanisms. An exception to the general statement is the case where $X = NH_3$: here no free-electron pairs are present on the ligand, so that a bridged complex cannot be formed; the result is that the electron transfer occurs very slowly by an outer-sphere mechanism.

Further insight into the inner-sphere mechanism is provided by the fact that if X^{-n} is OH^-, $^{16}OH^-$ is favored over $^{18}OH^-$ in the bridging position. This suggests that stretching of the Co^{3+}—OH bond is part of the activation process. Furthermore, the reaction is considerably slower in D_2O compared to H_2O, which suggests that stretching of the OH bond also takes place in the activation process [22].

In the light of what has just been discussed, the two mechanisms can be depicted formally as

$$\text{I.}\quad (NH_3)_5CoX^{3-n} + Cr(H_2O)_6^{2+} \rightleftharpoons (NH_3)_5CoXCr(H_2O)_5^{5-n} + H_2O$$
$$\rightleftharpoons Co(H_2O)_6^{2+} + 5NH_3 + Cr(H_2O)_5X^{3-n}$$

Inner-sphere mechanism

$$\text{II.}\quad (NH_3)_5CoX^{3-n} + Cr(bipy)_3^{2+} \rightleftharpoons (NH_3)_5CoX(bipy)Cr(bipy)_2^{5-n}$$
$$\rightleftharpoons Co^{2+}(aq) + 5NH_3 + Cr(bipy)_3^{3+}$$

Outer-sphere mechanism

(The exchange of water for ammonia in the inner coordination shell of Co^{2+} is a rapid process and is not included in the mechanisms.) The intermediate compound actually represents two species, one with the electron to be transferred associated with Cr^{2+} and the other with the electron associated with Co^{2+}. The role of the solvent molecules in outer-sphere electron transfers is not known, although it has been suggested that solvent may play a role as a bridging ligand.

A theory of outer-sphere electron transfer reactions has been developed by Marcus [27] and is briefly presented here. Electron-transfer reactions are usually quite slow, which at first glance is surprising. However, the motion of an electron is so rapid that the solvent and nuclei do not have time to orient themselves during a single-electron transfer, and in fact it is the adjustment of the solvent and reactants to a nonequilibrium state permitting electron transfer that is rate determining. This is basically a manifestation of the Franck–Condon principle which states that internuclear distances do not change during an electronic transition. The solvent dielectric constant is a measure of the orientation of the solvent molecular dipole in an electric field. If the electric field changes at a frequency faster than about 10^{11} sec^{-1}, the dipole orientation of water lags behind the field, lowering the effective dielectric constant. Electronic motions occur in times shorter than 10^{-11} sec, and solvent molecules are unable to respond to the rapidly varying electric field created by the electronic motion. Therefore, the effective dielectric constant for an electron transfer event is the high frequency dielectric constant (about 2 for water).

A quantitative analysis of electron transfer can be made in terms of the electron-transfer reaction

$$A^{z_A} + B^{z_B} \rightleftharpoons A^{z_A+\Delta z} + B^{z_B-\Delta z} \tag{7-39}$$

The rate constant for the forward reaction can be expressed as

$$k_{AB} = Z_{AB} e^{-\Delta G^*_{AB}/RT} \tag{7-40}$$

where Z_{AB} is a collision frequency in solution and ΔG^*_{AB} is the standard free energy necessary to produce an activated complex from the initial state. (This is not quite the free energy of activation because of the use of Z_{AB} rather than kT/h as a preexponential factor.) The free energy ΔG^*_{AB} consists of two terms:

(1) the work w_{AB} required to bring the reactants to within the reaction distance r_{AB},

$$w_{AB} = \frac{z_A z_B e^2}{\varepsilon_s r_{AB}}$$

where ε_s is the normal static dielectric constant; and

(2) the free energy needed to form the transition state with the solvent orientation adjusted to the nonequilibrium state characterized by the high frequency dielectric constant.

If the transition state is assumed to be formed by the transfer of a fractional charge $m\,\Delta ze$, the second term is simply the free-energy difference in forming the charges $m\,\Delta ze$ and $-m\,\Delta ze$ in a solvent with the high-frequency dielectric constant ε_0 and a solvent with the static dielectric constant ε_s. According to the principles we developed earlier in the chapter, this free-energy difference is

$$\left[\frac{(m\,\Delta ze)^2}{2r_A} + \frac{(-m\,\Delta ze)^2}{2r_B} + \frac{(m\,\Delta ze)(-m\,\Delta ze)}{r_{AB}}\right]\left[\frac{1}{\varepsilon_0} - \frac{1}{\varepsilon_s}\right]$$

where the first two terms are the terms associated with the individual charges formed and the third term is the free energy required to bring the charges together. Therefore,

$$\Delta G^*_{AB} = w_{AB} + m^2\lambda_0 \tag{7-41}$$

with

$$\lambda_0 = (\Delta ze)^2\left(\frac{1}{\varepsilon_0} - \frac{1}{\varepsilon_s}\right)\left(\frac{1}{2r_A} + \frac{1}{2r_B} - \frac{1}{r_{AB}}\right)$$

The value of m is found by deriving a similar expression for the standard free-energy change ΔG^*_{BA} associated with the reverse reaction. Since the charge to be transferred in the reverse direction is $-\Delta z$, the fractional charge transferred is $-\Delta ze - m\,\Delta ze = -(1 + m)\,\Delta ze$ so that

$$\Delta G^*_{BA} = w_{BA} + (1 + m)^2\lambda_0 \tag{7-42}$$

Furthermore, the standard free energy change for the reaction is

$$\Delta G^0_{AB} = \Delta G^*_{AB} - \Delta G^*_{BA} \tag{7-43}$$

Equations (7-41)–(7-43) can now be solved simultaneously for m with the result

$$m = -\left(\frac{1}{2} + \frac{\Delta G^0_{AB} + w_{BA} - w_{AB}}{2}\right)$$

Finally substituting this back into Eq. (7-41) gives

$$\Delta G^*_{AB} = \tfrac{1}{2}(w_{AB} + w_{BA}) + \frac{\lambda_0}{4} + \frac{\Delta G^0_{AB}}{2} + \frac{(\Delta G^0_{AB} + w_{BA} - w_{AB})^2}{4\lambda_0} \tag{7-44}$$

Thus an explicit expression for the rate constants can be obtained. The above treatment, however, neglects changes in the bond distances and bond angles in the inner coordination shells of the reactants during electron transfer. Calculation of the free energy for this process requires a detailed knowledge of the potential energy for vibrational displacements, i.e., the force constants and displacements associated with vibrational modes. Inclusion of this factor adds to λ_0 a term, λ_i, which is a function of force constants and vibrational displacements.

Because of the great difficulty in estimating λ_i, the absolute values of rate constants cannot be calculated very well from this theory. However, rate constants of related reactions can be correlated very successfully [28]. For example, the rate of an outer-sphere electron-transfer reaction can be related to the rates of isotope exchange reactions of the two reactants and the standard free energy change for the overall reaction. This theory is also useful for interpreting the kinetics of electrochemical reactions.

Electron-transfer reactions have been studied in considerably more detail than given here (cf. [22, 28–30]), but this discussion should illustrate that the determination of the kinetic rate law is just the beginning of the determination of a reaction mechanism.

Problems

7-1 Derive the expressions for the angle-averaged potential energy of interaction between an ion and a dipole given by Eqs. (7-7) and (7-8) from Eq. (7-6) and the Boltzmann expression for the equilibrium energy distribution.

7-2 Consider the reaction between an ion and an induced dipole of polarizability α, where the interaction potential is given by

$$U = -\frac{\alpha z^2 e^2}{2\varepsilon^2 r^4}$$

Derive an equation describing the dependence of the rate constant on the dielectric constant of the medium. Do you think this dependence could be verified by experiment? Why?

7-3 Derive the relation, valid in the limit of zero ionic strength, for the dependence of the specific rate constant k upon ionic strength for the reaction

$$A^{z_A} + H^+ \rightarrow \text{reaction products}$$

for which the rate equation is

$$-\frac{d(A)}{dt} = k(A)(H^+)$$

A^{z_A} is an ion of a strong electrolyte, but H^+ refers to the hydrogen ion in equilibrium with the weak acid HX according to the equation

$$HX \rightleftharpoons H^+ + X^-$$

Assume that a buffer system is used such that $(X^-)/(HX) = \text{const}$. The appropriate Debye–Hückel limiting law for the activity coefficients is

$$\log \gamma_i = 0.5 z_i^2 s^{1/2}$$

7-4 The solvolysis of *t*-butyl chloride has been studied in a variety of solvents, and the activation parameters in Table P7-4 have been obtained at 25°C [31].

TABLE P7-4

Solvent	$\Delta H^{0\ddagger}$ (kcal/mole)	$\Delta S^{0\ddagger}$ (eu)[a]
Water	23.2	+12.2
Methanol	24.8	−3.1
Ethanol	26.1	−3.2
Acetic acid	25.8	−2.5
90% Dioxane	21.5	−18.5
90% Acetone	21.8	−16.8

[a] Standard state is 1 *M*.

Explain why the enthalpy of activation is essentially constant, although the entropy of activation is not. Also account for the observed variation of the entropy of activation for the various solvents. What conclusions can be drawn concerning the nature of the transition state?

7-5 The kinetics of the reaction $2Fe^{3+} + Sn^{2+} \rightarrow 2Fe^{2+} + Sn^{4+}$ in aqueous solution has been studied by several workers. In approximately 2 *M* $HClO_4$ acid solutions and when $(Fe^{3+})/(Sn^{2+}) \gg 1$, the rate law is [32]

$$R = k(Cl^{-1})^3(Fe^{3+})(Sn^{2+})$$

for a range of Cl^- concentration from about 0.05 to 0.2 *M*. When Fe^{2+} is added initially at relatively high concentrations, the rate law is [33]

$$R = \frac{k'(Fe^{3+})^2(Sn^{2+})}{(Fe^{2+})}$$

(The dependence of the rate law on chloride-ion concentration has not been quantitatively measured in this case.)

Postulate a mechanism (or mechanisms) consistent with these facts. Show that the rate law for the proposed mechanism(s) behaves in the correct fashion.

7-6 The rate law for the reaction of hypochlorite ion and iodide ion

$$I^- + OCl^- = OI^- + Cl^-$$

in basic aqueous media is [34]

$$\frac{d}{dt}(OI^-) = k\frac{(I^-)(OCl^-)}{(OH^-)}$$

where $k = 60\ \text{sec}^{-1}$ at 25°C.

(a) Postulate a mechanism and show that it is consistent with the observed rate law. Calculate the rate constants of as many of the individual steps in the mechanism as possible.

(b) What does the rate law imply about the composition of the transition state for the rate limiting step?

Hint: Hypochlorous acid, HOCl, is a weak acid with an ionization constant of 3.4×10^{-8} M, so that OCl^- will hydrolyze slightly; the ionization constant for hypoiodous acid is about 10^{-11} M at 25°C.

7-7 The formation of an imine between glycine and pyridine-4-aldehyde proceeds at pH 10, 25°C, in aqueous solution according to the overall reaction shown in Fig. P7-7.

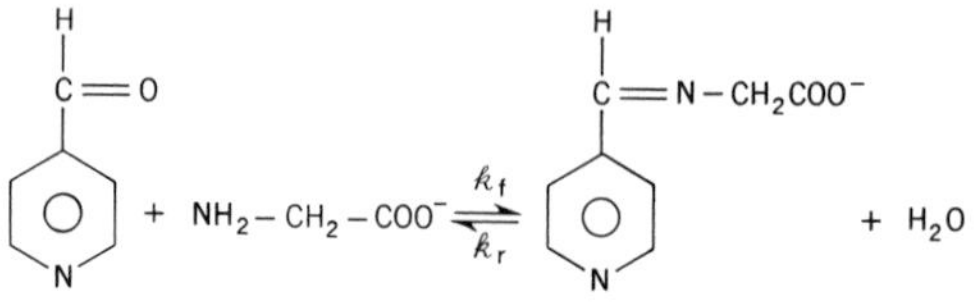

$$C_5H_4N\text{-}CHO + NH_2\text{-}CH_2\text{-}COO^- \underset{k_r}{\overset{k_f}{\rightleftharpoons}} C_5H_4N\text{-}CH{=}N\text{-}CH_2COO^- + H_2O$$

Fig. P7-7

The reaction was followed by observing the imine formation spectrophotometrically under conditions where the total glycine concentration was much greater than that of pyridinealdehyde. In this case, the reaction could be described by the rate law for a first-order reversible reaction. (This is an example of a *pseudo* first-order reaction.) The assumption was then made that the reaction in the forward direction was also first order with respect to glycine, and a second-order rate constant was obtained by dividing the first-order constant by the total glycine concentration. This second-order rate constant varied with the glycine concentration in the manner shown in Table P7-7 [35].

TABLE P7-7

(Glycine) (M)	k_f (M^{-1} sec^{-1})	k_r(sec^{-1})
0.00[a]	0.409	1.3×10^{-3}
0.05	0.355	1.3×10^{-3}
0.1	0.310	1.3×10^{-3}
0.2	0.257	1.3×10^{-3}
0.3	0.218	1.3×10^{-3}
0.4	0.190	1.3×10^{-3}
0.5	0.165	1.3×10^{-3}

[a] Extrapolated.

(a) Postulate a formal mechanism that is consistent with the above behavior and determine the value of as many of the individual rate constants as possible.

(b) Postulate a chemical mechanism for this reaction.

Hint: At equilibrium the imine formed is known to exist in equilibrium with a relatively small amount of carbinol amine (I), K being about 0.01 at 25°C.

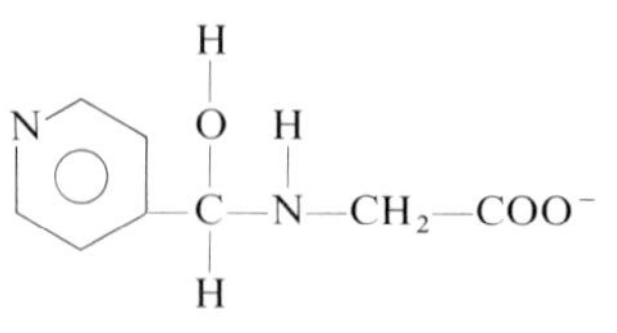

$$C_5H_4N\text{-}CH(OH)\text{-}NH\text{-}CH_2\text{-}COO^- \qquad \text{I}$$

References

1. R. A. Robinson and R. H. Stokes, "Electrolytic Solutions," p. 73. Butterworth, London, 1959.
2. J. F. Kincaid, H. Eyring, and A. E. Stearn, *Chem. Rev.* **28**, 301 (1941).
3. R. B. Bell, *Ann. Rep. Prog. Chem. Chem. Soc. London* **52**, 1472 (1930); J. H. Raley, F. F. Rust, W. E. Vaughan, and F. H. Seubold, *J. Am. Chem. Soc.* **70**, 88, 95, 1336 (1948); D. F. Smith, *ibid.* **49**, 43 (1927); H. Eyring and F. Daniels, *ibid.* **52**, 1472 (1930); A. Rembaum and M. Szwarc, *ibid.* **76**, 5975, 5978 (1954).
4. J. G. Kirkwood, *J. Chem. Phys.* **2**, 351 (1934).
5. W. Hecht and M. Conrad, *Z. Phys. Chem. (Leipzig)* **3**, 450 (1889).
6. G. H. Grant and C. N. Hinshelwood, *J. Chem. Soc.* **258** (1933).
7. W. Hecht, M. Conrad, and C. Bruchner, *Z. Phys. Chem. (Leipzig)* **4**, 273 (1889).
8. D. Segaller, *J. Chem. Soc.* **105**, 106 (1914).
9. E. S. Amis and V. K. LaMer, *J. Am. Chem. Soc.* **61**, 905 (1939); C. V. King and J. J. Josephs, *ibid.* **66**, 767 (1944).
10. J. N. Bronsted and R. Livingston, *J. Am. Chem. Soc.* **49**, 435 (1927).
11. D. P. Ames and J. E. Willard, *J. Am. Chem. Soc.* **73**, 164 (1951).
12. C. Postmus and E. L. King, *J. Phys. Chem.* **59**, 1217 (1955).
13. A. A. Frost and R. G. Pearson, "Kinetics and Mechanism," 2nd ed. p. 138. Wiley, New York, 1961.
14. J. N. Bronsted and R. Livingston, *J. Am. Chem. Soc.* **49**, 435 (1927), V. K. LaMer, *ibid.* **51**, 334 (1929); C. V. King and M. B. Jacobs, *ibid.* **53**, 1704 (1931); J. Barret and J. H. Baxandale, *Trans. Faraday Soc.* **52**, 210 (1956).
15. G. Scatchard, *Nat. Bur. Stand. US Circ.* 524, p. 185 (1953).
16. J. Walker and F. J. Hambly, *J. Chem. Soc.* **67**, 646 (1895).
17. F. D. Chattaway, *J. Chem. Soc.* **101**, 170 (1912).
18. J. C. Warner and F. B. Stitt, *J. Am. Chem. Soc.* **55**, 4807 (1933).
19. J. C. Warner and E. L. Warrick, *J. Am. Chem. Soc.* **57**, 1491 (1935); W. J. Svirbely and J. C. Warner, *ibid.* **57**, 1883 (1935); W. J. Svirbely and A. Schramm, *ibid.* **60**, 330 (1938); J. Lander and W. J. Svirbely, *ibid.* **60**, 1613 (1938); C. C. Miller, *Proc. Roy. Soc. London Ser. A* **145**, 288 (1934), **151**, 188 (1936).
20. H. Taube, H. Myers, and R. L. Rich, *J. Am. Chem. Soc.* **75**, 4118 (1953).
21. A. Zwickel and H. Taube, *Discuss. Faraday Soc.* **29**, 42 (1960).
22. H. Taube, *Adv. Inorg. Chem. Radiochem.* **1**, 1 (1959).
23. A. Zwickel and H. Taube, *J. Am. Chem. Soc.* **83**, 793 (1961).
24. A. Zwickel and H. Taube, *J. Am. Chem. Soc.* **83**, 1288 (1961).
25. J. D. Candlin, J. Halpern, and D. L. Trimim, *J. Am. Chem. Soc.* **86**, 1019 (1964).
26. D. K. Sebera and H. Taube, *J. Am. Chem. Soc. 83*, 1785 (1961).
27. R. A. Marcus, *Discuss. Faraday Soc.* **29**, 21 (1960); *J. Chem. Phys.* **44**, 679 (1965).
28. N. Sutin, *Acc. Chem. Res.* **1**, 225 (1968).
29. N. Sutin, *Ann. Rev. Phys. Chem.* **17**, 119 (1966).
30. A. Haim, *Acc. Chem. Res.* **8**, 264 (1975).
31. S. Winstein and A. H. Fainberg, *J. Am. Chem. Soc.* **79**, 5937 (1957).
32. F. R. Duke and P. C. Pinkerton, *J. Am. Chem. Soc.* **73**, 3045 (1951).
33. W. A. Noyes, *Z. Phys. Chem. (Leipzig)* **16**, 576 (1895).
34. Y. T. Chia and R. E. Connick, *J. Phys. Chem.* **63**, 1518 (1959).
35. T. C. French and T. C. Bruice, *Biochemistry* **3**, 1589 (1964).

CHAPTER 8

FAST REACTIONS IN LIQUIDS

8-1 INTRODUCTION

Although detailed theories of reaction rates in liquids for the experimentalist to use are clearly lacking, so many interesting reactions occur only in liquids (especially in aqueous media) that experimental investigations in this field have been numerous and will continue to be so. These studies have been concerned mainly with the elucidation of reaction mechanisms. Unfortunately, classical kinetic studies usually permit the determination only of an overall rate constant, which may contain equilibrium constants and several rate constants, depending on events prior to the rate-determining step. Such studies are thus extremely difficult to interpret reliably in terms of elementary mechanistic steps. What is desired is information about the entire time course of the reaction. In order to obtain such information, methods are needed for the kinetic study of very fast reactions.

The periods of molecular vibrations, which are physical phenomena rather than chemical transformations, are often in the range 10^{-12} to 10^{-13} sec. It is clear that intramolecular chemical transformations cannot occur in times shorter than a single molecular vibration. It is exceedingly difficult to distinguish between "physical" and "chemical" transformations at very short times; however, energy barriers generally become large enough for a process to be considered chemical at times of approximately 10^{-10} sec or longer. The characteristic time constant for an association reaction is dependent on the concentrations of reactants; from our previous consideration of the maximum values of association and dissociation rates, however, it is clear that in most systems these events also occur in times longer than 10^{-10} sec. The development of new techniques in recent years now permits reactions with characteristic time constants as small as picoseconds (10^{-12} sec) to be studied. In this chapter we present an introduction to fast-reaction techniques and their uses in the elucidation of reaction mechanisms.

8-2 EXPERIMENTAL TECHNIQUES

A summary of currently important techniques for studying fast reactions and their time range of application is given in Table 8-1. A complete discussion of all of these methods is clearly not within the scope of this book; interested readers should consult comprehensive review volumes [1, 2]. Instead, we will concentrate primarily on relaxation and magnetic resonance methods.

Flow techniques, as extensions of classical methods, were introduced in 1923 by Hartridge and Roughton. Since then, many technological improvements have been made, but the time resolution of the method is inherently limited by the time required for mixing, which cannot be made much less than about 1 msec without great difficulty. The technology of flow methods will not be discussed here, but extensive reviews are available [3]. For complex mechanisms, the interpretation of data obtained from rapid mixing techniques can be exceedingly difficult since coupled rate equations often must be solved. If all of the rate equations are linear first-order (or pseudo first-order) differential equations, analytical solutions can be obtained (see Section 8-3). However, very often this is not the case; analog and digital computers are frequently utilized to overcome this difficulty.

Chemical relaxation techniques were developed in the 1950s by Eigen and have had a great impact on the field of reaction kinetics. One basic principle is behind all the relaxation techniques. Rather than trying to mix the reactants, a reaction mixture already at equilibrium is perturbed by varying an external parameter. This causes a shift to new equilibrium concentrations,

TABLE 8-1

Fast Reactions

Method	Time range
Flow techniques	Minutes–10^{-4} sec
Relaxation techniques	
Temperature jump	1–10^{-8} sec
Pressure jump	1–10^{-6} sec
Electric pulses and waves	10^{-2}–10^{-10} sec
Ultrasonic absorption	10^{-5}–10^{-11} sec
Nuclear magnetic resonance	10^{-1}–10^{-6} sec
Electron paramagnetic resonance	10^{-5}–10^{-10} sec
Flash photolysis	Seconds–10^{-12} sec
Pulsed radiolysis	Seconds–10^{-11} sec
Photostationary methods	1–10^{-10} sec
Electrochemical methods	1–10^{-8} sec

and determination of the rate of this shift permits the kinetics of the reactions in question to be studied. The external parameters to be considered here are temperature, pressure, and electric field. They can be varied in several ways, e.g., periodically, by a rectangular pulse, or by a single-step impulse. Because the reaction rates are not infinite, the actual shift of the concentrations does not follow the perturbation exactly; i.e., the equilibrium shift lags behind the perturbation. Examples of this phenomenon are illustrated in Figs. 8-1 and 8-2 for systems with a characteristic relaxation time τ.

Before considering the special forms of rate equations near equilibrium, we consider how equilibrium perturbations can be introduced. Since we know from thermodynamics that

$$\frac{\partial \ln K_P}{\partial T} = \frac{\Delta H^0}{RT^2} \tag{8-1}$$

where K_P is an equilibrium constant and ΔH^0 is the corresponding enthalpy of reaction, it is clear that for reaction mechanisms where any one step in the reaction sequence has $\Delta H^0 \neq 0$, a temperature perturbation causes a change in the concentrations of reactants. In practice, the temperature jump is generally brought about by discharging a capacitor, initially charged to 10,000 to 100,000 volts, through the reaction mixture. A schematic diagram of a temperature jump apparatus is shown in Fig. 8-3. Uniform heating of the sample ($\sim$0.1–1 cc) is assured by a cell design which permits only parallel electric-field lines to do the heating. The energy used for heating is $\frac{1}{2}CV^2$ (C is the capacitance and V the voltage), and it can be shown that the temperature rise is an exponential function of time characterized by a first-order rate constant $2/RC$ (R is the resistance between cell electrodes). For cases in which the reactions occur in times much longer than $RC/2$, the heating can be considered to occur instantaneously. A typical apparatus produces 45 joules in 1 μsec (4.5×10^7 watts!). In order to keep R small, an inert electrolyte is usually added to the reaction mixture. Concentration changes are observed by optical techniques: absorption spectroscopy, fluorimetry, or polarimetry. After about 1 sec, convective mixing obscures the concentration changes, placing an effective limit on the slowness of reactions which can be studied by this method. Flash lamps and laser pulses also have been utilized to produce temperature jumps but are not widely used at the present time.

Clearly a pressure jump can also perturb chemical equilibria since

$$\frac{\partial \ln K_P}{\partial P} = -\frac{\Delta V^0}{RT} \tag{8-2}$$

where ΔV^0 is the volume change occurring in the reaction under consideration. In general, pressure perturbations, even in favorable cases, produce

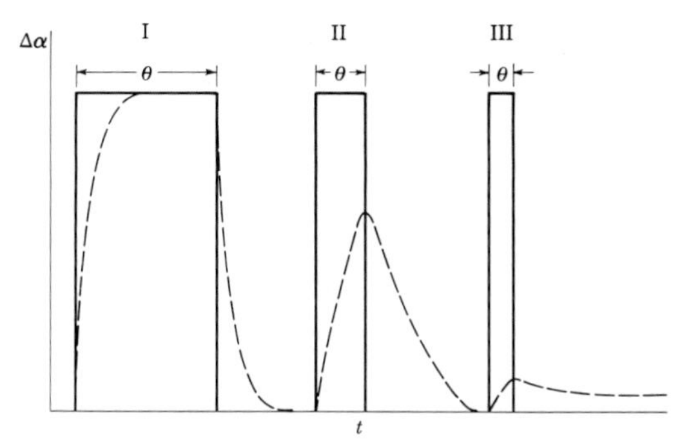

Fig. 8-1 Rectangular impulses; θ = impulse duration. I: $\tau/\theta = 0.1$; II: $\tau/\theta = 1$; III: $\tau/\theta = 10$. $\Delta\alpha$ is a measure of the equilibrium shift.—, external parameter; ---, relaxation curves.

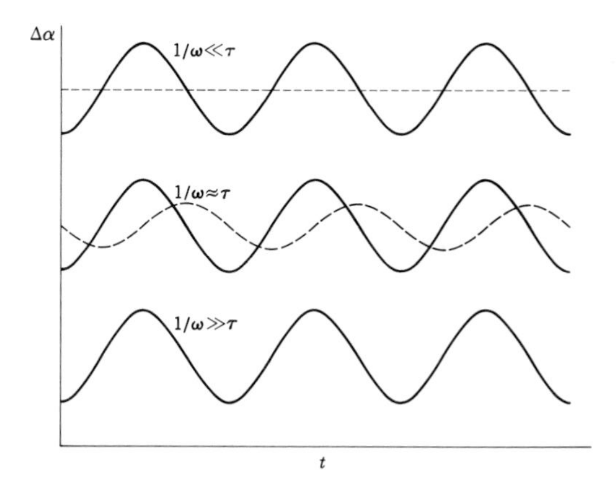

Fig. 8-2 Continuous wave perturbation, ω = frequency. $\Delta\alpha$ is a measure of the equilibrium shift. —, external parameter; ---, $\Delta\alpha$. (For $1/\omega \gg \tau$, the two curves are identical.)

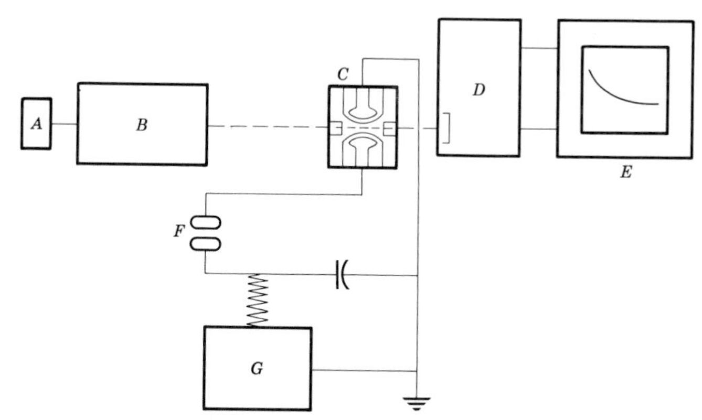

Fig. 8-3 Schematic drawing of a temperature jump apparatus. (A) Light source; (B) monochromator; (C) observation cell; (D) photomultiplier, emitter follower; (E) oscilloscope; (F) spark gap; (G) high voltage.

quite small concentration changes, so that only a sensitive differential-conductance method has been used routinely for detection. The actual apparatus usually transmits a high pressure to the solution (~ 10 to 100 atm) through a diaphragm, which is then ruptured to produce an effective step-function return to 1 atm. The time constant of the apparatus is largely a matter of design at present, but an absolute limit is set by the speed of sound, i.e., the rate of escape of the shock wave from the apparatus.

Electric-field-density changes perturb chemical equilibrium involving reactions where a change in the number of ions occurs. The effect is commonly called the second Wien effect, and a detailed theory has been worked out by Onsager [4]. For a 1:1 electrolyte,

$$\frac{\partial \ln K_P}{\partial |E|} = \frac{|z_A U_A| + |z_B U_B|}{U_A + U_B} |z_A z_B| \frac{e^3}{2\varepsilon k^2 T^2} + \cdots \tag{8-3}$$

where z_i is the ionic valence, U_i the mobility, e the electronic charge, ε the dielectric constant, k the Boltzmann constant, and T the absolute temperature. According to the Onsager theory, a field of 100 kV/cm would produce approximately a 6% conductance change. Electric-field perturbations can be applied as rectangular pulses (see Fig. 8-1). The actual perturbation is produced by the discharge of a condenser through the reaction mixture. The fields produced are of the order of 100 kV/cm. In this case the resistivity of the reaction mixture must be large to prevent heating effects. The changes in concentration are followed through the corresponding conductance changes. In general, electric fields can be used to perturb any chemical equilibria in which the polarizability of the reactants and products differ greatly. Thus equilibria in which ions are generated (or destroyed), or in which the dipole moments of the reactants are markedly changed, can be shifted by the use of electric fields. In practice, electric-field methods are difficult so that they have been extensively utilized only in a few laboratories.

Ultrasonic techniques are so numerous that a comprehensive discussion is not possible. Since an ultrasonic wave is an adiabatic pressure wave, in general both a temperature and a pressure perturbation of the system occurs. In most nonaqueous solvents, the temperature perturbation is of primary importance, because chemical equilibria are generally much more sensitive to temperature changes than to pressure changes. In aqueous solutions, however, the pressure perturbation is usually of primary importance, because the thermal-expansion coefficient of water is very small, so that the pressure wave is almost isothermal. A serious disadvantage of ultrasonic methods is that rather large volumes of solution are required for low-frequency measurements and relatively high concentrations ($> \sim 10^{-2}$ M) of reactants are required at all frequencies. Recent experimental innovations have alleviated these problems to some extent. The most common ultrasonic technique consists of a periodic perturbation, the subsequent chemical changes usually being detected by a direct measurement of the attenuation (or energy dissipation) of the ultrasonic wave. If a chemical reaction is occurring, a maximum in the ultrasonic absorption per wavelength occurs at a definite frequency. The relationship between this maximum and chemical reactions is presented in Section 8-3.

8-3 RATE EQUATIONS NEAR EQUILIBRIUM

In order to simplify our discussion, we first consider step-function perturbations. In this case, the conventional rate equation describes the system after the perturbation is complete. For example, the mechanism

$$\mathrm{A} + \mathrm{B} \underset{k_{21}}{\overset{k_{12}}{\rightleftharpoons}} \mathrm{AB} \tag{8-4}$$

can be described by the usual rate equation

$$\frac{d}{dt}(\mathrm{AB}) = -\frac{d}{dt}(\mathrm{A}) = -\frac{d}{dt}(\mathrm{B}) = k_{12}(\mathrm{A})(\mathrm{B}) - k_{21}(\mathrm{AB}) \tag{8-5}$$

Let us now define new concentration variables such that

$$(\mathrm{A}) = (\mathrm{A_e}) + \Delta(\mathrm{A}), \qquad (\mathrm{B}) = (\mathrm{B_e}) + \Delta(\mathrm{B}), \qquad (\mathrm{AB}) = (\mathrm{AB_e}) + \Delta(\mathrm{AB}) \tag{8-6}$$

where the subscript e designates equilibrium concentration and the Δ's deviations from the equilibrium concentrations. Also, from the stoichiometry we know that

$$\Delta C = \Delta(\mathrm{AB}) = -\Delta(\mathrm{A}) = -\Delta(\mathrm{B}) \tag{8-7}$$

Using Eqs. (8-6) and (8-7), we can rewrite Eq. (8-5) as

$$\begin{aligned} -\frac{d\Delta C}{dt} &= \{k_{12}[(\mathrm{A_e}) + (\mathrm{B_e})] + k_{21}\}\,\Delta C \\ &\quad - k_{12}(\mathrm{A_e})(\mathrm{B_e}) + k_{21}(\mathrm{AB_e}) - k_{12}(\Delta C)^2 \end{aligned} \tag{8-8}$$

However,

$$\frac{k_{12}}{k_{21}} = \frac{(\mathrm{AB_e})}{(\mathrm{A_e})(\mathrm{B_e})} \tag{8-9}$$

Furthermore, in the neighborhood of equilibrium, $(\Delta C)^2$ is so small that the last term in Eq. (8-8) can be neglected. Therefore, this equation can be written as

$$\frac{d\Delta C}{dt} = -\{k_{12}[(\mathrm{A_e}) + (\mathrm{B_e})] + k_{21}\}\,\Delta C = -\frac{\Delta C}{\tau} \tag{8-10}$$

which defines the relaxation time τ. For a single reaction, it is clear that all rate equations near equilibrium are linear first-order differential equations and that a relaxation time can be defined as in Eq. (8-10). This follows simply by neglecting all terms higher than the first-order concentration deviations from equilibrium, that is, $(\Delta C)^2$, etc. Integration of Eq. (8-10) gives

$$\Delta C = \Delta C_0 e^{-t/\tau} \tag{8-11}$$

where ΔC_0 is the concentration deviation at $t = 0$. Thus in any perturbation experiment, a reciprocal first-order rate constant τ is measured. Clearly, if the equilibrium concentrations are varied or if the equilibrium constant is known, both k_{12} and k_{21} can be obtained.

Most mechanisms are described by a series of coupled reactions. The number of independent concentration variables is simply the number of concentration variables minus the number of mass conservation equations. If there are n independent concentration variables in the mechanism, n independent rate equations are obtained; furthermore, near equilibrium all these are linear first-order differential equations of the form

$$-\frac{d\Delta C_i}{dt} = \sum_{j=1}^{n} a_{ij} \Delta C_j \tag{8-12}$$

where the a_{ij}'s are functions of the rate constants and equilibrium concentrations which follow directly from conventional rate equations in the neighborhood of equilibrium (just as in the simple example presented earlier). This system of equations is analogous to that encountered in vibration spectroscopy where the equations of motion of the individual atoms in a molecule are described by coupled first-order linear differential equations. This coupled set of equations of motion can be transformed to an uncoupled set of equations of motion in which the dependent variables are linear combinations of the coordinates characterizing the motions of the individual atoms, and the time dependence of the motion is described by normal mode frequencies which are functions of the individual bond frequencies. Similarly, the coupled rate equations, Eq. (8-12), can be transformed to a set of independent linear first-order rate equations which can be represented as

$$-\frac{d\Delta y_i}{dt} = \frac{\Delta y_i}{\tau_i} \tag{8-13}$$

The concentration variable Δy_i is a linear combination of the ΔC_i, and the τ_i are normal mode relaxation times that are functions of the a_{ij}. More explicitly, the time dependence of the individual concentrations can be represented as sums of exponentials or first-order rate equation terms:

$$\Delta C_i = \sum_{j} A_{ij} e^{-t/\tau_j} \tag{8-14}$$

where the sum has n terms and the τ_j are obtained by solving the determinant

$$\begin{vmatrix} a_{11} - 1/\tau & a_{12} & \cdots & a_{1n} \\ a_{21} & a_{22} - 1/\tau & \cdots & a_{2n} \\ \vdots & \vdots & & \vdots \\ a_{n1} & & \cdots & a_{nn} - 1/\tau \end{vmatrix} = 0 \tag{8-15}$$

A more exact solution of this problem can be handled best using elementary matrix methods. The transformation between the ΔC_i and the y_i can be written as

$$\{y\} = \{M\}\{C\} \tag{8-16}$$

where $\{y\}$ is the column matrix $(\Delta y_1, \Delta y_2, \ldots, \Delta y_n)$, $\{C\}$ a column matrix $(\Delta C_1, \Delta C_2, \ldots, \Delta C_n)$, and $\{M\}$ an n by n transformation matrix. The matrix $\{M\}$ must be such that

$$\{C\} = \{M^{-1}\}\{y\} \tag{8-17}$$

$$\{\Lambda\} = \{M\}\{A\}\{M^{-1}\} \tag{8-18}$$

where $\{M^{-1}\}$ is the inverse of $\{M\}$, $\{A\}$ the n by n matrix of a_{ij}, and $\{\Lambda\}$ a diagonal matrix of the normal mode reciprocal relaxation times obtained from Eq. (8-15). The transformation matrix and its inverse can be found by well-known mathematical procedures. More comprehensive mathematical accounts of chemical relaxation are available [2, 5]. In general, it is possible to obtain all of the rate constants by studying the relaxation times as a function of concentration.

As a particular example of a coupled system, we shall now calculate the relaxation spectrum for the mechanism

$$\mathrm{A + B} \underset{k_{-1}}{\overset{k_1}{\rightleftharpoons}} \mathrm{C} \underset{k_{-2}}{\overset{k_2}{\rightleftharpoons}} \mathrm{D}$$

The rate equations describing this mechanism in the neighborhood of equilibrium are obtained by the procedure previously described:

$$-\frac{d}{dt}(\mathrm{A}) = k_1(\mathrm{A})(\mathrm{B}) - k_{-1}(\mathrm{C}) = k_1[(\mathrm{A_e}) + (\mathrm{B_e})]\,\Delta(\mathrm{A}) - k_{-1}\,\Delta(\mathrm{C}) \tag{8-19}$$

$$-\frac{d}{dt}(\mathrm{D}) = -k_2(\mathrm{C}) + k_{-2}(\mathrm{D}) = -k_2\,\Delta(\mathrm{C}) + k_{-2}\,\Delta(\mathrm{D}) \tag{8-20}$$

Also, mass conservation (at constant volume) requires that

$$\Delta(\mathrm{A}) + \Delta(\mathrm{C}) + \Delta(\mathrm{D}) = 0 \tag{8-21}$$

Thus

$$-\frac{d}{dt}\Delta(\mathrm{A}) = \{k_1[(\mathrm{A_e}) + (\mathrm{B_e})] + k_{-1}\}\,\Delta(\mathrm{A}) + k_{-1}\,\Delta(\mathrm{D})$$

$$= a_{11}\,\Delta(\mathrm{A}) + a_{12}\,\Delta(\mathrm{D}) \tag{8-22}$$

$$-\frac{d}{dt}\Delta(\mathrm{D}) = k_2\,\Delta(\mathrm{A}) + (k_{-2} + k_2)\,\Delta(\mathrm{D})$$

$$= a_{21}\,\Delta(\mathrm{A}) + a_{22}\,\Delta(\mathrm{D}) \tag{8-23}$$

The solutions of these equations are given by Eq. (8-14), and the relaxation times are found by solving the determinant

$$\begin{vmatrix} a_{11} - (1/\tau) & a_{12} \\ a_{21} & a_{22} - (1/\tau) \end{vmatrix} = 0$$

The result is a quadratic equation with the solutions

$$\frac{1}{\tau_{1,2}} = \tfrac{1}{2}[(a_{11} + a_{22}) \pm \sqrt{(a_{11} + a_{22})^2 + 4(a_{12}a_{21} - a_{11}a_{22})}]$$

$$= \tfrac{1}{2}\{k_1[(\mathrm{A_e}) + (\mathrm{B_e})] + k_{-1} + k_2 + k_{-2}\}$$

$$\times\left(1 \pm \sqrt{1 - 4\frac{k_1k_2[(\mathrm{A_e}) + (\mathrm{B_e})] + k_1k_{-2}[(\mathrm{A_e}) + (\mathrm{B_e})] + k_{-1}k_{-2}}{\{k_1[(\mathrm{A_e}) + (\mathrm{B_e})] + k_{-1} + k_2 + k_{-2}\}^2}}\right) \tag{8-24}$$

One of the relaxation times corresponds to the plus sign of the square root, the other to the negative sign.

A useful procedure is to consider a limiting form of this equation. Assume the bimolecular step to be rapid compared to the unimolecular step; that is, $k_1[(\mathrm{A_e}) + (\mathrm{B_e})] + k_{-1} \gg k_{-2} + k_2$, or, as we shall see, $1/\tau_1 \gg 1/\tau_2$. In this case, the square-root term can be expanded in a series, that is, $(1 - x)^{1/2} \approx 1 - x/2$, and only the first nonvanishing term need be retained. The results are

$$\frac{1}{\tau_1} = k_1[(\mathrm{A_e}) + (\mathrm{B_e})] + k_{-1} \tag{8-25}$$

$$\frac{1}{\tau_2} = \frac{k_1k_2[(\mathrm{A_e}) + (\mathrm{B_e})] + k_1k_{-2}[(\mathrm{A_e}) + (\mathrm{B_e})] + k_{-1}k_{-2}}{k_1[(\mathrm{A_e}) + (\mathrm{B_e})] + k_{-1}}$$

$$= k_{-2} + \frac{k_2}{1 + k_{-1}/k_1[(\mathrm{A_e}) + (\mathrm{B_e})]} \tag{8-26}$$

Equations (8-25) and (8-26) can be obtained in a more straightforward manner. Since the first step equilibrates much more rapidly than the second step, it can be considered as an independent reaction, so that the relaxation time obtained in the simple example presented earlier [Eq. (8-10)] applies and is, in fact, identical to that given in Eq. (8-25). Also, while the second step is relaxing, the first step is essentially always at equilibrium. Therefore, recalling $\Delta(A) = \Delta(B)$ and Eq. (8-21), we obtain

$$\frac{k_1}{k_{-1}} = \frac{(C)}{(A)(B)} \qquad \text{or} \qquad \frac{k_1}{k_{-1}}[(A_e) + (B_e)]\,\Delta(A) = \Delta(C)$$

and

$$\Delta(D) = -\Delta(C)\left\{1 + \frac{k_{-1}}{k_1[(A_e) + (B_e)]}\right\}$$

Now Eq. (8-20) can be rewritten as

$$-\frac{d}{dt}[\Delta(D)] = \left\{k_{-2} + \frac{k_2}{1 + k_{-1}/k_1[(A_e) + (B_e)]}\right\}\Delta(D) = \frac{1}{\tau_2}\,\Delta(D)$$

Application of these principles to more complex mechanisms is laborious but straightforward.

For the more general case in which the equilibrium concentration is time-dependent, a time-independent reference concentration C_i^0 must be chosen, e.g., the stoichiometric concentration. The concentrations can then be written as

$$C_{i,e} = C_i^0 + \Delta C_{i,e}, \qquad C_i = C_i^0 + \Delta C_i$$

If the rate equations are now linearized using exactly the same procedure as before, a set of nonhomogeneous coupled rate equations is obtained. For a one-step mechanism, the appropriate equation is

$$\tau\frac{d\,\Delta C}{dt} + \Delta C = \Delta C_e(t) \tag{8-27}$$

Here ΔC_e is given by the time dependence of the perturbing function. For a periodic perturbation of radial frequency ω, $\Delta C_e = ae^{i\omega t}$ and

$$\Delta C = \left(\frac{ae^{i\omega t}}{1 + i\omega\tau}\right) \tag{8-28}$$

In the case of ultrasonic attenuation measurements, the chemical part of the absorption coefficient α_c at the wavelength λ is described by the equation

$$\alpha_c \lambda = \sum_i \frac{B\omega\tau_i}{1 + \omega^2 \tau_i^2} \tag{8-29}$$

Here α_c is defined by the equation

$$I = I_0 e^{-(\alpha_c + \alpha_0)x} \tag{8-30}$$

where I is the pressure amplitude of the ultrasonic wave after it has passed a distance x through the sample, I_0 is the pressure amplitude at $x = 0$, and α_0 is the absorption coefficient of the pure solvent. According to Eq. (8-29), $\alpha_c \lambda$ has a series of maxima at $\omega = 1/\tau_i$. The most important feature of these more complicated situations is that, regardless of the method employed, essentially the same relaxation time is always measured. (Strictly speaking, the relaxation time is dependent on the thermodynamic variables, which are held constant [6]. For example, the temperature jump determines an isothermal relaxation time, while with ultrasonics an adiabatic relaxation time is measured. Usually, however, the differences between the various possible relaxation times is orders of magnitude smaller than the experimental errors.)

The amplitudes of chemical relaxation processes are determined by the equilibrium concentrations (and strictly speaking, associated activity coefficients) and by thermodynamic variables appropriate for the particular perturbation method used. Thus, for example, an analysis of the amplitudes of relaxation processes associated with temperature jump measurements can lead to determination of the equilibrium constants and enthalpies associated with the mechanism under study. As might be anticipated from our previous discussion, the relaxation amplitudes are determined by normal mode thermodynamic variables which are linear combinations of the thermodynamic variables associated with the individual steps in the mechanism. The formal analysis of relaxation amplitudes has been developed in considerable detail [2, 5, 7].

Thus far we have considered only perturbations of equilibrium states. This generally requires that the equilibrium constants be such that appreciable concentrations of both reactants and products are present. However, perturbations of steady states also can be realized. The mathematical analysis is quite similar to that already discussed for equilibrium systems except that steady-state concentrations are utilized rather than equilibrium concentrations and the principle of detailed balance cannot be used. For example, a rapid mixing apparatus might be used to establish a steady state which is then perturbed by a temperature jump. While steady-state perturbations have not yet been extensively used, they represent a potentially important application of relaxation methods.

8-4 FLASH PHOTOLYSIS AND ELECTRON-PULSE RADIOLYSIS

Flash photolysis and electron-pulse techniques may be considered as cases of extreme perturbing functions, in the first case an extremely intense flash of light, in the other case a beam of electrons. Such perturbations cause extreme deviations from equilibrium concentrations, so that linear first-order rate equations no longer describe the time behavior of the system. In fact, molecules are often promoted to higher electronic states. With the advent of the laser, the time resolution of flash photolysis has been reduced to picoseconds. This permits the study of processes not previously accessible to kineticists. The types of systems studied include radiationless transitions, the solvated electron and chlorophyll. For example, the "cage" effect in liquids has been demonstrated by a study of the recombination of iodine atoms at very short times [8]. Although these methods are of considerable interest, we do not discuss them in further detail here (cf. Hammes [1]).

8-5 NUCLEAR MAGNETIC RESONANCE AND ELECTRON PARAMAGNETIC RESONANCE

Before discussing the application of nuclear magnetic resonance (NMR) to kinetics, let us briefly consider the salient features of NMR. If nuclei which possess a magnetic moment (i.e., a nonzero nuclear spin) are placed in a magnetic field, they precess around the direction of the field with a frequency ω which is proportional to the strength of the magnetic field H_{local} at the nucleus:

$$\omega = \gamma H_{\mathrm{local}} \tag{8-31}$$

Here γ is the gyromagnetic ratio of the particular nucleus in question. In an ordinary NMR experiment, a magnetic field of the order of kilogauss is applied to the sample. Simultaneously, a much smaller oscillating magnetic field of constant frequency is applied perpendicular to the large field through a coil wound around the sample. The kilogauss field is then varied until the nuclei are precessing at the same frequency as the oscillating field, i.e., until Eq. (8-31) is satisfied. A signal is induced in a receiver coil perpendicular to both fields when this resonance condition is satisfied. Usual NMR spectra consist of plots of receiver signal strength versus applied field.

In liquids, very sharp lines are found, and considerable chemical information can be obtained. The magnetic field at each particular nucleus depends strongly on the exact chemical environment and is in general different from the applied field because of diamagnetic shielding by orbital electrons. This means that nuclei with different electronic shielding exhibit resonance at slightly different applied fields. This difference in applied fields, measured with respect to some arbitrary standard, is called the chemical shift. Chemical shifts are sometimes expressed in frequency units [see Eq. (8-31)]. For example, acetaldehyde has one series of peaks for the CH_3 protons and another for the CHO protons at a slightly lower applied field.

A second feature which must be considered is the coupling of spins of neighboring nuclei through orbital electrons. This causes a splitting of the resonance lines. The number of components depends on the number and spins of neighboring nuclei in the molecule. In the case of acetaldehyde, the resonance due to the CH_3 protons rather than being a single line consists of a doublet. This splitting is due to the neighboring CHO proton (spin $= \frac{1}{2}$), which has two possible orientations of its magnetic moment in the magnetic field. Similarly, the CHO proton resonance consists of a symmetrical quadruplet with relative intensities 1:3:3:1, corresponding to the four possible spin states of the three CH_3 protons and their statistical weights.

These considerations apply as long as there is no rapid chemical exchange of the atom being studied between magnetically different positions. The effects of rapid exchange are shown in the following two examples. The OH proton of CH_3OH has a chemical shift (measured with respect to some external standard) which is different from that of the protons of H_2O. However, a mixture of CH_3OH and H_2O has only a single proton-resonance line for the OH and H_2O protons. Its chemical shift is a weighted average of the chemical shifts for protons in these two environments. This phenomenon is attributed to rapid exchange of the protons between water and methanol. A single resonance line is observed for the CH_3OH–H_2O mixture because the frequency of the exchange reaction is much greater than the chemical shift of the OH protons relative to the H_2O protons. During a sweep of the magnetic field through resonance, the protons make many round trips between the H_2O and the OH group of methanol. The effective shielding which determines H_{local} is, in this case, an average of that shown when there is no exchange.

Similarly, spin–spin splitting can be eliminated by a rapid exchange of an atom being studied or of the atoms causing the splitting. For example, the proton spectrum of pure, dry, liquid NH_3 consists of a triplet due to splitting by the ^{14}N nucleus (spin $= 1$), which has three possible orientations in the magnetic field. As shown in Fig. 8-4, which is from Ogg's work [9], trace amounts of $NaNH_2$ (or H_2O) catalyze the rapid exchange of the protons between NH_3 molecules, with the result that a single proton peak is obtained. When the exchange frequency is greater than the frequency difference between outer components of the spin–spin multiplet, the proton "sees" several different N nuclei during the time required to pass through resonance, and the splitting is eliminated.

The transition between a complex spectrum, in which peaks with different chemical shifts or with spin–spin splittings are observed, and the simpler

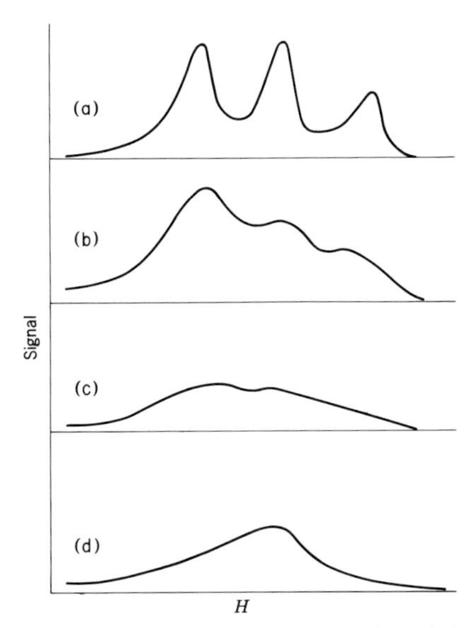

Fig. 8-4 Proton resonance in liquid NH_3. The concentration of $NaNH_2$ monotonically increases from (a) to (d); (c) corresponds to a $NaNH_2$ concentration of 10^{-7} *M*. (Adapted from Ogg [9].)

spectrum, produced by rapid exchange, occurs with a relaxation time given by

$$\tau \cong 1/\delta\omega \tag{8-32}$$

[This is essentially a manifestation of the uncertainty principle: $(h\omega/2\pi)(\tau) = h/2\pi$.] Here τ, which is simply related to ordinary rate constants, is the mean lifetime for the chemical exchange; $\delta\omega$ is the frequency separation (in radians per second) between lines obtained when there is no exchange. For the two cases just considered, $\delta\omega$ is the chemical shift of the OH proton of CH_3OH relative to the H_2O protons and the separation of the outer components of the NH_3 proton spin–spin triplet.

We are now in a position to consider the use of NMR to determine exchange rates in solution. It is necessary, first of all, that the system being studied have an NMR spectrum in which particular peaks can be identified with particular types of nuclei. By varying some parameter, such as temperature or catalyst concentration, the rate of a chemical-exchange reaction must be increased to a point where peaks associated with nuclei having different chemical shifts coalesce or, alternatively, spin–spin splitting is eliminated. Since $\delta\omega$ is typically of the order of 10 to 10^3 radians sec^{-1}, this technique is applicable to reactions with mean lifetimes of the order of 10^{-1} to 10^{-3} sec. It should be clear that the exchange process, even if it is always so rapid that only one resonance line is seen, may contribute to the linewidth. Thus rates can be measured by studying linewidths at various concentrations, a procedure that extends the time range of the method somewhat. Similar comments apply to the case involving spin–spin splitting, although this situation is somewhat more complex.

To illustrate how this approach can be used to determine the rate constants for a particular chemical reaction, Ogg's work on the NH_3–$NaNH_2$ system may be considered [9]. Assume that the mechanism is

$$NH_3 + NH_2^- \rightarrow NH_2^- + NH_3 \tag{8-33}$$

Here the concentration of NH_2^- is much smaller than that of NH_3, so that the process is pseudo first order, and

$$\tau = 1/k(NH_2^-) \tag{8-34}$$

In general, the relaxation time is given by

$$\frac{1}{\tau_i} = \frac{\text{rate of removal of molecules from } i\text{th state by exchange}}{\text{number of molecules in } i\text{th state}}$$

In the experiments, the frequency separation of the proton triplet components $\delta\omega$ was found to be 290 radians sec^{-1}. The concentration of $NaNH_2$ was then increased until the spin–spin splitting was just eliminated. From Fig. 8-4 it can be seen that the critical exchange frequency, where the triplet structure just disappears, occurs at 10^{-7} M $NaNH_2$. At this concentration, $\tau \cong 1/\delta\omega$. Since $\delta\omega = 290\ sec^{-1}$,

$$k = 3 \times 10^9 \quad M^{-1}\ sec^{-1}$$

Although the preceding analysis is qualitatively correct, it is much too simplistic for quantitative kinetic studies. A proper treatment involves solution of the equations describing the coupling of the motion of nuclei in a magnetic field with the pertinent kinetic processes (the Bloch equations). The behavior of nuclei in a magnetic field is governed by two relaxation times, the spin–lattice relaxation time T_1, and the spin–spin relaxation time T_2. The former is a measure of the rate of equilibration of the nuclei with their environment, while the latter also includes relaxation due to the interaction of nuclear spins. Because T_2 includes additional possible modes of relaxation, $1/T_2 \geq 1/T_1$. Both relaxation times can be measured: T_1 usually requires the use of pulsed magnetic fields, while T_2 is related to the linewidth of conventional continuous wave spectra. Both relaxation times are altered by kinetic processes.

Although a complete discussion of these phenomena is beyond the scope of this book, the following example illustrates the relationship between the magnetic resonance relaxation times and the kinetic relaxation time. We consider the case in which a ligand can bind to another molecule M, and the concentration of the bound ligand is much smaller than that of the unbound ligand. The magnetic resonance relaxation times of each magnetically distinct nucleus in the ligand are given by the equations

$$\frac{1}{T_{1,\mathrm{obs}}} = \frac{1}{T_{1,\mathrm{A}}} + \frac{P_{\mathrm{M}}}{T_{1,\mathrm{M}} + \tau} \tag{8-35}$$

$$\frac{1}{T_{2,\mathrm{obs}}} = \frac{1}{T_{2,\mathrm{A}}} + \frac{P_{\mathrm{M}}}{\tau}\left[\frac{\dfrac{1}{T_{2,\mathrm{M}}}\left(\dfrac{1}{T_{2,\mathrm{M}}} + \dfrac{1}{\tau}\right) + \Delta\omega^2}{\left(\dfrac{1}{T_{2,\mathrm{M}}} + \dfrac{1}{\tau}\right)^2 + \Delta\omega^2}\right] \tag{8-36}$$

Here $T_{i,\mathrm{obs}}$ is the measured relaxation time, $T_{i,\mathrm{A}}$ the relaxation time of the nucleus in the unbound ligand, P_{M} the fraction of the ligand nucleus bound to M, $T_{i,\mathrm{M}}$ the relaxation time of the ligand nucleus bound to M, τ the relaxation time for ligand dissociation, and $\Delta\omega$ the chemical shift of the bound nucleus relative to the unbound nucleus. Two limiting cases can be readily discerned for T_1. In the slow exchange limit, $1/T_{1,\mathrm{M}} \gg 1/\tau$ and $1/T_{1,\mathrm{obs}} - 1/T_{1,\mathrm{A}} = P_{\mathrm{M}}/\tau_{\mathrm{M}}$, or in other words, the measured spin–lattice relaxation time directly measures the kinetics of ligand exchange. In the fast exchange limit, $1/\tau \gg 1/T_{1,\mathrm{M}}$, and $1/T_{1,\mathrm{obs}} - 1/T_{1,\mathrm{A}} = P_{\mathrm{M}}/T_{1,\mathrm{M}}$, or the measured spin–lattice relaxation time directly measures $T_{1,\mathrm{M}}$. These two limiting cases can be readily distinguished experimentally by determining the temperature dependence of the spin–lattice relaxation time: $1/\tau$ increases with increasing temperature, whereas $1/T_{1,\mathrm{M}}$ generally decreases with increasing temperature. A schematic representation of the temperature dependence of $1/T_{1,\mathrm{obs}} - 1/T_{1,\mathrm{M}}$ is shown in Fig. 8–5. At sufficiently low temperatures, it should be possible to determine τ. The situation is somewhat more complex for T_2. If $\Delta\omega = 0$, equations analogous to those describing T_1 are obtained. If $\Delta\omega \neq 0$, fast and slow exchange regions still exist, but an intermediate region also is found. However, the temperature dependence of T_2 is similar in that at low temperatures the ligand exchange rate is directly measured, whereas at high temperatures $T_{2,\mathrm{M}}$ can be determined.

The analysis of continuous wave spectra modulated by kinetic processes is usually carried out by computer simulation of spectra, which of course involves solving the equations for relaxation of nuclei in the magnetic field. The magnetic resonance method is best suited for studying fairly simple kinetic processes in which only a single chemical relaxation time is of importance. More specialized treatises and the original literature should be consulted for a detailed description of the application of nuclear magnetic resonance to kinetic investigations [10, 11].

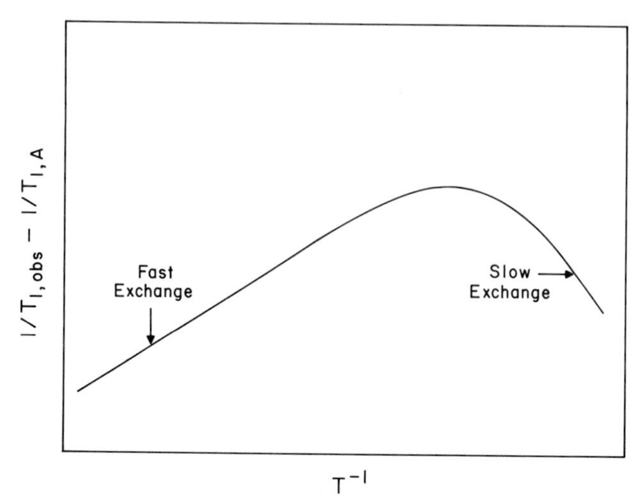

Fig. 8-5 Schematic plot of $1/T_{1,\mathrm{obs}} - 1/T_{1,\mathrm{A}}$ versus the reciprocal absolute temperature.

Electron paramagnetic resonance (EPR) essentially measures the effect of a magnetic field on unpaired electrons. The spectra obtained are generally much more complex than NMR spectra, and only a few kinetic applications have been attempted. Potentially this method is capable of yielding a considerable amount of information concerning fast free-radical and electron-transfer reactions.

8-6 PROTOLYTIC REACTIONS

As an example of an important class of reactions studied by relaxation and NMR techniques, we consider protolytic reactions, which have been studied extensively, especially by Eigen and co-workers, and are now quite well understood. For normal acids and bases, protonation and deprotonation are diffusion-controlled processes with rate constants of the order of magnitude predicted by Eq. (2-90). These reactions can be thought of as occurring in two steps: diffusion of the solvated proton (or hydroxyl ion) toward the base (or acid) followed by a rapid transfer of the proton through the hydrogen-bonded complex by some type of jump mechanism. In a highly ordered structure, e.g., ice, this transfer rate has a specific rate constant of approximately 10^{12} sec^{-1}. This fast transfer rate explains why these reactions are usually diffusion-controlled if the partners are able to form hydrogen bonds with the solvent. Rate constants for typical protonation and deprotonation reactions in aqueous solution are summarized in Table 8-2. Most of the second-order rate constants are in the range 10^{10}–10^{11} M^{-1} sec^{-1}. On the

TABLE 8-2

Protonation–Deprotonation Rates

	Reaction[a]	k_f (M^{-1} sec^{-1})	k_r (sec^{-1})	Reference
Normal	$H^+ + OH^-$	1.3×10^{11}	2.6×10^{-5}	13
	$H^+ + CH_3COO^-$	4.5×10^{10}	8×10^5	14
	$H^+ + NH_3$	4.3×10^{10}	24	15
	H^+ + imidazole	1.5×10^{10}	1.7×10^3	16
	$H^+ + CuOH^+$	$\sim 1 \times 10^{10}$	$\sim 1 \times 10^2$	17
	$OH^- + HCO_3^-$	6×10^9	1.4×10^2	17
	$OH^- + NH_4^+$	3.4×10^{10}	5×10^5	14, 18
	OH^- + imidazole$^+$	2.3×10^{10}	2.5×10^3	16
	$OH^- + HPO_4^{2-}$	$\sim 2 \times 10^9$	$\sim 3 \times 10^2$	12
Hindered due to	$OH^- + EDTA^{3-}$	3.8×10^7	6.9×10^3	12
H-bond perturbation	$OH^- + NTA^{2-}$	1.3×10^7	7.0×10^2	12
Hindered by	OH^- + DMA	1.2×10^7	32	19
internal chelation	OH^- + PAS	3×10^7	3×10^4	12

[a] EDTA = ethylenediaminotetraacetic acid, NTA = nitrilotriacetic acid, DMA = dimethylanthranilic acid, PAS = *p*-aminosalicyclic acid.

other hand, dissociation rate constants k_r vary considerably, reflecting large differences in the strength of the chemical bond broken.

As might be expected, several factors can contribute to deviations from "normal" diffusion-controlled rates. Charge effects are of some importance, but they generally do not cause deviations from "normal" rate constants of more than a factor of 10. Similarly, molecular shape and steric hindrance can limit the accessibility of the site, but such effects usually cause changes in the rate constants of less than one order of magnitude. Appreciable deviations do occur, however, if the hydrogen-bonding solvent structure is disrupted at the acceptor or donor. Examples given in Table 8-2 are reactions of aminopolycarboxylic acids such as $EDTA^{3-}$ with OH^-, where the negatively charged carboxylic groups disrupt the hydrogen-bonded structure between the amine proton and water. Also, if the donor proton is involved in an internal hydrogen bond, the solvent hydrogen link can be completely blocked. For many systems this internal chelation can be extremely stable, e.g., salicylic acid and acetylacetone enol, causing reductions from the normal rates of combination by factors greater than 10^3.

In the case of pseudoacids, for example, C acids, the resulting ions are stabilized by changes in electronic structure, so that the reactions are generally not diffusion-controlled. A very good example of this phenomenon is the protolysis and hydrolysis of acetylacetone, which has been studied in detail by Eigen and Kruse [12]. This acid can exist in keto (85%) and enol forms (15%); the conjugate base can have two resonance structures (I).

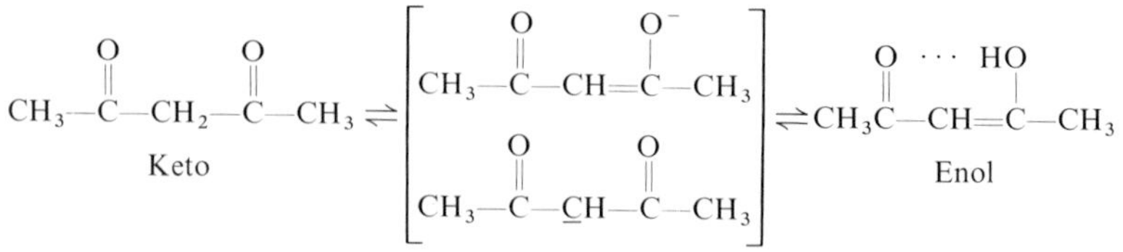

$$\mathrm{CH_3{-}\overset{\displaystyle O}{\overset{\|}{C}}{-}CH_2{-}\overset{\displaystyle O}{\overset{\|}{C}}{-}CH_3} \rightleftharpoons \left[\begin{array}{c}\mathrm{CH_3{-}\overset{\displaystyle O}{\overset{\|}{C}}{-}CH{=}\overset{\displaystyle O^-}{\overset{|}{C}}{-}CH_3} \\ \mathrm{CH_3{-}\overset{\displaystyle O}{\overset{\|}{C}}{-}\underline{C}H{-}\overset{\displaystyle O}{\overset{\|}{C}}{-}CH_3}\end{array}\right] \rightleftharpoons \mathrm{CH_3\overset{\displaystyle O}{\overset{\|}{C}}{-}CH{=}\overset{\displaystyle HO}{\overset{|}{C}}{-}CH_3}$$

Keto — Conjugate base — Enol

I

All the possible protolytic rate constants for this system have been obtained and are recorded in Table 8-3. Note that the protonation at the C^- atom is slower than protonation at the O^- atom by a factor of 10^3. A similar effect is exhibited by the deprotonation rates. In this case the enol form is stabilized by an internal H bond, so that its rate of reaction with OH^- is slower than a diffusion-controlled reaction by a factor of 10^3. These relatively slow protonation and deprotonation rates are characteristic of pseudoacids in general.

Many intermolecular proton-transfer reactions have now been studied; furthermore, Eigen has proposed a theory which correlates all the data and is of fundamental significance in understanding acid–base catalysis.

Let us first consider the simplest possible mechanism for proton transfer:

$$HD^+ + A \underset{k_{-1}}{\overset{k_1}{\rightleftharpoons}} D\overset{+}{H}A \underset{k_{-2}}{\overset{k_2}{\rightleftharpoons}} HA^+ + D \tag{8-37}$$

where A is a proton acceptor and D a proton donor. In this mechanism, k_1 and k_{-2} are assumed to have the values characteristic of a diffusion-controlled process. (Although this mechanism obviously is oversimplified, in that more intermediate states are probably involved, its salient features will prove useful as limiting cases.) If $D\overset{+}{H}A$ is assumed to be in a steady state, the experimental rate constants are

$$k_f = \frac{k_1}{1 + k_{-1}/k_2}, \qquad k_r = \frac{k_{-2}}{1 + k_2/k_{-1}}$$

TABLE 8-3

Protolysis and Hydrolysis of Acetylacetone[a]

Reaction	k_f (M^{-1} sec^{-1})	k_r (sec^{-1})
H^+ + enolate ⇋ enol	3×10^{10}	1.7×10^2
H^+ + enolate ⇋ keto	1.2×10^7	1.4×10^{-2}
OH^- + enol ⇋ enolate + H_2O	1.6×10^7	28
OH^- + keto ⇋ enolate + H_2O	4×10^4	3.5×10^{-1}

[a] 298°K, $s = 0.1$ M (Eigen and Kruse [12]).

Now when $k_{-1} \ll k_2$, the rate in the forward reaction is diffusion-controlled, while if $k_{-1} \gg k_2$, the rate in the reverse direction is diffusion-controlled. This idea can be expressed in another way: the rate of proton transfer for the forward reaction is diffusion-controlled if the acceptor binds the proton more tightly than the donor, that is, $pK_A \gg pK_D$. As long as this inequality prevails, k_f is constant. By definition, $\log k_f - \log k_r \equiv pK_A - pK_D \equiv \Delta(pK)$ so that $\log k_r$ is linearly related to $\Delta(pK)$. This can be expressed in still another way as

$$\frac{\partial \log k_f}{\partial \Delta(pK)} = \alpha \tag{8-38}$$

where α is a constant equal to zero if $\Delta(pK)$ is positive and equal to 1 if $\Delta(pK)$ is negative. When $pK_A = pK_D$, there is a fifty–fifty chance for the proton to go to either A or D, and a transition region occurs. This behavior is depicted in Fig. 8-6. For cases in which the proton transfer is not symmetric with respect to charge, for example, $A^- + DH^+ \rightleftharpoons AH + D$, the diffusion-controlled rate is different in both directions, so that an asymmetry in the $\log k_f - \Delta(pK)$ relationship occurs. Any other type of process whereby one species is stabilized over the other produces a similar effect effect, e.g., internal H bonding.

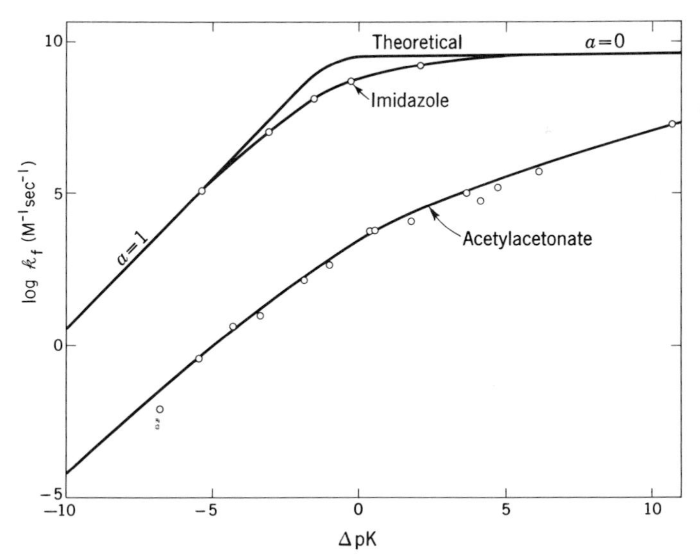

Fig. 8-6 $\text{Log}\, k_f - \Delta pK$ plot for "ideal" case, imidazole ($pK_A = 6.95$, donors; glucose, phenol, $HP_2O_7^{3-}$, *p*-nitrophenol, acetic acid) and acetylacetonate ($pK_A = 9.5$, donors: H_2O, glycerine, mannose, glucose, trimethylphenol, phenol, *o*-chlorophenol, *m*-nitrophenol, *p*-nitrophenol, dimedone, acetic acid, benzoic acid, chloracetic acid, H_3O^+. (Data from Eigen and Kruse [12], Eigen [30], and Maass [31].)

In practice, this ideal case is only approached (see Fig. 8-6). Instead, the transition from $\alpha = 1$ to 0 occurs smoothly in the range $pK_A \approx pK_D$. In terms of the mechanism in Eq. (8-37), this means that more intermediate states are kinetically significant. This transition becomes broader for resonance-stabilized donors, until finally in the case of pseudoacids the transition is extremely broad. As a case in point, an actual experimental curve obtained with acetylacetone is included in Fig. 8-6.

8-7 ACID–BASE CATALYSIS

The nature of catalysis in homogeneous systems has been the subject of a considerable amount of research. A catalyst is any substance which affects the rate of reaction but is not consumed in the overall reaction. From thermodynamic principles we know that the equilibrium constant for the overall reaction must be independent of the mechanism, so that a catalyst for the forward reaction must also be one for the reverse reaction. In aqueous solution, a large number of reactions are catalyzed by acids and bases; for our purposes we shall employ the Brönsted definition of acids and bases as proton donors and acceptors, respectively. Catalysis by acids and bases involves proton transfer either to or from the substrate. For example, the dehydration of acetaldehyde hydrate is subject to acid catalysis [20], probably by the mechanism (II).

$$\begin{array}{l}\quad\mathrm{OH}\\ \mathrm{H}/\\ \mathrm{RC}\\ \quad\backslash\\ \quad\mathrm{OH}\end{array} + \mathrm{HA} \overset{\text{fast}}{\rightleftharpoons} \begin{array}{c}\mathrm{H\quad H}\\ \mathrm{RC{-}OH}\\ \mathrm{OH+}\end{array} + \mathrm{A^-} \xrightarrow{\text{slow}} \begin{array}{c}\mathrm{HH}\\ \mathrm{RCOH}\\ \mathrm{O}\end{array} + \mathrm{HA} \xrightarrow{\text{fast}} \begin{array}{c}\mathrm{H}\\ \mathrm{RC{=}O}\end{array} + \mathrm{H_2O} + \mathrm{HA}$$

II

Clearly, a detailed discussion of the equilibrium and kinetic properties of proton-transfer reactions is necessary in order to establish a molecular mechanism of acid–base catalysis. Only a few specific aspects of acid–base catalysis are discussed here; more detailed accounts are available elsewhere [21, 22].

A distinction is often made between specific catalysis by hydrogen and hydroxyl ions (*specific* hydrogen and hydroxide-ion catalysis) and catalysis by acids and bases in general (*general* acid and base catalysis). Actually such a distinction is somewhat artificial, as we shall see. As an example of general base catalysis, consider the simple mechanism

$$\mathrm{SH} + \sum_i \mathrm{B}_i \underset{\Sigma k_{-i}}{\overset{\Sigma k_i}{\rightleftharpoons}} \mathrm{S'} + \sum_i \mathrm{HB}_i, \qquad \mathrm{S} \xrightarrow{k_2} \text{products} \tag{8-39}$$

(The necessary deprotonation of HB_i is assumed to be fast relative to and after the rate-determining step.) Here S designates the substrate and B_i a base. If S′ is assumed to be in a steady state, the rate of the reaction is equal to

$$R = \frac{(\mathrm{SH}) \sum_i k_i(\mathrm{B}_i)}{1 + \sum_i k_{-i}(\mathrm{HB}_i)/k_2}$$

If $k_2 \gg \sum_i k_{-i}(\mathrm{HB}_i)$, that is, proton transfer is rate-limiting,

$$R = (\mathrm{SH}) \sum_i k_i(\mathrm{B}_i) \tag{8-40}$$

This can be regarded as a normal rate equation for general base catalysis. If the first step is at equilibrium,

$$R = (\mathrm{SH})k_2 \sum_i \frac{k_i}{k_{-i}} \frac{(\mathrm{B}_i)}{(\mathrm{HB}_i)} = (\mathrm{SH})k_2 \sum_i \frac{k_i}{k_{-i}} \frac{\mathrm{K}_i}{\mathrm{K}_w}(\mathrm{OH}^-) \tag{8-41}$$

where K_w is the ionization constant of water and K_i is the ionization constant of HB_i. Thus, even though the reaction mechanism is a case of general base catalysis, the rate equation is the same as that found for specific hydroxide-ion catalysis. Therefore, extreme caution must be used in interpreting these rate equations in terms of mechanism. Analogous mechanisms for general acid catalysis are obvious.

In some cases, two proton transfers are involved in the mechanism, and in addition both acids and bases are effective catalysts. Two general mechanisms are possible, stepwise proton transfers

$$\begin{aligned} &\mathrm{SH} + \mathrm{B}_j \rightleftharpoons \mathrm{S} + \mathrm{HB}_j \rightarrow \mathrm{SH}' + \mathrm{B}_j \\ &\mathrm{SH} + \mathrm{HB}_i \rightleftharpoons \mathrm{HSH} + \mathrm{B}_i \rightarrow \mathrm{SH}' + \mathrm{HB}_i \end{aligned} \tag{8-42}$$

or a *concerted* mechanism, in which the transition state contains one acid and one base molecule

$$\mathrm{B}_j + \mathrm{SH} + \mathrm{HB}_i \rightleftharpoons \mathrm{B}_i\mathrm{HSHB}_j \rightarrow \mathrm{B}_i + \mathrm{S}' + \mathrm{HB}_j \tag{8-43}$$

If the intermediates are assumed to be in a steady state, the rate laws are

$$R = (\mathrm{SH})\left[\sum_i k_i(\mathrm{HB}_i) + \sum_j k_j(\mathrm{B}_j)\right] \tag{8-44}$$

$$R = (\mathrm{SH}) \sum_i \sum_j k_{ij}(\mathrm{HB}_i)(\mathrm{B}_j) \tag{8-45}$$

If the catalytic power of a given acid or base is independent of its catalytic partner, Eq. (8-45) becomes

$$R = (\mathrm{SH}) \sum_i k_i(\mathrm{HB}_i) \sum_j k_j(\mathrm{B}_j) \tag{8-46}$$

In practice, the rate laws for these two mechanisms are extremely difficult to distinguish in aqueous solution. This is because terms of the form $k(X)(H_2O)$ are indistinguishable from $k(X)$. In addition, cross terms in Eq. (8-46) of the form $k_i(B_i)(HB_i)$ are usually small and difficult to detect. This product term has been definitely observed for the iodination of acetone in aqueous solution [23–25]. The mutarotation of glucose also probably occurs via a concerted mechanism [25].

The catalytic behavior of a large number of structurally similar acids and bases is correlated remarkably well by a relationship known as the Brönsted equation, which states that

$$k = G K_A{}^{\alpha} \tag{8-47}$$

or

$$\log k = \log G - \alpha \mathrm{p}K_A$$

where k is the catalytic rate constant for the reaction, K_A the ionization constant for the acid catalyst, and G and α are constants, α being restricted to values between zero and one. A more exact formulation of the Brönsted equation considers statistical corrections to Eq. (8-47). If, for a given catalyst, p is the number of equivalent protons on the acid and q is the number of equivalent positions on the conjugate base at which a proton can be accepted, then Eq. (8-47) should be written as [26]

$$\frac{k}{p} = G\left(\frac{q}{p} K_A\right)^{\alpha} \tag{8-48}$$

for acid catalysis. For the case of base catalysis, a similar relationship is valid:

$$\frac{k}{q} = G'\left(\frac{q}{p} K_A\right)^{-\beta}$$

where β is a constant lying between 0 and 1, and for a given reaction, $\alpha + \beta = 1$. Usually p and q are taken to refer to different atoms in the same molecule, so that, for example, p is taken as 2 for fumaric acid but as 1 for ammonium ion (even though 4 would seem appropriate). Equations (8-47) and (8-48) are applicable only if structurally and electrostatically similar acid and base catalysts are compared. Note that Eq. (8-38) can be cast into an essentially identical form since the pK of the proton acceptor, the substrate, is constant for acid catalysis, and the pK of the proton donor, the substrate, is constant for base catalysis. This is not too surprising, since a proton-transfer reaction is often rate-determining in acid–base catalysis.

In most cases, the Brönsted coefficient, α or β, approximates the slope of an intermediate linear portion of a plot such as Fig. 8-6. This is usually because the range of pK_A values used is too small for observation of the limiting slopes. Also, since substrates are usually extreme pseudoacids, the transition from $\alpha = 0$ to $\alpha = 1$ will occur so gradually that α often will appear to have a constant value even if $\Delta(pK)$ varies considerably. Curvature of Brönsted plots, however, has been noted [21]. In some instances, several intermediates may occur with proton transfer only being partially rate determining. Occasionally, the limiting value of 0 or 1 is approached. An interesting reaction in which $\alpha \approx 0$ is the transfer of an acyl group from sulfur to nitrogen in S acetylmercaptoethanolamine above pH 2.3 [27]. The postulated mechanism is (8-49).

$$CH_3C(=O)\text{–}S\text{–}CH_2\text{–}CH_2\text{–}NH_2 \underset{}{\overset{K}{\rightleftharpoons}} CH_3C(O^-)(S\text{–}CH_2\text{–}CH_2\text{–}NH_2^+) \xrightarrow{k_{HA}(BH)} CH_3C(OH)(S\text{–}CH_2\text{–}CH_2\text{–}NH_2^+) \rightleftharpoons CH_3C(OH)(S\text{–}CH_2\text{–}CH_2\text{–}NH) \rightleftharpoons HS\text{–}CH_2\text{–}CH_2\text{–}NH\text{–}C(=O)CH_3 \quad (8\text{-}49)$$

If the rate-determining step is assumed to be the proton transfer in the second step, the rate of the overall reaction is

$$R = Kk_{HA}(BH)(S) \tag{8-50}$$

A plot of $\log k_{HA}$ versus pK_A is presented in Fig. 8-7. (The value of K was determined by assuming that k_{HA} is diffusion-controlled when BH is H_3O^+.) Note the similarity to Fig. 8-6! These results can be explained if the rate controlling step is the protonation of the second species in Eq. (8-49) by a diffusion-controlled reaction with BH when pK_A is less than the pK of the second species ($\alpha = 0$). A change in slope of the plot occurs approximately at the pK of the intermediate (~ 7.4), and at higher values of pK_A, the rate becomes diffusion controlled in the reverse direction ($\alpha = 1$). At first glance the idea that a diffusion-controlled proton transfer is rate limiting may seem unreasonable. However, the rate is a product of concentrations and a rate constant, and in this instance the concentration of the second species in Eq. (8-49) is extremely low. While this mechanism provides a consistent representation of available data, a kinetically indistinguishable possibility is general base catalysis with a protonated substrate

$$R = k_B(B^-)(SH^+) = k_B \frac{K_A}{K_{SH}}(BH)(S) \tag{8-51}$$

[K_{SH} is the ionization constant of the second species in Eq. (8-49).] This type of ambiguity is quite general in acid–base reactions, and a distinction between general acid and general base catalysis cannot be made through the rate law

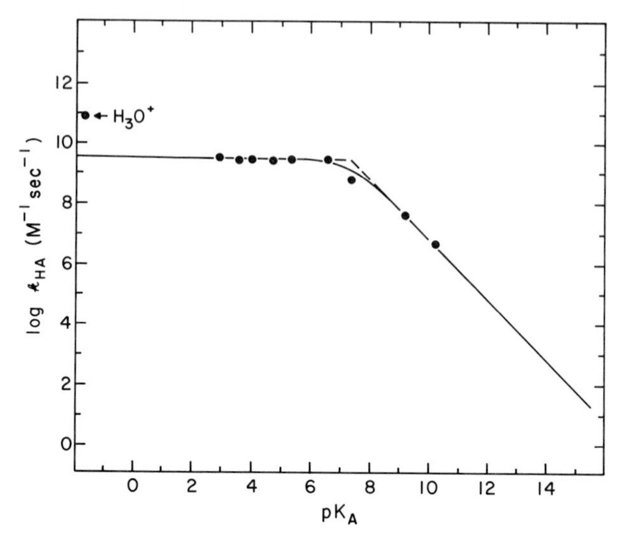

Fig. 8-7 Brönsted plot for general acid catalysis of the acetyl transfer reaction of S-acetylmercaptoethylamine at 50° and ionic strength 1.0 M, based on a value of $k_{H_3O^+} = 6.5 \times 10^{10}$ M^{-1} sec^{-1}. (Adapted from R. E. Barnett and W. P. Jencks, *J. Am. Chem. Soc.* **91**, 2358 (1969), Fig. 5. Copyright 1969 by the American Chemical Society. Reproduced by permission of the copyright owner.)

[Eqs. (8-50) and (8-51)]. Other information, however, may cause one mechanism to be favored over another. In the present case, general acid catalysis is preferred because with a variety of acids k_{HA} is found to have the value expected for a diffusion-controlled reaction.

The Brönsted relationships are a general manifestation of the concept of linear free energies. The basic idea is that only a constant fraction of the change in the overall free-energy change is found in the free energy of activation:

$$\Delta G^{0\ddagger} = RT \ln(k_i'/k_i) = \beta \Delta G^0 \tag{8-52}$$

That β should be less than one is reasonable, since ionization involves complete removal of the proton, whereas the activation step involves only partial removal of the proton. A molecular basis can be established in a qualitative way [21, 28] if we consider the potential-energy curves of a proton-transfer reaction as a function of the distance of the proton from the two centers X and Y in the reaction

$$\text{X—H} + \text{Y} \rightarrow \text{X} + \text{Y—H} \tag{8-53}$$

A qualitative description is shown in Fig. 8-8. If a change is made to Y′, the dashed curve results. If the curves have the same shape and the displacement ΔE_a is along a linear portion of the curve, the geometry of the system requires

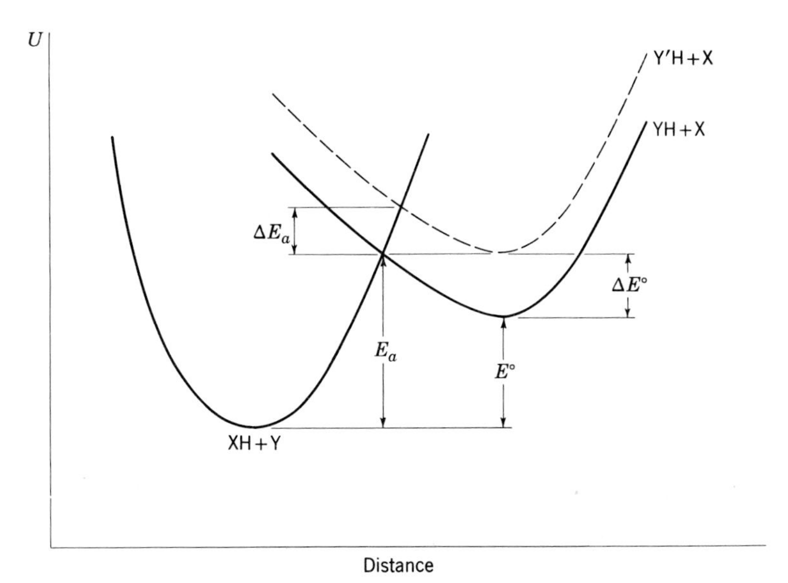

Fig. 8-8 Schematic potential-energy curve for proton transfer.

that

$$\Delta E = \alpha \, \Delta E^0 \tag{8-54}$$

If the entropy factors are the same,

$$\Delta(\log k_i) = \alpha \, \Delta(\log K_{\mathrm{D}})$$

Since such an argument involves several critical assumptions about the shape of the potential-energy curves and the entropy of activation, it is not too surprising that the Brönsted catalysis law has not been found to be completely adequate [21]. This simple argument also predicts that different acids and bases should have rate constants which differ only because their activation energies differ. This prediction is not consistent with available experimental results.

In principle, similar types of relationships are also applicable to other reaction types, such as electron-transfer and substitution reactions. Hammett, for example, has had considerable success in correlating the rate of benzene substitution reactions [29]. For an interesting treatise on this general subject, see Leffler and Grunwald [32].

From this discussion, it should be apparent that the study of fast reactions can provide an intimate knowledge of reaction mechanisms not available with conventional techniques. The development of techniques for studying fast reactions has caused a renaissance in kinetic studies of reactions in solution.

Problems

8-1 The formation of a hydrogen-bonded dimer between α-pyridone molecules proceeds according to the reaction shown in Fig. P8-1. The kinetics of this reaction has

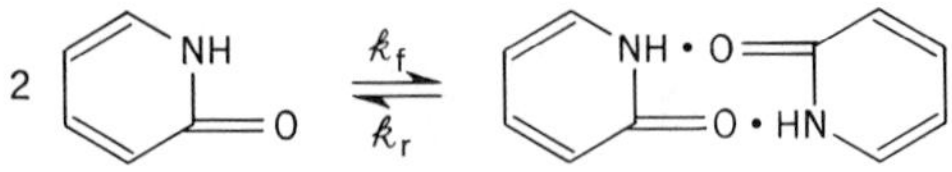

Fig. P8-1

been studied by using ultrasonic techniques, and the results shown in Table P8-1 have been obtained at 13°C in dioxane [33].

TABLE P8-1

$f\,(=\omega/2\pi) \times 10^{-6}$ (cps)	$2\alpha_c\lambda \times 10^3$	$f\,(=\omega/2\pi) \times 10^{-6}$ (cps)	$2\alpha_c\lambda \times 10^3$
15.1	2.24	55.1	4.62
25.1	3.58	74.9	4.40
35.1	4.22	95.2	4.00
45.1	4.54	114.9	3.86

Evaluate the relaxation time for this system and determine the rate constants k_f and k_r. The total concentration of α-pyridone is 0.1 M and the equilibrium constant for the dimerization in dioxane is 19 M^{-1} at 13°C.

8-2 Consider the following reaction mechanism, where ATP = adenosine 5′-triphosphate:

$$Mg^{2+} + ATP^{4-} \underset{k_{21}}{\overset{k_{12}}{\rightleftharpoons}} MgATP^{2-}, \qquad ATP^{4-} + H^+ \underset{k_{31}}{\overset{k_{13}}{\rightleftharpoons}} ATPH^{3-}$$

(a) Derive expressions for the relaxation times of this system (1) in the general case, (2) assuming that the protolytic equilibrium is adjusted rapidly compared to that for metal-complex formation, and (3) assuming that there is buffering of (H^+), that is, $\Delta C_{H+} = 0$ and that the protolytic equilibrium is rapidly adjusted.

TABLE P8-2

C_{H^+} (M)	C_{ATP} (M)	C_{ATPH} (M)	C_{Mg} (M)	τ (μsec)
$10^{-7.11}$	1.03×10^{-4}	2.52×10^{-5}	1.28×10^{-4}	240
$10^{-7.48}$	1.12×10^{-4}	1.17×10^{-5}	1.24×10^{-4}	250
$10^{-7.11}$	6.4×10^{-5}	1.57×10^{-5}	7.97×10^{-5}	370
$10^{-7.11}$	1.6×10^{-4}	3.92×10^{-5}	1.99×10^{-4}	180
$10^{-7.11}$	2.78×10^{-4}	6.82×10^{-5}	3.46×10^{-4}	110

(b) The values given in Table P8-2 have been obtained for the slow relaxation time in buffered solutions (case 3 above). Also $k_{31}/k_{13} = 10^{-6.5}$ M and $k_{12}/k_{21} = 10^4$ M^{-1}. Obtain values for the rate constants k_{12} and k_{21}.

Hint: A graphical procedure is recommended.

8-3 The following results have been obtained in a nuclear magnetic resonance study of methanol at 22°C [34].

Pure methanol has roughly the spectrum shown in Fig. P8-3a and when HCl is added to methanol, the spectrum becomes that shown in Fig. P8-3b.

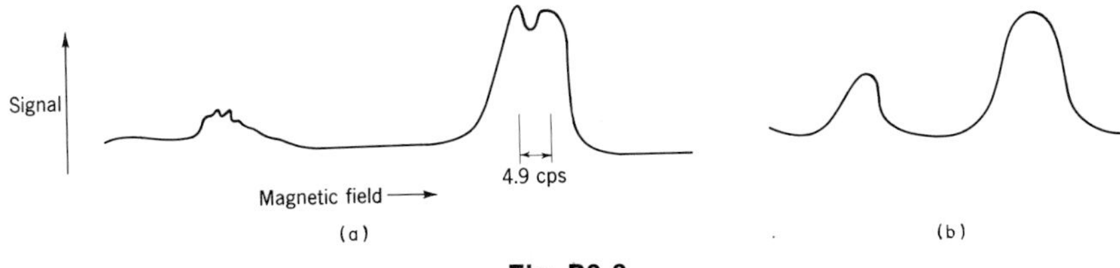

Fig. P8-3

(a) Assume this change is entirely due to an exchange reaction between CH_3OH and $CH_3OH_2^+$ and postulate two possible one-step mechanisms. The specific rate constant for the exchange reaction is about 10^8 M^{-1} sec^{-1}.

(b) On the basis of your general knowledge of proton-transfer reactions, cite any evidence which favors one of the mechanisms.

(c) On the basis of the preceding data, at what $CH_3OH_2^+$ concentration would you predict coalescence of the methyl doublet? Is this concentration consistent with the fact that the autoprotolysis constant for methanol is

$$(CH_3O^-)(CH_3OH_2^+) = 2 \times 10^{-17} \quad M^2?$$

8-4 The kinetics of the protolysis of methylammonium ion in aqueous solution has been studied with nuclear magnetic resonance. Relaxation times were obtained by studying the CH_3, NH_3^+, and H_2O resonances as a function of pH and methylammonium ion concentration. The experimental results were consistent with the mechanism

$$CH_3NH_3^+ + OH^- \xrightarrow{k_5} CH_3NH_2 + HOH$$

$$CH_3NH_3^+ + NH_2CH_3 \xrightarrow{k_6} CH_3NH_2 + NH_3CH_3^+$$

$$CH_3\overset{H}{\underset{H}{N}}\text{—}H^+ + \overset{H}{O}H + \overset{H}{\underset{H}{N}}CH_3 \xrightarrow{k_7} CH_3\overset{H}{\underset{H}{N}} + H\text{—}\overset{H}{O} + {}^+H\text{—}\overset{H}{\underset{H}{N}}\text{—}CH_3$$

At 19°C the relaxation times given in Table P8-4 were measured from changes in the CH_3 quadruplet with concentrations [35].

Derive an expression for the relaxation time of the given mechanism and determine as many of the individual rate constants as possible. The ionization constant of methylammonia is 1.22×10^{-11} M and that of water is 0.63×10^{-14} M at 19°C.

TABLE P8-4

$(CH_3NH_3^+)$ (M)	$(H^+) \times 10^3$ (M)	$1/\tau$ (sec^{-1})
1.03	1.078	8.0
	0.710	11.6
	0.526	14.5
	0.416	19.7
0.83	0.558	14.3
0.545	0.258	15.3
0.272	0.267	9.3

8-5 The halogenation of acetone

$$CH_3-\overset{\overset{\displaystyle O}{\|}}{C}-CH_3 + X_2 \rightleftharpoons XCH_2-\overset{\overset{\displaystyle O}{\|}}{C}-CH_3 + HX$$

is first order in acetone and zero order in the halogen. Moreover, the bromination and iodination have identical rates. With different ketones, it has also been observed that the rate of deuterium exchange is identical to the rate of halogenation. This reaction is subject to general acid and base catalysis. In acetate buffers ($K_A = 3.1 \times 10^{-5}$ M), the data presented in Table P8-5 have been obtained at 25°C and constant ionic strength [25]. The contribution to the rate from the uncatalyzed reaction and specific hydrogen and hydroxide-ion catalysis ($k_0 + k_{H^+}(H_3O^+) + k_{OH^-}(OH^-)$) is 9.7, 16.5, and 23.4 $\times$ 10^{-8} min^{-1}, respectively, for the three ratios of acid to base.

(a) Determine the dependence of the rate constant on the concentration of buffer components, and calculate as many of the specific catalytic rate constants as possible.

(b) Postulate a mechanism and show it is consistent with all of the given experimental facts.

TABLE P8-5

$(HA)/(A^-)^a$	$(A^-) \times 10^2$	$10^8 k$ (min^{-1})	$(HA)/(A^-)^a$	$(A^-) \times 10^2$	$10^8 k$ (min^{-1})
1	2.59	63.0	2	7.41	231
1	5.74	132	3	2.60	103
1	13.10	306	3	5.31	201
2	2.06	70.5	3	10.03	393

[a] HA = acetic acid, A^- = acetate ion.

8-6 Bell and Higginson [20] have studied the acid catalysis of the dehydration of acetaldehyde in aqueous acetone at 25°C; some of their results are presented in Table P8-6.

Using the Brönsted equation, make a graphical correlation of the tabulated data. Be sure to include appropriate statistical corrections and calculate the parameter α in

TABLE P8-6

Acid	k (M^{-1} min^{-1})[a]	K_A (M)[b]
Diethylketoxime	0.104	2.5×10^{-13}
Hydroquinone	0.013	4.5×10^{-11}
Resorcinol	0.026	1.55×10^{-10}
Nitromethane	0.00084	5.8×10^{-10}
o-Chlorophenol	0.112	3.2×10^{-9}
Benzoylacetone (enol)	0.0088	5.9×10^{-9}
p-Nitrophenol	0.52	6.75×10^{-8}
2,4,6-Trichlorophenol	1.53	3.9×10^{-7}
Propionic	18.0	1.35×10^{-5}
Acetic	19.2	1.76×10^{-5}
Phenylacetic	33.0	4.88×10^{-5}
Diphenylacetic	44.2	1.15×10^{-4}
Glycollic	56.6	1.54×10^{-4}
Dichloracetic	773	5.0×10^{-2}

[a] Catalytic specific rate constant.
[b] Acid ionization constant in water.

Eq. (8-48). Explain any marked deviations from the behavior predicted by the Brönsted equation.

8-7 Hammet [36] has compiled the data given in Table P8-7 for the alkaline hydrolysis of ethyl esters in 85% ethanol at 25°.

(a) Using the concept of linear free-energy changes, derive an equation relating the rate and ionization constants given in the table.

(b) Plot the tabulated data according to the equation obtained in (a) and determine any empirical parameters involved in the equation. What explanation can you give for any extreme deviations of the data from the derived equation?

TABLE P8-7

Ester	k (M^{-1} sec^{-1} $\times 10^4$)	pK_A[a]
Benzoic	5.50	4.21
p-Aminobenzoic	0.127	4.92
p-Methoxybenzoic	1.15	4.47
p-Chlorobenzoic	23.7	3.99
p-Nitrobenzoic	720	3.44
m-Chlorobenzoic	47.7	3.83
m-Nitrobenzoic	429	3.45
Acetic	69.5	4.76
Isobutyric	8.01	4.85

[a] Acid ionization constant of the substituted benzoic acid in water at 25°.

References

1. G. G. Hammes, ed., "Techniques of Chemistry," Vol. 6, part 2. Wiley (Interscience), New York, 1974.
2. C. F. Bernasconi, "Relaxation Kinetics." Academic Press, New York, 1976.
3. B. Chance, *in* Ref. 1, pp. 5–62.
4. L. Onsager, *J. Chem. Phys.* **2**, 599 (1933).
5. M. Eigen and L. de Maeyer, *in* Ref. 1, pp. 63–146; G. W. Castellan, *Ber. Bunsenges. Phys. Chem.* **67**, 898 (1963).
6. J. Meixner, *Kolloid-Z.* **134**, 3 (1953).
7. D. Thusius, *Biochemie* **55**, 277 (1973).
8. T. J. Chuang, G. W. Hoffman, and K. B. Eisenthal, *Chem. Phys. Lett.* **25**, 201 (1974).
9. R. A. Ogg, *J. Chem. Phys.* **22**, 560 (1954); *Discuss. Faraday Soc.* **17**, 215 (1954).
10. T. J. Swift, *in* Ref. 1, pp. 521–564.
11. R. A. Dwek, "Nuclear Magnetic Resonance in Chemistry." Oxford Univ. (Clarendon) Press, London and New York, 1973.
12. M. Eigen and W. Kruse, *Z. Naturforsch.* **186**, 857 (1963), and unpublished results.
13. M. Eigen and L. de Maeyer, *Z. Elektrochem.* **59**, 986 (1955).
14. M. Eigen and J. Schoen, *Z. Elektrochem.* **59**, 483 (1955).
15. M. T. Emerson, E. Grunwald, and R. A. Kromhout, *J. Chem. Phys.* **33**, 547 (1960).
16. M. Eigen, G. G. Hammes, and K. Kustin, *J. Am. Chem. Soc. 82*, 3482 (1960).
17. M. Eigen and G. Schwarz, unpublished results.
18. M. Eigen, *Z. Phys. Chem. (Leipzig)* **NF1**, 176 (1954).
19. M. Eigen and E. M. Eyring, *J. Am. Chem. Soc.* **84**, 3254 (1962).
20. R. P. Bell and W. C. E. Higginson, *Proc. Roy. Soc. London Ser. A* **197**, 141 (1949).
21. R. P. Bell, "The Proton in Chemistry." Cornell Univ. Press, Ithaca, New York, 1st ed, 1959; 2nd ed, 1973.
22. W. P. Jencks, *in* "Catalysis in Chemistry and Enzymology," pp. 163–242. McGraw-Hill, New York, 1969; W. P. Jencks, *Acc. Chem. Res.* **9**, 425 (1976).
23. H. M. Dawson and E. Spivey, *J. Chem. Soc.* p. 2180 (1930).
24. C. G. Swain, *J. Am. Chem. Soc.* **72**, 4578 (1950).
25. R. P. Bell and P. Jones, *J. Chem. Soc.* p. 88 (1953).
26. J. N. Brönsted and K. Pederson, *Z. Phys. Chem. (Leipzig)* **108**, 185 (1924).
27. R. E. Barnett and W. P. Jencks, *J. Am. Chem. Soc.* **91**, 2358 (1969).
28. J. Horuiti and M. Polanyi, *Acta Physiochim. URSS* **2**, 505 (1935).
29. L. P. Hammett, "Physical Organic Chemistry," Chapter 7. McGraw-Hill, New York, 1940.
30. M. Eigen, *Angew. Chem.* **75**, 489 (1963); *Angew. Chem. Int. Ed. Engl.* **3**, 1 (1964).
31. G. Maass, Dissertation, Göttingen, 1962.
32. J. E. Leffler and E. Grunwald, "Rates and Equilibria of Organic Reactions." Wiley, New York, 1963.
33. G. G. Hammes and H. O. Spivey, *J. Am. Chem. Soc.* **88**, 1621 (1966).
34. Z. Luz, D. Gill, and S. Meiboom, *J. Chem. Phys.* **30**, 1540 (1959).
35. E. Grunwald, A. Loewenstein, and S. Meiboom, *J. Chem. Phys.* **27**, 630 (1957).
36. L. Hammett, Ref. 29, p. 121.

CHAPTER 9

ENZYME KINETICS

9-1 INTRODUCTION

The molecular details of enzymatic reactions are currently of great interest and particularly so to kineticists because of the special kinetic problems involved. First, the complex mechanisms of enzymatic reactions require rather special phenomenological rate equations; second, since enzymes are catalysts, the nature of catalysis in general must be considered; and finally, the complex nature of even the simplest enzymatic mechanism makes its clarification a challenge.

As catalysts, enzymes are interesting not only because of their obvious biological importance, but also because of their remarkable efficiency and specificity. Under ordinary conditions, almost all biological reactions fail to proceed at an appreciable rate in the absence of enzymes. In fact, on a molar basis other types of catalysts generally do not come within a factor of a million of enzymatic efficiency for a given reaction.

A number of enzymes have been obtained in highly purified form, even crystalline, and their chemical and physical properties have been extensively studied. In general, enzymes operate most efficiently in aqueous media. All enzymes are macromolecules, with molecular weights ranging from about 10^4 to 10^6. More precisely, a portion of all enzymes is *protein* in nature. Proteins are polymers of α-amino acids (polypeptides) and have the general formula

$$\mathrm{NH_2\overset{R}{C}H{-}\overset{\overset{O}{\|}}{C}{-}(NH{-}\overset{R}{C}H{-}\overset{\overset{O}{\|}}{C}{-})_nNH{-}\overset{R}{C}H{-}COOH}$$

The carbon–nitrogen bond between the amino acids is called a peptide bond. Here R can be a number of different groups, either lyophobic or lyophilic in nature. An important characteristic of enzymes is that these side chains can contain ionizable groups, and at physiological pH values proteins contain a large number of charged groups. There are many different amino acids within a single protein molecule, and it is possible to determine the order in which the acids are arranged if the polypeptide chain is not too long. An even more important and difficult structural problem, i.e., the three-dimensional protein configuration, has been studied in quite a few cases, and for some enzymes it is possible to correlate the enzymatic mechanism with the enzyme structure on a molecular basis.

Many enzymes have been shown to be entirely protein in nature, but others contain a protein part, called the apoenzyme, and a nonprotein part, called the prosthetic group. The prosthetic group is usually a complex organic molecule and is generally known as a *coenzyme*. However, certain enzymes also have ions strongly attached to the protein, and removal of the ions from the protein results in a complete loss of catalytic activity. Such ions (often metals) are called *activators*. Many prosthetic groups cannot be split from the protein without considerably altering the protein molecule, whereas others dissociate quite readily. All these types of prosthetic groups are similar, differing only in the firmness and mode of the attachment to the apoenzyme, and too sharp a distinction should not be made between the types.

Enzymes catalyze such a large array of reaction types that a complete compilation is prohibitive. We present a partial but by no means inclusive classification of major reaction types. The substrate molecules range from small organic molecules, e.g., urea, to macromolecules, e.g., proteins.

Hydrolytic enzymes catalyze the general reaction

$$\mathrm{AB + H_2O \rightleftharpoons AOH + BH} \tag{9-1}$$

Three main subclasses of hydrolytic enzymes are

(1) the proteolytic enzymes, which catalyze the hydrolysis of the peptide linkage;

(2) the esterases, which hydrolyze esters to the acid and the alcohol, e.g., lipases, which hydrolyze fats to fatty acids and glycerol, and phosphatases, which hydrolyze phosphate esters to phosphoric acid and the alcohol; and

(3) the glycosidases, which catalyze the hydrolysis of glycosidic linkages in polysaccharides, e.g.,

$$\mathrm{sugar^1{-}O{-}sugar^2 + H_2O \rightleftharpoons sugar^1{-}OH + sugar^2{-}OH}$$

Phosphorylases are similar to hydrolytic enzymes except that a molecule of phosphoric acid replaces water in the reaction.

Oxidative enzymes are concerned with various oxidation–reduction processes. Major subdivisions are dehydrogenases, which remove two hydrogen atoms from a substrate and transfer it to a coenzyme, and oxidases, which transfer hydrogen to molecular oxygen.

Transferring enzymes catalyze reactions in which groups are interchanged between two molecules:

$$AB + CD \rightleftharpoons AC + BD \tag{9-2}$$

Important examples are phosphokinases, which catalyze the transfer of a phosphoryl group, and transaminases, which bring about the exchange of amino and keto groups between amino and keto acids.

Addition reactions such as

$$A + B \rightleftharpoons AB \tag{9-3}$$

are also catalyzed; e.g., carbonic anhydrase catalyzes the hydration of CO_2.

Finally, isomerizing enzymes should be mentioned. An example is phosphoglucomutase, which catalyzes the interconversion of glucose 1-phosphate to glucose 6-phosphate.

Another remarkable feature of enzymes is their specificity. Some enzymes work only with a single substrate. For example, fumarase catalyzes only the hydration of fumaric acid to *l*-malic acid, and even the *d* isomer of malic acid and maleic acid (the stereoisomer of fumaric acid) are completely unaffected, as are other molecules quite similar to the substrates. (However, it has been possible to substitute fluorine for some of the hydrogens.) A second class of enzymes exhibits group specificity; i.e., certain groupings must be present on the substrate molecule. For example, carboxypeptidase will catalyze peptide hydrolysis if, among other things, a carboxyl group is adjacent to the peptide bond. The lowest type of specificity is shown by enzymes which catalyze a certain type of reaction irrespective of adjacent groups to the linkage being acted upon.

This introduction to enzyme chemistry is far from complete, and interested readers can find further details in standard texts [1].

9-2 STEADY-STATE KINETICS

In most cases, enzymes are such efficient catalysts that concentrations must be very low in order to make the reaction occur during a convenient time interval, i.e., greater than seconds. Typical enzyme concentrations range from approximately 10^{-8} to $10^{-10}\,M$, while substrate concentrations are usually greater than $10^{-6}\,M$. Under these conditions, all reaction intermediates are present at much smaller concentrations than the substrates, so that they can be considered to be in a steady state (after a short induction period). A majority of enzyme reactions have been investigated under these conditions, and such experiments are appropriately called steady-state kinetic studies. In view of the large number of possible mechanisms for the great variety of reactions catalyzed by enzymes, no attempt at a complete discussion is made. Instead, the simplest possible mechanism, i.e., that involving a single substrate and product, is discussed in considerable detail. A prototype reaction of this type is the reaction (9-4) catalyzed by the enzyme fumarase.

$$\underset{\text{Fumarate}}{\ce{^-O_2C-CH=CH-CO_2^-}} + H_2O \rightleftharpoons \underset{l\text{-Malate}}{\ce{^-O_2C-CH(OH)-CH_2-CO_2^-}} \tag{9-4}$$

When the initial velocity (rate) v of our prototype reaction is studied as a function of substrate concentration (S) at constant enzyme concentration, the results depicted in Fig. 9-1 are obtained. The maximum initial velocity reached is called the maximum velocity V_s and is directly proportional to the total enzyme concentration (E_0). The ratio $V_s/(E_0)$ is often called the turnover number and is a direct measure of the catalytic efficiency of the enzyme. The substrate concentration at which the initial velocity reaches one half the maximum velocity is called the Michaelis constant. A simple possible mechanism consistent with these findings is the Michaelis–Menten mechanism

$$E + S \underset{k_{-1}}{\overset{k_1}{\rightleftharpoons}} X \underset{k_{-2}}{\overset{k_2}{\rightleftharpoons}} E + P$$

where S and P represent the substrate and product, E is the free enzyme, and X is a reaction intermediate. Since X is present in a steady state,

$$-\frac{dX}{dt} = (k_{-1} + k_2)(X) - k_1(E)(S) - k_{-2}(E)(P) = 0 \tag{9-5}$$

Also, conservation of mass and the restriction that $(S_0) \gg (E_0)$ require that

$$(E_0) = (E) + (X) \tag{9-6}$$

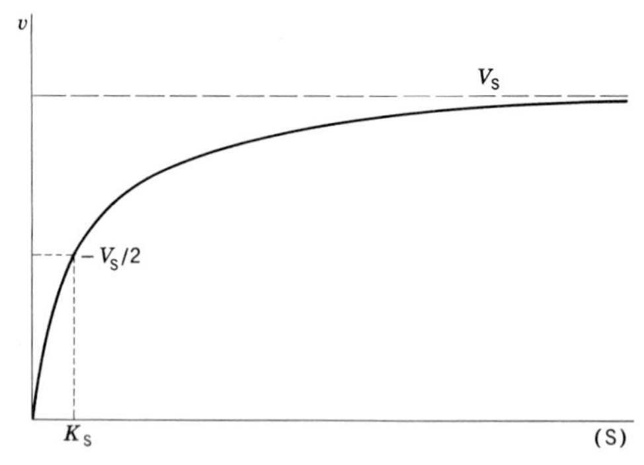

Fig. 9-1 Schematic plot of initial velocity versus substrate concentration for a mechanism involving one substrate.

and

$$(S_0) = (S) + (P) \tag{9-7}$$

so that

$$-\frac{d}{dt}(S) = \frac{d}{dt}(P) = k_1(E)(S) - k_{-1}(X) \tag{9-8}$$

Combining Eqs. (9-5)–(9-8), we obtain

$$-\frac{d}{dt}(S) = \frac{d}{dt}(P) = \frac{[k_1 k_2(S) - k_{-1}k_{-2}(P)](E_0)}{k_1(S) + k_{-2}(P) + k_{-1} + k_2} \tag{9-9}$$

This equation is usually written as

$$-\frac{d}{dt}(S) = \frac{[V_S/K_S](S) - [V_P/K_P](P)}{1 + (S)/K_S + (P)/K_P} \tag{9-10}$$

where

$$V_S = k_2(E_0), \qquad V_P = k_{-1}(E_0),$$
$$K_S = (k_{-1} + k_2)/k_1, \qquad K_P = (k_{-1} + k_2)/k_{-2} \tag{9-11}$$

For initial-velocity measurements we can set (P) = 0 and obtain

$$v = -\frac{d}{dt}(S) = \frac{k_2(E_0)}{1 + (k_{-1} + k_2)/k_1(S)} = \frac{V_S}{1 + K_S/(S)} \tag{9-12}$$

This equation displays the dependence on substrate concentration, shown in Fig. 9-1, since as $(S) \to \infty$, $v \to V_S$, and when $(S) = K_S$, $v = V_S/2$. The Michaelis constant K_S, in general, is not an equilibrium constant but a "steady-state" constant and measures the ratio of steady-state concentrations (E)(S)/(X). Note that the rate equation obtained is identical with the Langmuir adsorption isotherm and is, in fact, valid for any catalytic reaction where only one type of noninteracting catalytic site is present and saturation of the catalyst is possible. Obviously V_S and K_S can be determined from initial velocities; one of the most convenient methods is to plot $(S)/v$ versus (S), which gives a straight line with an intercept equal to K_S/V_S and a slope equal to $1/V_S$. If the reaction is reversible, the same procedure can be carried out for the reverse reaction. Since four kinetic parameters are obtained, all four rate constants can be obtained. Alternatively, Eq. (9-10) can be integrated so that the entire time course of the reaction is studied. Usually the integrated rate equation is not so convenient to use as the initial-velocity equation. It is important to remember that all these constants are not independent but are related through the equilibrium constant K:

$$K = \frac{k_1 k_2}{k_{-1}k_{-2}} = \frac{V_S K_P}{V_P K_S} = \frac{(P)}{(S)} \tag{9-13}$$

This relationship is obtained by setting $d(\mathrm{P})/dt$ equal to zero and is known as the Haldane equation.

The case we have discussed so far is unrealistic in the sense that more than one intermediate must be involved in such a complicated reaction, but the form of the rate equation is independent of the number of intermediates inserted into the mechanism! For example, if the mechanism is written as

$$\mathrm{E} + \mathrm{S} \underset{k_{-1}}{\overset{k_1}{\rightleftharpoons}} \mathrm{X}_1 \underset{k_{-2}}{\overset{k_2}{\rightleftharpoons}} \mathrm{X}_2 \underset{k_{-3}}{\overset{k_3}{\rightleftharpoons}} \mathrm{P} + \mathrm{E} \tag{9-14}$$

the rate equation can again be written as in Eq. (9-10), with the following definitions of the Michaelis constants and maximum velocities:

$$K_{\mathrm{S}} = \frac{k_2 k_3 + k_{-1} k_3 + k_{-1} k_{-2}}{k_1(k_{-2} + k_2 + k_3)}, \qquad K_{\mathrm{P}} = \frac{k_{-1} k_{-2} + k_{-1} k_3 + k_2 k_3}{k_{-3}(k_{-2} + k_2 + k_{-1})}$$
$$V_{\mathrm{S}} = \frac{k_2 k_3 (\mathrm{E}_0)}{k_3 + k_{-2} + k_2}, \qquad V_{\mathrm{P}} = \frac{k_{-1} k_{-2} (\mathrm{E}_0)}{k_{-1} + k_{-2} + k_2} \tag{9-15}$$

Of course now it is clearly impossible to calculate all six of the rate constants, since only four independent experimental parameters can be obtained. Moreover, treatment of the general case of an arbitrary number of isomeric intermediates yields a rate law of the same form [2].

Actually we should not be surprised at the fact that steady-state rate studies are not decisive for determination of the number (or nature) of the intermediates in the mechanism. By the very nature of the steady-state assumption, the intermediates are virtually impossible to detect experimentally. What then is the value of carrying out steady-state rate experiments? For one thing, information about the structural specificity of the enzyme can often be obtained by varying the substrate; for another, in more complicated mechanisms, possible reaction pathways can be inferred from the form of the rate law. Also of considerable interest to the kineticist is the fact that knowledge of the steady-state kinetic constants allows the determination of a lower bound for *all* the rate constants in the mechanism. For example, in the case of a reaction mechanism of the type we are considering with n reaction intermediates (where n is an arbitrary number), the following inequalities can be shown to prevail [2]:

$$k_{i+1} \geq \frac{V_{\mathrm{S}}}{(\mathrm{E}_0)} \quad (i \neq 0), \qquad k_1 \geq \frac{V_{\mathrm{S}} + V_{\mathrm{P}}}{(\mathrm{E}_0) K_{\mathrm{S}}}$$
$$k_{-i} \geq \frac{V_{\mathrm{P}}}{(\mathrm{E}_0)} \quad (i \neq n+1), \qquad k_{-(n+1)} \geq \frac{V_{\mathrm{S}} + V_{\mathrm{P}}}{(\mathrm{E}_0) K_{\mathrm{P}}} \tag{9-16}$$

The equalities are applicable if $n = 1$; it is easy to verify that the inequalities are correct for the case of $n = 2$ by rearrangement of Eqs. (9-15). The extension to the general case of n intermediates can be made but requires rather tedious algebraic manipulations.

A very useful general method of deriving the steady-state rate equations for enzyme reactions has been presented by King and Altman [3]. It is unique in that the rate law can be written down without having to solve simultaneous equations. The proof underlying this method involves the theory of determinants and is not presented here. The method is applicable to any catalytic process in which the concentration of substrate(s) is much greater than that of catalyst. The mechanisms previously discussed are now reconsidered in the light of the method of King and Altman. First the mechanisms are written in terms of the enzyme species (I and II).

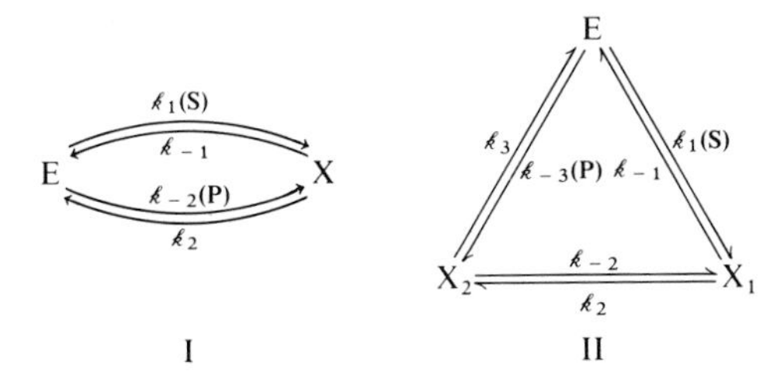

A schematic way of calculating the ratio of any enzyme species to that of the total enzyme concentration was derived by King and Altman. Consider all possible paths leading directly to a given species within the given reaction mechanism; block out in succession one individual step in the pathway and omit the reverse pathway. The results for the species E in the two mechanisms being considered are shown in III and IV.

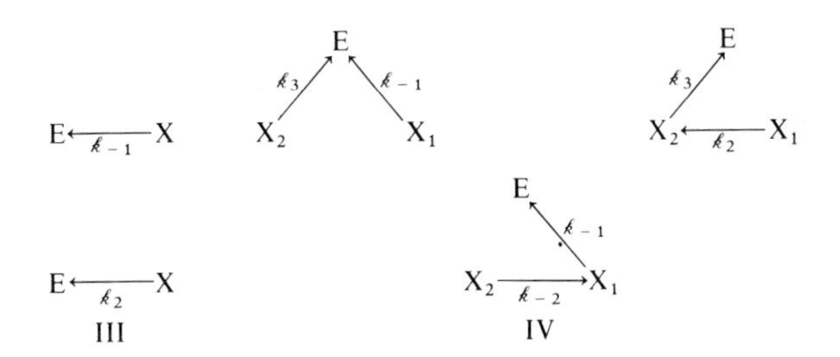

The ratio $(E)/(E_0)$ is equal to the sum of the product of the rate constants involved in each of the possible paths divided by a term D, which will be defined shortly. Thus

$$\text{I.}\quad \frac{(\mathrm{E})}{(\mathrm{E}_0)} = \frac{k_{-1} + k_2}{D}$$

$$\text{II.}\quad \frac{(\mathrm{E})}{(\mathrm{E}_0)} = \frac{k_{-1}k_3 + k_2k_3 + k_{-1}k_{-2}}{D}$$

In general, each term of the numerator involves rate constants (and concentrations) associated with reaction steps which individually or in sequence lead to the species in question. If the number of enzyme species is n, $n - 1$ rate constants are found in each term in the numerator and are associated with $n - 1$ different enzyme-containing species. All the possible combinations of $n - 1$ rate constants which conform to this requirement are present as numerator terms. In a similar manner, an expression can be obtained for other enzyme species:

$$\text{I.}\quad \frac{(\mathrm{X})}{(\mathrm{E}_0)} = \frac{k_1(\mathrm{S}) + k_{-2}(\mathrm{P})}{D}$$

$$\text{II.}\quad \frac{(\mathrm{X}_1)}{(\mathrm{E}_0)} = \frac{k_1k_3(\mathrm{S}) + k_1k_{-2}(\mathrm{S}) + k_{-2}k_{-3}(\mathrm{P})}{D}$$

$$\frac{(\mathrm{X}_2)}{(\mathrm{E}_0)} = \frac{k_1k_2(\mathrm{S}) + k_2k_{-3}(\mathrm{P}) + k_{-1}k_{-3}(\mathrm{P})}{D}$$

The denominator term for a given mechanism is simply the sum of all of the numerator terms; thus

$$\text{I.}\quad D = k_{-1} + k_2 + k_1(\mathrm{S}) + k_{-2}(\mathrm{P})$$

$$\text{II.}\quad D = k_{-1}k_3 + k_2k_3 + k_{-1}k_{-2} + k_1k_3(\mathrm{S}) + k_1k_{-2}(\mathrm{S}) + k_{-2}k_{-3}(\mathrm{P}) + k_1k_2(\mathrm{S}) + k_2k_{-3}(\mathrm{P}) + k_{-1}k_{-3}(\mathrm{P})$$

The rate law for each of the mechanisms is

$$\text{I.}\quad -\frac{dS}{dt} = k_1(\mathrm{E})(\mathrm{S}) - k_{-1}(\mathrm{X}) = (\mathrm{E}_0)\left[\frac{k_1(\mathrm{S})(\mathrm{E})}{(\mathrm{E}_0)} - \frac{k_{-1}(\mathrm{X})}{(\mathrm{E}_0)}\right]$$

$$\text{II.}\quad -\frac{dS}{dt} = k_1(\mathrm{E})(\mathrm{S}) - k_{-1}(\mathrm{X}_1) = (\mathrm{E}_0)\left[\frac{k_1(\mathrm{S})(\mathrm{E})}{(\mathrm{E}_0)} - \frac{k_{-1}(\mathrm{X}_1)}{(\mathrm{E}_0)}\right]$$

If appropriate substitutions are now made, the same rate equation as Eq. (9-10) is obtained. An additional simplifying rule is that all parallel steps in a mechanism can be replaced by a single step with the effective rate constants being the sum of the rate constants for all paths in a given direction.

Thus the one intermediate Michaelis–Menten mechanism can be written as

$$E \underset{k_{-1}+k_2}{\overset{k_1(S)+k_{-2}(P)}{\rightleftharpoons}} X$$

The numerator for the ratio $(E)/(E_0)$ is simply the rate constant for the backward reaction, while the numerator for the ratio $(X)/(E_0)$ is the rate constant for the forward reaction.

This schematic method is applicable to all mechanisms, including multisubstrate ones, and its use saves considerable time and effort for all but the very simplest mechanisms.

An important method of investigating substrate specificity is to study the effect on the rate of substances which are structurally similar to the substrate. In general, the rate is decreased by the presence of such substances, and this phenomenon is called inhibition. Several kinetically distinguishable types of inhibition are observed, depending on whether the inhibitor in question reacts with the substrate-binding site on the free enzyme, on a neighboring site, or with a reaction intermediate. Probably the most commonly studied type of inhibition is that which occurs when the inhibitor reacts with the same binding site on the free enzyme as the substrate. This is termed *competitive* inhibition. The phenomenon can be fitted into the reaction scheme under consideration by simply adding the reaction

$$E + I \rightleftharpoons EI \tag{9-17}$$

where I is the inhibitor. The reaction in Eq. (9-17) has an equilibrium dissociation constant

$$K_I = \frac{(E)(I)}{(EI)} \tag{9-18}$$

Straightforward application of the steady-state assumption leads to the following equation for the initial velocity:

$$v = \frac{V_S}{1 + [K_S/(S)][1 + (I)/K_I]} \tag{9-19}$$

Obviously K_I can be readily determined experimentally by studying the initial velocity as a function of inhibitor concentration. By studying the differences in binding of the structurally similar inhibitors, the structural specificity of the enzyme can be determined.

9.3 pH AND TEMPERATURE DEPENDENCE OF KINETIC PARAMETERS

Reaction rates in enzyme systems are usually extremely sensitive to variation of experimental conditions, i.e., to pH, temperature, ionic strength, specific ions, solvent, etc. We now consider in some detail the influence of pH and temperature on enzymatic reactions.

In general, enzymes exhibit maximum catalytic activity at a definite pH. This optimum pH is generally in the vicinity of pH 7 (± 1), although exceptions are well known. The dependence of the kinetic parameters of the fumarase reaction on pH is illustrated in Figs. 9-2 and 9-3. Several different possible effects of pH on the reaction must be distinguished. In the first place, many substrates may have ionizable groups and only one of the ionized forms of the substrate may be acted upon by the enzyme. Since substrate ionization constants can be determined quite easily and precisely,

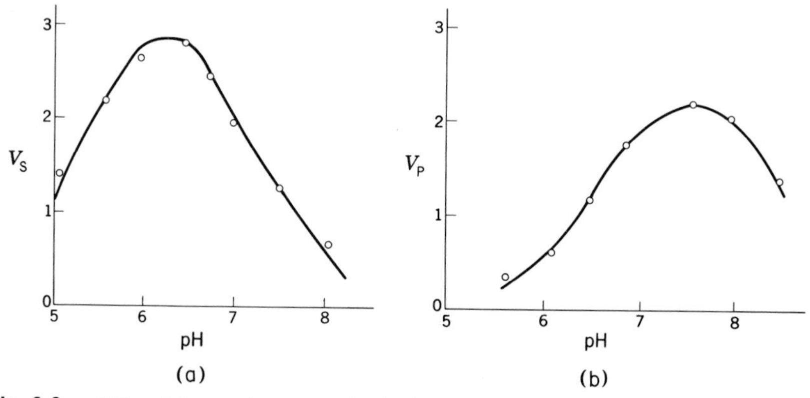

Fig. 9-2 (a) Plot of the maximum velocity for fumarate as a function of pH in 0.01 M acetate buffer at 25°; the solid line is a theoretical curve. (b) Plot of the maximum velocity for l-malate as a function of pH in 0.01 M acetate buffer at 25°; the solid line is a theoretical curve. (Adapted from Frieden and Alberty [4].)

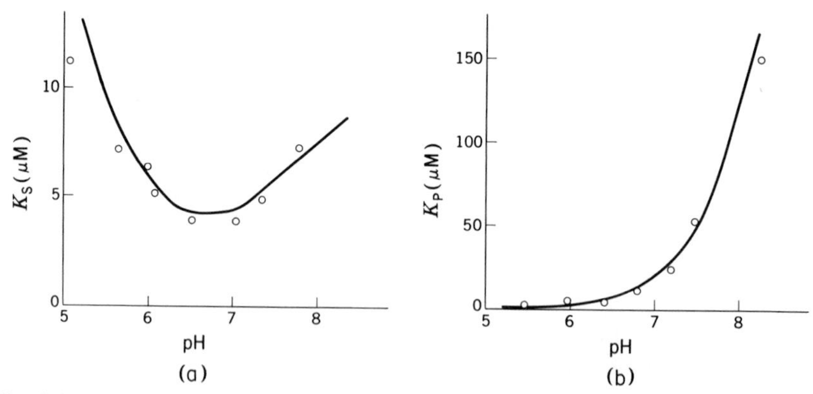

Fig. 9-3 (a) Plot of the Michaelis constant for fumarate as a function of pH in 0.01 M acetate buffer at 25°; the solid line is a theoretical curve. (b) Plot of the Michaelis constant for l-malate as a function of pH in 0.01 M acetate buffer at 25°; the solid line is a theoretical curve. (Adapted from Frieden and Alberty [4].)

it is generally (but not always) fairly easy to determine which form of the substrate undergoes reaction. For example, in the case of fumarase the doubly charged acid anions are the reactive species. In some cases, difficulties may arise from a close coupling of substrate ionizations with enzyme ionizations.

Two reasons are most often given for the effect on the enzyme of varying pH. Changes in pH can produce substantial structure changes in the enzyme, and this partial denaturation, i.e., loss of native structure, can produce a sharp decline in the specific activity of the enzyme. Usually such effects occur only at extreme pH's and can be recognized by a general lack of reproducibility of the experimental results and variation of the rate with past history of the enzyme, e.g., the length of time the enzyme has been exposed to the extreme pH's. The type of variation of rate parameters with pH illustrated for fumarase is usually explained as being due to ionizations at the "active site" of the enzyme, with only one of the ionized forms of the enzyme being catalytically active. The active site of the protein is considered to be the binding site for the substrate, where catalytic activity occurs. Since substrates are usually small molecules compared to the enzymes, this site can involve only a relatively small portion of the enzyme. This behavior can be incorporated into the simple one-intermediate mechanism by scheme (9-20)

$$\begin{array}{ccccc} EH_2 & & XH_2 & EH_2 & \\ K_a \upharpoonleft\downharpoonright & & K_{xa} \upharpoonleft\downharpoonright & \upharpoonleft\downharpoonright & \\ EH + S & \rightleftharpoons & XH & \rightleftharpoons EH + P & \\ K_b \upharpoonleft\downharpoonright & & K_{xb} \upharpoonleft\downharpoonright & \upharpoonleft\downharpoonright & \\ E & & X & E & \end{array} \tag{9-20}$$

Here hydrogen ions have been omitted for the sake of simplicity and the K_i's define ionization constants in the usual fashion; e.g.,

$$K_a = \frac{(EH)(H^+)}{(EH_2)}$$

Since protolytic reactions are usually very fast (especially in the presence of buffers, where proton-transfer reactions occur readily), all the protolytic steps can be assumed to be in equilibrium throughout the course of the reaction. The derivation of the rate equation is exactly as before except that the enzyme-conservation equation is given by

$$(E_0) = (EH) + (E) + (EH_2) + (XH) + (X) + (XH_2) \tag{9-21}$$

or

$$(E_0) = (EH)\left[1 + \frac{(H^+)}{K_a} + \frac{K_b}{(H^+)}\right] + (XH)\left[1 + \frac{(H^+)}{K_{xa}} + \frac{K_{xa}}{(H^+)}\right] \tag{9-22}$$

and

$$\frac{d}{dt}[(X) + (XH) + (XH_2)] = 0$$

$$= -(k_{-1} + k_2)(XH) + k_1(EH)(S) + k_{-2}(EH)(P) \tag{9-23}$$

The resulting rate equation is readily seen to have the same form as Eq. (9-10), but now

$$K_S = K_S' \frac{1 + (H^+)/K_a + K_b/(H^+)}{1 + (H^+)/K_{xa} + K_{xb}/(H^+)}, \qquad K_P = K_P' \frac{1 + (H^+)/K_a + K_b/(H^+)}{1 + (H^+)/K_{xa} + K_{xb}/(H^+)}$$

$$V_S = V_S' \frac{1}{1 + (H^+)/K_{xa} + K_{xb}/(H^+)}, \qquad V_P = V_P' \frac{1}{1 + (H^+)/K_{xa} + K_{xb}/(H^+)} \tag{9-24}$$

where now the primed quantities are the same functions of the rate constants as given by Eq. (9-11). Note that the ratio $K_S(E_0)/V_S$ [and $K_P(E_0)/V_P$] is dependent only on the ionization constants of the free enzyme. This function exhibits a maximum at a definite pH, which should be the same for both the forward and reverse reactions. This type of behavior is illustrated in Fig. 9-4 for the fumarase reaction. Both pK_a and pK_b can be determined and have

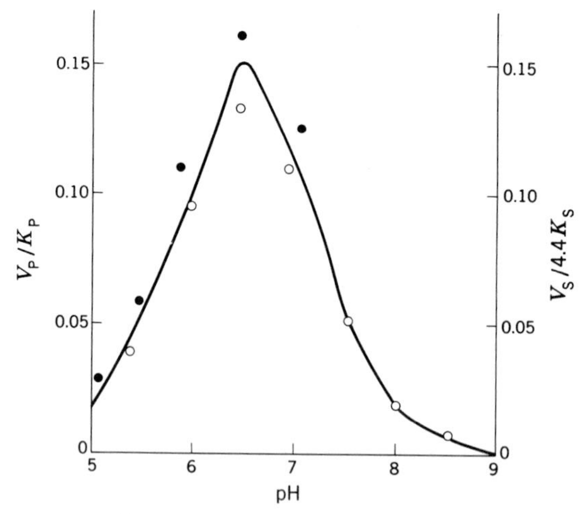

Fig. 9-4 Plot of $V_S/4.4\,K_S$ (●) and V_P/K_P (○) versus pH in 0.01 *M* acetate buffer at 25°. The factor 4.4 is the equilibrium constant for the overall reaction and normalizes the ordinates. The solid line is the theoretical curve calculated according to the equations given in the text. (Adapted from Frieden and Alberty [4].)

been found to be 6.2 and 6.8 at 25° for the case illustrated. Moreover, this interpretation is independent of the number of intermediates in the mechanism. Of course, in some cases more or less than two ionization constants for the free enzyme are required to explain the data. The individual kinetic parameters, however, depend on pH and the number of intermediates in such a manner that ionization constants for the reaction intermediates cannot be reliably obtained from steady-state data (although this fact is often ignored). This simple mechanism also predicts that V_S and V_P have maxima at the same pH. This is not found for fumarase, indicating that at least one more reaction intermediate must be added to the mechanism.

A critical evaluation of this explanation of the pH dependence for enzymatic reactions reveals that this mechanism is only one relatively simple type consistent with the data. For example, if the steps $E + S \rightleftharpoons X$, $S + EH_2 \rightleftharpoons XH_2$, etc., were included, exactly the same rate law would result, although the interpretation of the data would be considerably changed [5]. Other types of explanations for the pH dependence of steady-state kinetic parameters have emerged from time to time but have not gained wide acceptance.

The temperature dependence of enzymatic reaction rates is even more difficult to interpret unequivocally. More often than not, plots of $\log V_S$, $\log K_S$, $\log K_S/V_S$, etc., versus the reciprocal temperature are straight lines over fairly extended temperature ranges. In some cases, these plots display a gradual curvature. In all cases, when the temperature becomes sufficiently high, thermal inactivation of the enzyme occurs, resulting in a sharp decrease in the reaction rate. The interpretation of the "activation energies" and "enthalpies" obtained from the data is not clear. The activation energy values are usually quite similar to those of ordinary solution reactions. The observed straight line could be due to the fact that one of the reaction steps is clearly rate-determining and that the activation energy of this step is being measured; alternatively, the experimental activation energy might be a composite of enthalpies and an activation energy. Curvature in the plots can be due to heat-capacity effects, a change in the rate-controlling step with temperature, or thermally induced configurational changes of the enzyme (reversible denaturation).

An excellent study has been made of the temperature dependence of the steady-state kinetic parameters of the fumarase reaction [6]. By studying the temperature dependence over a wide range of pH, the apparent activation energies and standard-enthalpy changes associated with the pH-independent steady-state parameters, the lower bounds of the rate constants, and the ionization constants of the groups at the active site were obtained. The results are summarized in Table 9-1. In this case the temperature dependence of all parameters appears quite normal. The standard-enthalpy changes of

TABLE 9-1

Apparent Activation Energies and Standard Enthalpy Changes for the Fumarase Reaction[a]

Parameter	E_a (kcal/mole) or ΔH^0 (kcal/mole)	Parameter	E_a (kcal/mole) or ΔH^0 (kcal/mole)
$V_S'/(E_0)$	8.2	$(V_S' + V_P')/K_S'(E_0)$	7.5
$V_P'/(E_0)$	13.9	$(V_S' + V_P')/K_P'(E_0)$	6.0
K_S'	2.8	pK_a	−1.7
K_P'	4.0	pK_b	7.7

[a] Brant *et al.* [6].

the apparent ionization constants suggest that a carboxyl and an imidazole group are involved in the catalytic process; however, an unequivocal interpretation of the activation parameters in terms of mechanism is not possible even in the case of such a comprehensive study.

9-4 MULTISUBSTRATE MECHANISMS

Our discussion of steady-state enzyme kinetics has been confined to a relatively simple mechanism. Most enzyme mechanisms are more complex in that more than one substrate or coenzyme is involved. The treatment of such mechanisms is exactly the same, however, and the critical comments are just as apt.

As an illustration of more complex reactions we consider the overall reaction

$$\mathrm{A + B \rightleftharpoons C + D} \tag{9-25}$$

Some possible mechanisms are:

I. $$\mathrm{A + E} \underset{k_{-1}}{\overset{k_1}{\rightleftharpoons}} \mathrm{EA}$$

$$\mathrm{EA + B} \underset{k_{-2}}{\overset{k_2}{\rightleftharpoons}} \mathrm{X_1} \underset{k_{-3}}{\overset{k_3}{\rightleftharpoons}} \mathrm{ED + C}$$

$$\mathrm{ED} \underset{k_{-4}}{\overset{k_4}{\rightleftharpoons}} \mathrm{E + D}$$

II. $$\mathrm{A + E} \underset{k_{-1}}{\overset{k_1}{\rightleftharpoons}} \mathrm{X_1} \underset{k_{-2}}{\overset{k_2}{\rightleftharpoons}} \mathrm{C + X_2}$$

$$\mathrm{X_2 + B} \underset{k_{-3}}{\overset{k_3}{\rightleftharpoons}} \mathrm{X_3} \underset{k_{-4}}{\overset{k_4}{\rightleftharpoons}} \mathrm{D + E}$$

III. $A + E \overset{K_1}{\rightleftharpoons} EA$

$E + B \overset{K_2}{\rightleftharpoons} EB$

$EA + B \overset{K_2}{\rightleftharpoons} X_1 \overset{k}{\rightleftharpoons} X_2 \rightleftharpoons EC + D$ (or ED + C)

$EB + A \overset{K_1}{\rightleftharpoons} X_1 \overset{k}{\rightleftharpoons} X_2 \rightleftharpoons EC + D$ (or ED + C)

$EC \overset{K_3}{\rightleftharpoons} E + C$

$ED \overset{K_4}{\rightleftharpoons} E + D$

Mechanism I is referred to as a *compulsory-pathway* mechanism since the order of addition of substrates to the enzyme is fixed; mechanism II is often called a *shuttle* or *ping pong* mechanism because part of a substrate is shuttled back and forth between substrates and enzyme; and mechanism III involves a *random* addition of substrates to the enzyme. Obviously a large number of additional mechanisms could be written by permuting the substrates and by combining two of the mechanisms. If (C) = (D) = 0, the initial velocities for the first two mechanisms can be easily obtained using the method of King and Altman:

$$\text{I.} \quad \frac{(E_0)}{v} = \phi_1 + \frac{\phi_2}{(A)} + \frac{\phi_3}{(B)} + \frac{\phi_4}{(A)(B)} \tag{9-26}$$

$$\text{II.} \quad \frac{(E_0)}{v} = \phi_1' + \frac{\phi_2'}{(A)} + \frac{\phi_3'}{(B)} \tag{9-27}$$

where

$$\phi_1 = \frac{k_3 + k_4}{k_3 k_4}, \qquad \phi_2 = \frac{1}{k_1}, \qquad \phi_3 = \frac{k_3 + k_{-2}}{k_2 k_3}, \qquad \phi_4 = \frac{k_{-1}}{k_1}\phi_3$$

$$\phi_1' = \frac{k_2 k_4}{k_2 + k_4}, \qquad \phi_2' = \frac{k_{-1} + k_2}{k_1 k_2}, \qquad \phi_3' = \frac{k_{-3} + k_4}{k_3 k_4}$$

Note that mechanisms I and II can be easily distinguished experimentally. Both rate laws predict that a plot of $(E_0)/v$ versus $1/(A)$ should be linear at a constant concentration of B, but the lines obtained at different concentrations of B should be intersecting for the first mechanism and parallel for the second mechanism as shown schematically in Fig. 9-5. In both cases, however, A and B appear in the rate laws in a symmetric manner, so that it is not possible to say which substrate is added to the enzyme first. Finally, let us consider mechanism III. The general rate law in this case is extremely complex. However, if the assumptions are made that A and B bind independently to the enzyme, e.g., the reactions $E + A \rightleftharpoons EA$ and $EB + A \rightleftharpoons X_1$ have the same equilibrium constants, and that all reaction steps prior to the breakdown of

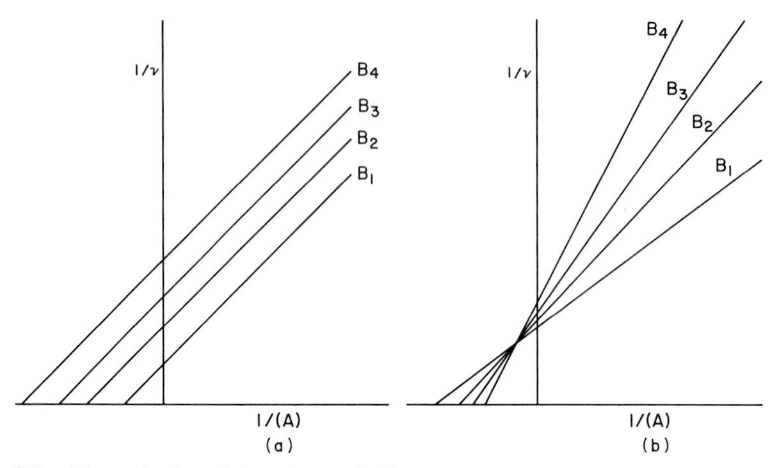

Fig. 9-5 Schematic plots of the reciprocal initial velocity, $1/v$, versus the reciprocal concentration of substrate A for a two-substrate reaction: (a) successive binary complex formation between enzyme and substrates, Eq. (9-27); (b) ternary complex formation between both substrates and enzyme, Eq. (9-26). The concentrations of the second substrate, B, are such that $B_1 > B_2 > B_3 > B_4$.

X_1 are so rapid as to be at equilibrium during the course of the reaction, a rate law for the initial velocity of the forward reaction identical to Eq. (9-26) is obtained, where

$$\phi_1 = \frac{1}{k}, \qquad \phi_2 = \frac{1}{K_1 k}, \qquad \phi_3 = \frac{1}{K_2 k}, \qquad \phi_4 = \frac{1}{K_1 K_2 k}$$

Thus mechanisms I and III cannot be distinguished by simply measuring initial velocities at various substrate concentrations.

A distinction between mechanisms I and III can be made by measuring the effect of product inhibition on the initial rate. (Note that for the random, rapid equilibration mechanism, $\phi_4 = \phi_2\phi_3/\phi_1$; unfortunately the experimental precision is generally not sufficient to utilize this relationship as a reliable mechanistic indicator.) Thus experiments can be carried out with varying concentrations of A, B, and C in the absence of D and with varying concentrations of A, B, and D in the absence of C. The steady-state initial velocities for mechanism I for these two cases are

$$\frac{E_0}{v} = \phi_1[1 + \theta_2(C)] + \frac{\phi_2}{(A)} + \frac{\phi_3}{(B)}[1 + \theta_3(C)] + \frac{\phi_4}{(A)(B)}[1 + \theta_3(C)] \quad (9\text{-}28)$$

$$\frac{E_0}{v} = \phi_1 + \frac{\phi_2}{(A)}[1 + \theta_1(D)] + \frac{\phi_3}{(B)} + \frac{\phi_4}{(A)(B)}[1 + \theta_1(D)] \quad (9\text{-}29)$$

In these equations the θ_i are constants that can be expressed in terms of rate constants. Since the substrates do not appear symmetrically in these equations, it is possible to determine which substrate binds first, which binds second, which product is released first, and which is released last. The corresponding initial velocity equations for mechanism III are

$$\frac{E_0}{v} = \phi_1 + \frac{\phi_2}{(A)} + \frac{\phi_3}{(B)} + \frac{\phi_4}{(A)(B)}[1 + K_3(C)] \tag{9-30}$$

$$\frac{E_0}{v} = \phi_1 + \frac{\phi_2}{(A)} + \frac{\phi_3}{(B)} + \frac{\phi_4}{(A)(B)}[1 + K_4(D)] \tag{9-31}$$

where K_3 and K_4 are the equilibrium binding constants for the binding of C and D to the enzyme. The forms of the rate laws for mechanisms I and III are clearly different and can be distinguished experimentally. In practice, this distinction is not always clear-cut because the rate laws are sufficiently complex and the experimental error sufficiently large to cause some ambiguity. Also, additional mechanistic steps often must be assumed to account for the data.

A steady-state analysis of mechanism III yields a complex rate law that is not of practical utility. More extensive discussions of complex reactions are available elsewhere [7, 8]. As might be expected from the previous discussion of the single substrate–single product reaction, the form of the rate law for two substrate–two product mechanisms is independent of the number of reaction intermediates, lower bounds for the rate constants can be obtained from steady-state parameters, and the pH dependence of the steady-state parameters can give information about the ionizable groups at the active site [9].

To summarize, steady-state kinetics gives information about the overall reaction mechanism, particularly about such things as substrate specificity and the substrate binding sequence. The rate data allow lower bounds for the rate constants to be determined, although almost nothing can be said about the number and nature of the intermediates. The effect of pH on the rate parameters allows qualitative statements to be made about the pK values of ionizable groups at the active site, although detailed interpretations of data must be viewed with caution. On the other hand, thermal data on the steady-state kinetic parameters generally cannot be reliably interpreted on a mechanistic basis.

9-5 TRANSIENT STATE KINETICS

From the preceding discussion it is evident that in order to obtain information about the intermediates in enzyme catalysis, experiments must be done at sufficiently high enzyme and substrate concentrations so that the intermediates can be detected with available experimental methods. However, because of the great catalytic efficiency of enzymes, the rates of enzymatic reactions at sufficiently high concentrations for detection of reaction intermediates are very fast, necessitating the use of fast reaction methods. In fact, the elementary steps associated with enzyme catalysis generally occur in times considerably shorter than 1 sec. Flow methods have been extensively used to study enzymatic reactions [10], but unfortunately the individual steps in many cases occur too rapidly for this technique. The temperature jump method has been used to extend the time range for investigating enzyme catalysis [11]. Unfortunately, other fast-reaction techniques have not proved to be of general utility for enzyme systems.

As an example of an enzyme reaction which has been studied with fast-reaction techniques, we consider the mechanism of action of pancreatic ribonuclease A. Ribonuclease catalyzes the breakdown of ribonucleic acid in two distinct steps as shown in Fig. 9-6. First the diester linkage is broken, and a pyrimidine 2′3′-cyclic phosphate is formed; then the cyclic phosphate is hydrolyzed to give the pyrimidine 3′-monophosphate and purine oligonucleotides with a terminal pyrimidine 3′-phosphate. Ribonuclease has been extensively studied with a variety of chemical and physical techniques, and its three-dimensional structure is known (cf. Richards and Wyckoff [12] for a comprehensive review).

Kinetic studies generally do not employ ribonucleic acid itself as a substrate because the system becomes inhomogeneous as ribonucleic acid is degraded, thus making a detailed kinetic analysis impossible. Instead, relatively simple substrates such as dinucleosides, pyrimidine 2′ : 3′-cyclic phos-

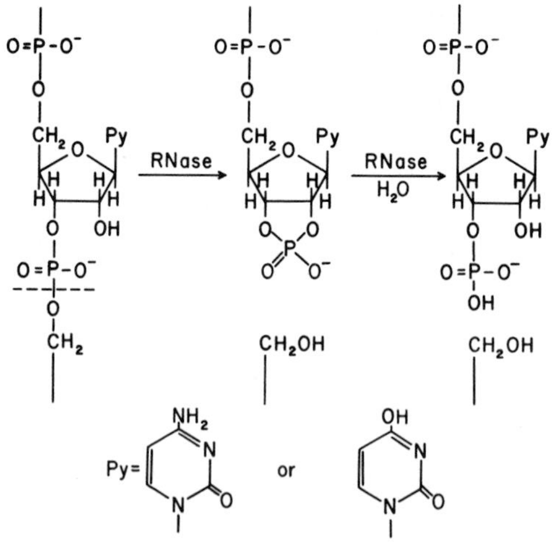

Fig. 9-6 The two-step hydrolysis of ribonucleic acid as catalyzed by bovine pancreatic ribonuclease A.

phates, and pyrimidine 3′-phosphates have been used. The reaction can be conveniently divided into three stages, corresponding to these three types of model compounds. The reactions separating these three types of compounds occur slowly relative to the rates characterizing the enzyme–substrate reactions. However, at equilibrium, essentially only pyrimidine 3′-phosphates are present so that the stopped flow–temperature jump method must be utilized for studying the interaction between ribonuclease and the other two types of substrates [13].

In the absence of substrates, a relaxation process is observed in solutions of ribonuclease with the temperature jump method having a relaxation time of 0.1 to 1 msec. This relaxation time is independent of the enzyme concentration but is pH dependent. It can be attributed to an isomerization or conformational change of the enzyme, and a simple mechanism consistent with the data is

$$\mathrm{E'H} \underset{k_{-1}}{\overset{k_1}{\rightleftharpoons}} \mathrm{EH} \overset{K_A}{\rightleftharpoons} \mathrm{E} + \mathrm{H}^+ \tag{9-32}$$

In this mechanism, the prime and unprimed states are different conformational states of the enzyme. If the protolytic reaction is assumed to equilibrate rapidly relative to the conformational change,

$$\frac{1}{\tau_1} = k_1 + \frac{k_{-1}}{1 + K_A/(\mathrm{H}^+)} \tag{9-33}$$

The data are quantitatively consistent with this equation and pK_A is approximately 6. This conformational change can be understood in terms of the three-dimensional structure of the enzyme. Ribonuclease is a compact kidney-shaped molecule with the active site located along a groove. Inhibitors are bound to the enzyme near two histidine residues (numbers 12 and 119 of the amino acid sequence), and a considerable amount of evidence suggests these residues are at the catalytic site. At the top of the "hinge" of the groove is a third histidine (number 48). The imidazole ring is partially buried, and its environment could be altered by an opening and closing of the groove associated with the site. The observed relaxation process may be associated with an opening and closing of the groove such that the imidazole residue is buried in E′H and has a pK of approximately 6 when exposed in the EH isomer.

The interaction of dinucleosides, pyrimidine 2′:3′-cyclic phosphates, or pyrimidine 3′-phosphates with the enzyme is characterized by two relaxation processes, in addition to the process associated with the unliganded enzyme. In all cases the results can be described by a two-step mechanism: a bimolecular combination of enzyme and substrate followed by an isomerization or conformational change of the enzyme–substrate complex:

$$\mathrm{E} + \mathrm{S} \underset{k_{-2}}{\overset{k_2}{\rightleftharpoons}} \mathrm{X}_1 \underset{k_{-3}}{\overset{k_3}{\rightleftharpoons}} \mathrm{X}_2 \tag{9-34}$$

TABLE 9-2

Representative Rate Constants Associated with Enzyme–Substrate Complex Formation

$$E + S \underset{k_{-2}}{\overset{k_2}{\rightleftharpoons}} X_1 \underset{k_{-3}}{\overset{k_3}{\rightleftharpoons}} X_2$$

	$10^{-7} k_2$ (M^{-1} sec^{-1})	$10^{-3} k_{-2}$ (sec^{-1})	$10^{-3} k_3$ (sec^{-1})	$10^{-3} k_{-3}$ (sec^{-1})	Reference
Ribonuclease A					
Cytidine 3′-phosphate	4.6	4.2	~1	~0.8	15
Uridine 3′-phosphate	7.8	11	0.97	0.77	15
Cytidine 2′3′-cyclic phosphate	2–4	10–20	11	2	16
Uridine 2′3′-cyclic phosphate	1.1	21	9	26	17
Cytidylyl 3′,5′-cytidine	1.4	7	($k_3 + k_{-3} = 8.6 \times 10^3$)		18
Creatine kinase					
Adenosine 5′-diphosphate (ADP)	2.3	18	1.67	0.24	19
MgADP	0.53	5.1	1.67	0.24	19
CaADP	0.17	1.2	1.67	0.24	19
MnADP	0.74	4.1	1.67	0.24	19
Chymotrypsin					
Furyloyl-L-tryptophanamide	0.6	10	1.5	30	20

Thus far it has not been possible to establish unequivocally whether the substrate reacts with the prime or unprimed conformation of Eq. (9-32), so this complication is neglected in Eq. (9-34). This uncertainty does not alter the basic nature of the mechanism. If the first step is assumed to be rapid relative to the second, the reciprocal relaxation times for this mechanism can be written as (see Chapter 8)

$$\frac{1}{\tau_2} = k_2[(E_e) + (S_e)] + k_{-2} \tag{9-35}$$

$$\frac{1}{\tau_3} = k_{-3} + \frac{k_3}{1 + k_{-1}/\{k_1[(E_e) + (S_e)]\}} \tag{9-36}$$

where the subscript e indicates the equilibrium concentration. Some of the rate constants obtained for a variety of substrates are given in Table 9-2. Many of the rate constants have been determined as a function of pH. The pH dependence of the second-order rate constant is similar to that of $V_S/K_S(E_0)$ observed with steady-state studies and can be used to infer the pK values of the ionizable groups on the free enzyme important for catalysis. The bell-shaped plot of k_2 versus pH can be analyzed in terms of two ionizable groups at the active site, one in its basic form, the other in its acid form, with approximate pK values of 5.4 and 6.4. These are undoubtedly the imidazole rings of histidines 12 and 119. The pH dependence of the relaxation time associated with the conformational change following the binding of substrates is similar (but not identical) to that of the relaxation time associated with the conformational change of the unliganded enzyme, and these two conformational changes are probably very similar in nature. Nuclear magnetic resonance measurements also support the occurrence of such a conformational change.

The minimal mechanism consistent with the data can be represented as in Fig. 9-7. Basically, each type of substrate reacts very rapidly with the enzyme, which exists in two conformations, a conformational change occurs,

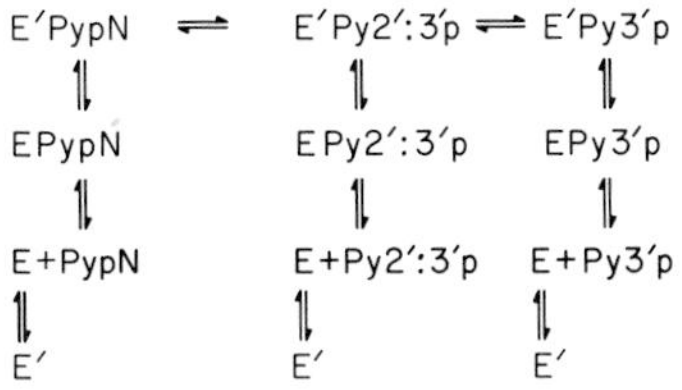

$$\begin{array}{ccccc}
E'PypN & \rightleftharpoons & E'Py2'{:}3'p & \rightleftharpoons & E'Py3'p \\
\updownarrow & & \updownarrow & & \updownarrow \\
EPypN & & EPy2'{:}3'p & & EPy3'p \\
\updownarrow & & \updownarrow & & \updownarrow \\
E+PypN & & E+Py2'{:}3'p & & E+Py3'p \\
\updownarrow & & \updownarrow & & \updownarrow \\
E' & & E' & & E'
\end{array}$$

Fig. 9-7 A schematic mechanism for the hydrolysis of a dinucleoside by ribonuclease A. The various conformational states inferred to be present from kinetic studies are shown. In this mechanism PypN is a pyrimidine 3′5′-nucleoside, Py 2′3′p is a pyrimidine 2′3′ cyclic phosphate and E is the enzyme. The primes designate different conformational states of the enzyme and enzyme–substrate complexes.

and the various intermediates are interconverted in rate-determining steps, one for transesterification and the other for cyclic phosphate hydrolysis. In terms of the three-dimensional structure, the enzymatic reaction can be envisaged as follows. The enzyme exists in dynamic equilibrium between two forms differing in the structure of the active site groove. When the enzyme binds a substrate, almost as fast as the two can diffuse together, the groove shape is altered and lysine 41 swings over to the enzyme to assist in

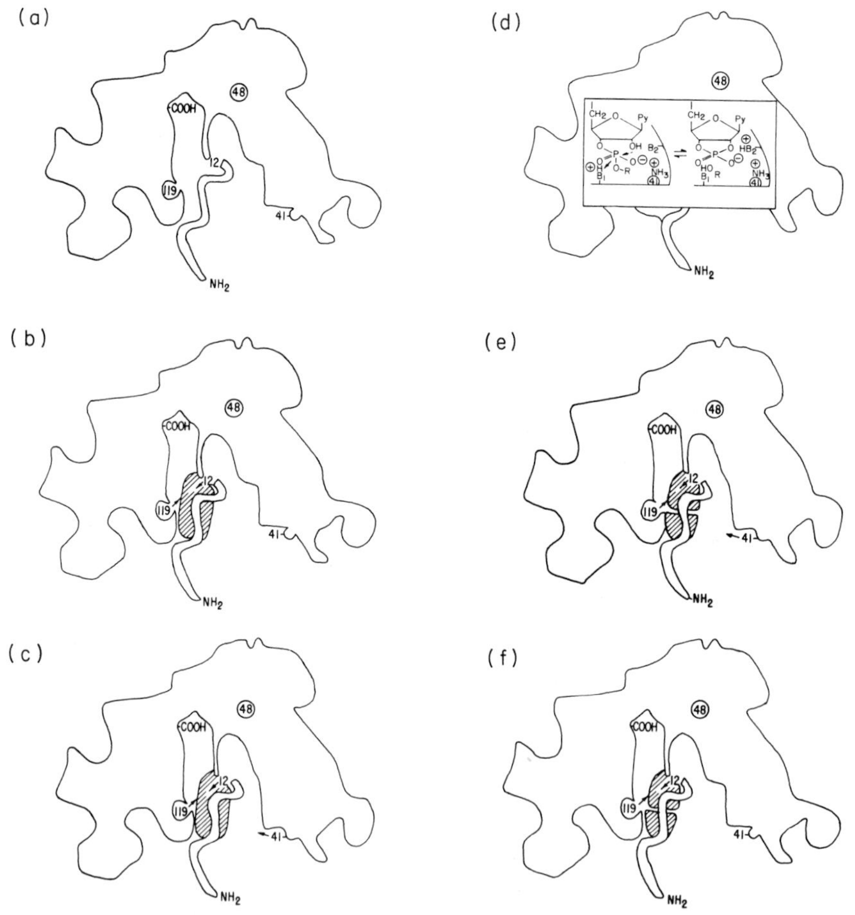

Fig. 9-8 A pictorial representation of the ribonuclease reaction. The free enzyme (a) exists in two conformational states differing by small movements of the hinge region joining the two halves of the molecule. The substrate is bound (b) and a conformational change occurs closing the hinge (c). Concerted acid–base catalysis then occurs (d); products are formed (e); the conformational change is reversed (f); and product(s) dissociate to give free enzyme.

the binding process. The substrate is protected from water and is oriented very precisely so that the imidazole residues (histidines 12 and 119) can catalyze the chemical reaction. The details of the proton transfer reactions cannot be studied directly. However, detailed stereochemical studies and consideration of the three-dimensional structure indicate that the mechanism probably involves an in-line concerted proton transfer between the two imidazole residues and the substrate [14]. The conformational change is then reversed and the product dissociates. The mechanism is shown schematically in Fig. 9-8 for the transesterification reaction. The hydrolysis step proceeds in a similar manner.

Thus a combination of detailed kinetic, chemical, and structural studies has led to a quite complete understanding of the catalytic process.

9-6 THE MECHANISM OF ENZYMATIC ACTION

The variety of enzymatic reactions is so large that a comprehensive discussion of enzymatic mechanisms is not appropriate here, but it is worthwhile considering the general enzymatic process in terms of some of the elementary steps commonly involved in enzyme catalysis. Three types of reactions are now considered in detail:

1. Enzyme–substrate complex formation
2. Conformational changes and isomerizations
3. Acid–base catalysis

All enzymatic reactions are initiated by the combination of enzyme and substrate, and many such reactions have been studied. In many cases, the two-step mechanism of Eq. (9-34) occurs. A few typical rate constants for the bimolecular reaction and the subsequent conformational change are presented in Table 9-2. The second-order rate constants generally are quite large, in spite of the great stereospecificity of enzymes, but they seem to be slightly smaller than might be expected for a diffusion-controlled process. The dissociation rate constants show great variation, primarily reflecting the stability of the enzyme–substrate complex.

The conformational change triggered by substrate binding appears to be a general phenomenon. In essence the binding energy is used to produce a macromolecular structure that is better optimized for catalysis. Probably an important function of this conformational change is to place the substrate in a hydrophobic pocket where the strength of hydrogen bonding and electrostatic interactions is enhanced; the exclusion of water also permits efficient acid–base catalysis to occur. Obviously the rate constants associated with the conformational change must be greater than the turnover number of the enzyme if the conformational transition is catalytically significant.

The most important noncovalent interactions in these transitions are hydrogen bonding, solvation, and hydrophobic interactions. The dynamics of these processes have been extensively studied in model systems, and the results obtained provide some insight into the corresponding processes in proteins [21].

The dynamics of hydrogen bonding are difficult to study in water because water itself is a potent hydrogen bond donor and acceptor. However, many kinetic studies of hydrogen bonding have been made in nonaqueous solvents, for example, the dimerization of 2-pyridone and the dimerization of benzoic acid. In weakly hydrogen bonding solvents, the formation of hydrogen bonded dimers of 2-pyridone and benzoic acid is diffusion controlled. As previously discussed for proton transfer reactions (Chapter 8), this implies that the specific rate constant for hydrogen bonding after the reactants have diffused together is about 10^{12} sec^{-1}. When the solvent itself is able to form relatively strong hydrogen bonds, the rate-determining step in dimer formation becomes desolvation of the monomeric species which typically is associated with a rate constant of about 10^8 sec^{-1}. Direct measurements of water dissociation from solutes have been made. The dissociation is diffusion controlled for ammonia (the rate constant is about 10^{11} sec^{-1}), but the rate constant decreases to a value of about 10^8 sec^{-1} as the solute is made more hydrophobic. This can be interpreted as being due to the formation of a sheath of strongly interacting molecules around the hydrophobic groups, thus making dissociation more difficult than for normally structured water.

Although the elementary step of hydrogen bond formation has a specific rate constant of about 10^{12} sec^{-1}, and desolvation of hydrophobic groups, which very likely is often rate limiting for hydrogen bonding processes, has a specific rate constant of about 10^8 sec^{-1}, conformational transitions in proteins are characterized by much smaller rate constants. Therefore, a crucial factor must be missing in the model systems considered thus far. The missing element is *cooperativity*. For example, polyglutamic acid can exist in either an α helical or random coil configuration. A cooperative transition between the two states can be triggered by changes in pH, provided the polymer contains more than six residues [22]. Although only changes in hydrogen bonding and solvation are involved in the transition, the rate constant for this process at the midpoint of the transition is only 10^6 sec^{-1} [23]. The overall rate is considerably slower than the elementary steps because a cooperative change occurs. This suggests that the conformational transitions in proteins are slow relative to the elementary steps because the conformational transitions are highly cooperative. Cooperative processes require a large number of interacting elements, such as found in enzymes.

Acid–base catalysis appears to be an important factor in virtually all enzymatic reactions. The rates of protolytic reactions have been discussed previously (Chapter 8). Both protonation and deprotonation must occur to complete a catalytic cycle. For water-mediated proton transfer, this cycle of reactions can be written as

$$B + H^+ \rightleftharpoons BH^+ \tag{9-37}$$

$$BH^+ + OH^- \rightleftharpoons B + H_2O \tag{9-38}$$

As previously mentioned, the rate constants in the forward direction are diffusion controlled and have typical values of about $10^{10}\ M^{-1}\ sec^{-1}$. Therefore, the rate constants for the reverse reactions are approximately $10^{10} K_A\ sec^{-1}$ and $10^{10} K_w/K_A\ sec^{-1}$ for the reactions of Eqs. (9-37) and (9-38), respectively, where K_A is the acid ionization constant of B and K_w is the ionization constant for water. Since for catalysis both of the reactions must occur, the maximum catalytic rate constant, $\sim 10^3\ sec^{-1}$, occurs when the pK_A is about 7. However, acid–base catalysis need not be mediated by water. A general formulation of *intramolecular* proton transfer is

$$DH + A \rightleftharpoons HA + D \tag{9-39}$$

where D and A denote a proton donor and acceptor, respectively. If the pK of the acceptor is much higher than that of the donor, the specific rate constant for proton transfer is approximately $10^{12}\ sec^{-1}$, while the specific rate constant for the reverse reaction is approximately $10^{12} K_A/K_D\ sec^{-1}$ where K_A and K_D represent the acid ionization constants of the acceptor and donor, respectively. Again for catalysis, a cycle of both the forward and reverse reactions must occur. The difference between pK values associated with protein ionizable groups involved in catalysis and common substrates is typically very large. For example, if the pK difference is 7 units, the maximum catalytic rate constant is about $10^5\ sec^{-1}$; this large pK difference also implies the concentration of one of the intermediates would be only 10^{-7} $(10^5/10^{12})$ of the total enzyme concentration; this in turn requires very large rate constants for further reactions if a high catalytic rate is to be maintained. Finally, the possibility of concerted acid–base catalysis should be considered. A simultaneous proton acceptance and donation by the substrate could readily be part of a cooperative conformational change of the macromolecule–substrate framework. Such a reaction would eliminate the necessity for forming a reaction intermediate in very low concentrations. An upper bound for the rate constant associated with such a process is the rate constant for the direct proton transfer between the acid and base groups on the enzyme involved in the catalysis. A typical pK difference is 2 units which gives an upper bound of $10^{10}\ sec^{-1}$; this is obviously too high an estimate since substrates are generally not good proton acceptors and donors and might be expected to reduce this maximum rate constant several orders of magnitude.

TABLE 9-3

Approximate Turnover Numbers of Representative Enzymes

Enzyme	$V_S/(E_0)$ (sec^{-1})	Reference
Chymotrypsin	10^2–10^3	26
Carboxypeptidase	10^2	26
Urease	10^4	26
Fumarase	10^3	27
Ribonuclease A	10^2–10^4	12
Transaminases	10^3	28
Carbonic anhydrase	10^6	29
Acetylcholinesterase	10^4	30

Some typical turnover numbers for enzymatic reactions are summarized in Table 9-3. The maximum turnover numbers are about 10^5 to 10^6 sec^{-1}, with typical values of 10^2 to 10^3 sec^{-1}. This is close to the maximum possible rate constant predicted from consideration of proton transfer reactions. However, the maximum rates are derived from consideration of proton transfer rates between good proton acceptors and donors. These rates are considerably decreased for poor acceptors and donors, such as typical substrates. Therefore, the protein must alter the effective pK values of the substrate. Several plausible mechanisms can be envisaged. The formation of a hydrophobic pocket for the substrate permits proton transfer reactions to occur without the competition of water; a small number of structured water molecules even might assist in proton transfer. Cooperative conformational transitions coordinated with proton transfer could lower the required activation energies. Such an effect is seen in simple reactions where the simultaneous breaking of one bond and formation of another has a lower activation energy than breaking of the same bond alone [24]. In a protein, the multitude of noncovalent interactions acting cooperatively could act in a similar manner to lower the activation energy for a process such as concerted proton transfer. Another way of viewing the role of the protein is that it distorts the structure of the substrate through conformational changes until it closely resembles the transition state, thus effectively lowering the activation energy for formation of the transition state of the rate-determining step [25]. The driving energy then is the binding energy which would be gained by the distortion of the substrate to a more tightly bound species.

In a number of enzymatic reactions, the substrate forms a covalent intermediate either with the enzyme or with a tightly, often covalently, bound coenzyme. The advantage of forming covalent intermediates generally is to break down a kinetically difficult reaction into kinetically easier steps. The enzyme is able to create a unique environment favorable for formation of the covalent intermediate and an equally unique environment favorable for its decomposition.

An enzymatic reaction has important advantages over a similar reaction in solution. Several different functional groups important in catalysis can simultaneously interact with substrates. This is a considerable entropic gain for the system, or in more simplistic terms raises the effective concentrations of the reactants. The enzyme also restricts the substrates to a specific structure and orientation with respect to the functional groups of the enzyme. This too is basically an entropic effect in that the number of possible conformations (or isomers) of the system is restricted. The catalytic activity of an enzyme then can be depicted as being derived from the creation of a specific polypeptide structure through cooperative interactions that binds the substrate (or substrates). This binding event initiates a series of cooperative conformational changes which optimize the polypeptide structure for catalysis of the reaction as already discussed. The enzyme appears to break down the overall reaction into a number of steps, with the enzyme adjusting its configuration for optimal catalysis of each step. This flexibility of structure, which only can be obtained with a macromolecule, seems to be an essential part of enzyme catalysis.

We have considered only one function of an enzyme, namely, catalysis. An equally important function is to regulate metabolic fluxes. The kinetic behavior of regulatory enzymes can be quite complex: the initial velocities are often not hyperbolic functions of substrate concentrations, and the binding of nonsubstrate metabolites can have profound kinetic effects. This fascinating subject is beyond the scope of this book, but many comprehensive reviews are available [31–34].

Problems

9-1 Derive the steady-state rate equation for the mechanism

$$E + S \rightleftharpoons X_1 \rightleftharpoons X_2 \rightleftharpoons X_3 \rightarrow E + P$$

9-2 A common type of enzyme inhibition occurs when an inhibitor combines with a reaction intermediate to form an inactive complex. This mechanism can be written as

$$E + S \rightleftharpoons ES \rightarrow E + P, \qquad ES + I \rightleftharpoons EIS$$

Derive the equation for the steady-state initial velocity predicted by the above mechanism.

9-3 The mechanism proposed for the hydrolysis of esters by the enzyme chymotrypsin can be depicted as

$$E + S \underset{k_{-1}}{\overset{k_1}{\rightleftharpoons}} X_1 \underset{k_{-2}}{\overset{k_2}{\rightleftharpoons}} X_2 + P_1, \qquad X_2 \underset{k_{-3}}{\overset{k_3}{\rightleftharpoons}} E + P_2$$

where E is the enzyme, S is the ester, P_1 is the alcohol formed, P_2 is the acid formed, and the X's are reaction intermediates. Derive an expression for the initial reaction velocity under steady-state conditions.

9-4 The hydration of CO_2

$$CO_2 + H_2O \rightarrow HCO_3^- + H^+$$

is catalyzed by the enzyme carbonic anhydrase. A stopped-flow apparatus has been used to study the steady-state kinetics of the forward and reverse reactions at pH 7.1, 0.5°C, and 2×10^{-3} *M* phosphate buffer with bovine carbonic anhydrase. Some typical data obtained are shown in Table P9-4 [35].

TABLE P9-4

Hydration		Dehydration	
$1/v$ $(M^{-1}\text{ sec} \times 10^{-3})^a$	(CO_2) $(M \times 10^3)$	$1/v$ $(M^{-1}\text{ sec} \times 10^{-3})^a$	(HCO_3^-) $(M \times 10^3)$
36	1.25	95	2
20	2.5	45	5
12	5	29	10
6	20	25	15

[a] Reciprocal initial velocity with a total enzyme concentration of 2.8×10^{-9} *M*.

(a) Calculate the steady-state kinetic parameters for the forward and reverse reactions.

(b) Calculate the equilibrium constant for the overall reaction.

(c) Calculate lower bounds of the rate constants for a mechanism with an arbitrary number of isomeric reaction intermediates.

9-5 The reaction catalyzed by the enzyme hexokinase is

$$\text{MgATP} + \text{G} \rightleftharpoons \text{MgADP} + \text{G6P}$$

TABLE P9-5

(MgATP) $(M \times 10^4)$	(G) $(M \times 10^4)$	$(E_0)/v \times 10^3$ (sec)	(MgATP) $(M \times 10^4)$	(G) $(M \times 10^4)$	$(E_0)/v \times 10^3$ (sec)
4.73	10	2.55	19.4	10	1.83
	5	2.90		5	2.17
	2	4.54		2	3.17
	1	7.48		1	4.87
9.20	10	2.21	40.0	10	1.72
	5	2.51		5	2.03
	2	3.90		2	2.93
	1	6.10		1	4.44

where MgATP is magnesium adenosine 5′-triphosphate, G is glucose, MgADP is magnesium adenosine 5′-diphosphate, and G6P is glucose 6-phosphate. What mechanisms are consistent with the initial-rate steady-state data given in Table P9-5 for yeast hexokinase at pH 8, 25°C, and 0.3 M $(CH_3)_4NCl$? [36].

9-6 The enzyme aspartate aminotransferase catalyzes the reaction

$$\text{As} + \text{Kg} \rightleftharpoons \text{Oa} + \text{Gm}$$

where As is aspartic acid, Kg is ketoglutaric acid, Gm is glutamic acid, and Oa is oxalacetic acid. Two mechanisms have been proposed: one involves formation of a ternary complex between the enzyme, the keto acid, and the amino acid; the other involves transfer of the amino group to a coenzyme (pyridoxal phosphate) on the enzyme, which in turn transfers it to the keto acid, so that only binary complexes are formed. On the basis of the kinetic data presented in Table P9-6, decide which of these mechanisms is operative [28].

TABLE P9-6

v^{-1} [a]	Kg ($M \times 10^3$)	As ($M \times 10^3$)	v^{-1} [a]	Kg ($M \times 10^3$)	As ($M \times 10^3$)
4.3	0.04	0.4	2.0	0.16	0.8
3.3	0.08		2.7	0.04	1.6
2.9	0.16		1.9	0.08	
2.6	0.32		1.45	0.16	
3.4	0.04	0.8	1.25	0.32	
2.5	0.08				

[a] Reciprocal initial velocity in arbitrary units at constant enzyme concentration, pH 7.3 in 0.04 M sodium arsenate and 26°C.

9-7 Consider the mechanism

$$\text{E} + \text{S} \rightleftharpoons \text{X}_1 \rightleftharpoons \text{X}_2 \rightleftharpoons \text{E} + \text{P}$$

(a) Derive the general form of the relaxation spectrum.

(b) At high concentrations of S and P, it can be assumed that the first and last steps in the mechanism equilibrate rapidly compared to the rate of interconversion of the intermediates. Derive explicit expressions for the relaxation times under these conditions.

(c) Derive an expression for the steady-state relaxation time predicted by the above mechanism in terms of Michaelis constants and maximum velocities.

References

1. For example, P. D. Boyer, ed., "The Enzymes." Academic Press, 1970ff. This series is a primary source for enzymology. The first two volumes cover basic properties of enzymes. Later volumes deal with specific enzymes.
2. L. Peller and R. A. Alberty, *J. Am. Chem. Soc.* **81**, 5907 (1959).
3. E. L. King and C. Altman, *J. Phys. Chem.* **60**, 1375 (1956).

4. C. Frieden and R. A. Alberty, *J. Biol. Chem.* **212**, 859 (1955).
5. R. A. Alberty and V. Bloomfield, *J. Biol. Chem.* **238**, 2804 (1963).
6. D. A. Brant, L. B. Barnett, and R. A. Alberty, *J. Am. Chem. Soc.* **85**, 2204 (1963).
7. A. Cornish-Bowden, "Principles of Enzyme Kinetics." Butterworth, London, 1976.
8. W. W. Cleland, *in* Ref. 1, Vol. 2, p. 1, 1970.
9. V. Bloomfield, L. Peller, and R. A. Alberty, *J. Am. Chem. Soc.* **84**, 4367 (1962).
10. B. Chance, *in* "Investigation of Rates and Mechanisms," Part II, Investigations of Elementary Reaction Steps in Solution and Very Fast Reactions (G. G. Hammes, ed.). Wiley (Interscience), New York, 1974.
11. G. G. Hammes and P. R. Schimmel, *in* Ref. 1, Vol. 2, p. 67, 1970.
12. F. M. Richards and H. W. Wyckoff, *in* Ref. 1, Vol. 4, p. 647, 1971.
13. G. G. Hammes, *Acc. Chem. Res.* **1**, 321 (1968).
14. D. A. Usher, E. S. Erenrich, F. Eckstein, *Proc. Nat. Acad. Sci. USA* **69**, 115 (1972).
15. G. G. Hammes and F. G. Walz, Jr., *J. Am. Chem. Soc.* **91**, 7179 (1969).
16. J. E. Erman and G. G. Hammes, *J. Am. Chem. Soc.* **88**, 5607 (1966).
17. E. J. del Rosario and G. G. Hammes, *J. Am. Chem. Soc.* **92**, 1750 (1970).
18. J. E. Erman and G. G. Hammes, *J. Am. Chem. Soc.* **88**, 5164 (1966).
19. G. G. Hammes and J. K. Hurst, *Biochemistry* **8**, 1083 (1969).
20. G. P. Hess, *in* Ref. 1, Vol. 3, p. 213, 1971.
21. G. G. Hammes, *Adv. Chem. Phys.* **39**, (1978).
22. P. Doty, A. Wade, J. T. Yang, and E. R. Blout, *J. Polym. Sci.* **23**, 851 (1957).
23. A. F. Barksdale and J. E. Steuhr, *J. Am. Chem. Soc.* **94**, 3334 (1972).
24. G. G. Hammes, *Nature* **204**, 342 (1964).
25. W. P. Jencks, *Adv. Enzymol.* **43**, 219 (1975).
26. K. J. Laidler, *Discuss. Faraday Soc.* **20**, 83 (1955).
27. R. A. Alberty and W. H. Peirce, *J. Am. Chem. Soc.* **79**, 1523 (1957).
28. S. F. Velick and J. Vavra, *J. Biol. Chem.* **237**, 2109 (1962).
29. S. Lindskog, L. E. Henderson, K. K. Kannon, A. Liljas, P. O. Nyman, and B. Strandberg, *in* Ref. 1, Vol. 5, p. 589, 1971.
30. H. C. Froede and I. B. Wilson, *in* Ref. 1, Vol. 5, p. 87, 1971.
31. D. E. Koshland, Jr., *in* Ref. 1, Vol. 1, p. 342, 1970.
32. E. R. Stadtman, *in* Ref. 1, Vol. 1, p. 398, 1970.
33. D. E. Atkinson, *in* Ref. 1, Vol. 1, p. 461, 1970.
34. G. G. Hammes and C.-W. Wu, *Ann. Rev. Biophys. Bioeng.* **3**, 1 (1974).
35. H. DeVoe and G. B. Kistiakowsky, *J. Am. Chem. Soc.* **83**, 274 (1961).
36. G. G. Hammes and D. Kochavi, *J. Am. Chem. Soc.* **84**, 2069, 2073, 2076 (1962).

APPENDIX A

KINETIC THEORY SUMMARY

The fraction of molecules dc/c with x components of velocity between v_x and $v_x + dv_x$ is given by the Maxwell–Boltzmann equation

$$\frac{dc}{c} = \left(\frac{m}{2\pi kT}\right)^{1/2} \exp\left(-\frac{mv_x^2}{2kT}\right) dv_x \qquad \text{(A-1)}$$

In three dimensions the corresponding expression is

$$\frac{dc}{c} = \left(\frac{m}{2\pi kT}\right)^{3/2} \exp\left(-\frac{mv^2}{2kT}\right) dv_x \, dv_y \, dv_z \qquad \text{(A-2)}$$

where v^2 $(= v_x^2 + v_y^2 + v_z^2)$ is the square of the magnitude of the velocity vector. The distribution of the magnitudes of the velocities, without regard to direction, namely, the scalar speeds, can be derived from Eq. (A-2) by writing the volume element $dv_x \, dv_y \, dv_z$ in polar coordinates $v^2 \, dv \sin\theta \, d\theta \, d\phi$ and integrating over θ and ϕ to obtain

$$\frac{dc}{c} = 4\pi \left(\frac{m}{2\pi kT}\right)^{3/2} v^2 \exp\left(-\frac{mv^2}{2kT}\right) dv \qquad \text{(A-3)}$$

The average value of a molecular property $\overline{P}$ can be found by using the appropriate distribution expression for dc and integrating over all phase space:

$$\overline{P} = \frac{\int_{\text{all phase space}} P \, dc}{\int_{\text{all phase space}} dc} \qquad \text{(A-4)}$$

For example, the average speed in one dimension is given by

$$\overline{v} = \left(\frac{2m}{\pi kT}\right)^{1/2} \int_0^{\infty} \exp\left(-\frac{mv_x^2}{2kT}\right) |v_x| \, dv_x = \left(\frac{2kT}{\pi m}\right)^{1/2}$$

where $|v_x|$ is the absolute value of the x component of velocity. The corresponding average speed in three dimensions is obtained by using Eq. (A-3) and is given by

$$\bar{v} = \left(\frac{8kT}{\pi m}\right)^{1/2}$$

APPENDIX B

STATISTICAL–MECHANICAL SUMMARY

The most important statistical–mechanical quantity for our purposes is the molecular partition function, defined as

$$f = \sum_i g_i e^{-\epsilon_i/kT} \tag{B-1}$$

where the ϵ_i and g_i are the energy levels and statistical weights (degeneracies), respectively. It can usually be assumed that the total energy may be written as a sum of translational, rotational, vibrational, and electronic energies

$$\epsilon = \epsilon_T + \epsilon_R + \epsilon_V + \epsilon_E \tag{B-2}$$

so that

$$f = f_T f_R f_V f_E \tag{B-3}$$

For translation, it is rarely necessary to consider discrete energy levels, and the translational partition function in one dimension is

$$f_T = \frac{(2\pi mkT)^{1/2}}{h} l \tag{B-4}$$

while in three dimensions it is

$$f_T = \frac{(2\pi mkT)^{3/2}}{h^3} V \tag{B-5}$$

where l is the distance that the particle can move along the single coordinate and V is the volume of the system. For rotation, the energy levels are discrete, but they are usually fully excited at room temperature. For a linear diatomic molecule,

$$f_R = \frac{8\pi^2 IkT}{h^2 \sigma} \tag{B-6}$$

where I is the moment of inertia and σ is the symmetry number. For nonlinear molecules, three different principal moments of inertia A, B, and C are necessary and

$$f_R = \frac{8\pi^2(8\pi^3 ABC)^{1/2}(kT)^{3/2}}{\sigma h^3} \tag{B-7}$$

In the case of vibration, the discrete energy levels must be taken into account. For a harmonic oscillator, $\epsilon_v = (v + \frac{1}{2})h\nu$, where ν is the characteristic frequency and v is the quantum number. The sum in Eq. (B-1) can be obtained in closed form, and for each vibrational degree of freedom

$$f_V = e^{-\epsilon_0/kT}(1 - e^{-h\nu/kT})^{-1} \tag{B-8}$$

where ϵ_0 $(= \frac{1}{2}h\nu)$ is the zero-point energy. For several vibrational degrees of freedom, f_V is simply the product of terms of the form of Eq. (B-8). The electronic partition function is obtained by taking the appropriate sum directly. However, electronic energy levels above the ground state are seldom excited at ordinary temperatures. The equilibrium constant for a chemical reaction can be written in terms of partition functions. For the reaction

$$a\text{A} + b\text{B} \rightleftharpoons e\text{E} + d\text{D}$$

we have

$$K = \frac{f_E{}^e f_D{}^d}{f_A{}^a f_B{}^b} e^{-\Delta E_0/RT} \tag{B-9}$$

where $f_A, f_B, \ldots$ are the total partition functions, as given by Eq. (B-3) for the reactants and products. In Eq. (B-9) the concentration unit is 1 molecule/cc, and the partition functions are associated with a volume of 1 cc. The energy ΔE_0 is the difference in spectroscopic dissociation energies between products and reactants where the products of dissociation are atoms in their ground states.

APPENDIX C

THEORY OF SHOCK TUBES

A useful one-dimensional qualitative model for illustrating the formation of a shock wave in a gas has been proposed by Becker [1] as follows. The high-pressure gas in the driver section is represented as a close-fitting piston, which is being accelerated into the expansion section in a series of small successive movements. Each of these small movements of the piston sends a pressure pulse at sonic velocity through the gas, which thereby undergoes adiabatic heating. The second pulse therefore travels through gas heated by the first pulse at a more rapid rate than the first pulse, since at the elevated temperature its velocity is the sum of its greater sonic velocity $(\gamma RT/M)^{1/2}$, where γ is the heat capacity ratio C_P/C_V, and a higher flow velocity imparted to the gas by the pulse itself. Because of the increase in the velocity, later pulses tend to overtake earlier ones, and since the pulses cannot pass each other, they eventually unite to form a single discontinuity. It is thus possible for a piston moving with uniform acceleration for a finite time to produce a pressure discontinuity in a time period which is essentially infinitesimal. Figure C-1 is a graphical representation of a one-dimensional shock wave resulting from the coalescence of pressure pulses.

When a stationary state has been set up in the shock front, we can calculate the change in state of the gas and the velocity of the shock. We consider a tube of uniform (unit) cross section and select a frame of coordinates moving with the shock front in the x direction, as shown in Fig. C-2. Relative to the front, the unshocked gas at density ρ_u and temperature T_u is moving into the front with linear velocity v_u and mass flux $\rho_u v_u$. In a similar manner, relative to the front, the shocked gas at density ρ_s and temperature T_s is moving away from the front at a linear velocity v_s and a mass flux $\rho_s v_s$.

Conservation of mass requires that the mass flux be constant, so that

$$\rho_u v_u = \rho_s v_s \tag{C-1}$$

across the shock front, while conservation of momentum gives for the

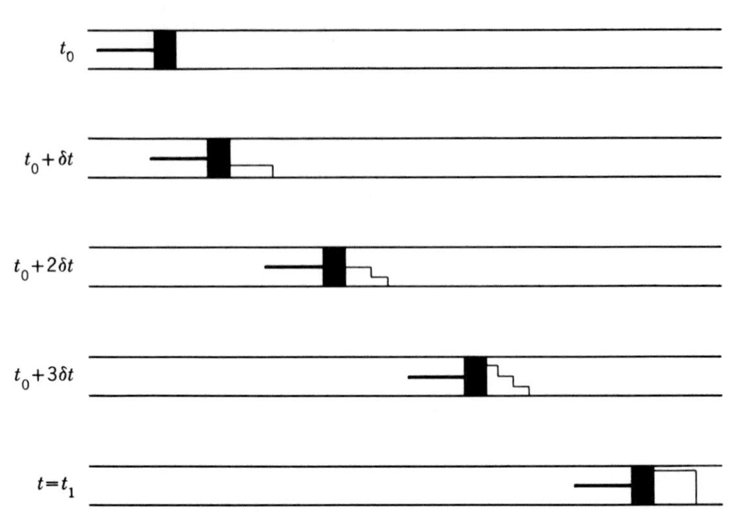

Fig. C-1 Pressure–distance profiles of Becker's model for the formation of a one-dimensional shock wave. (Adapted from Becker [1].)

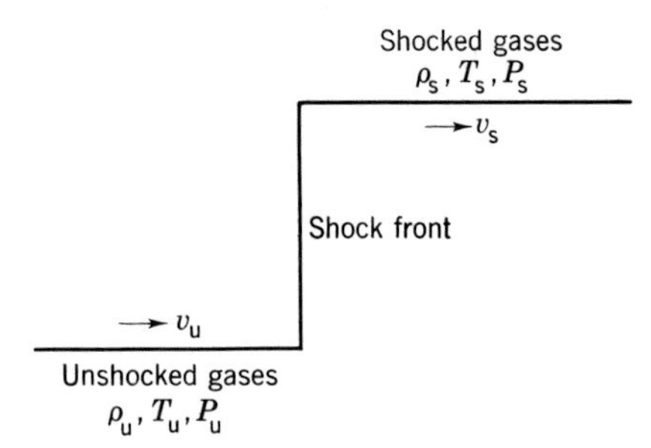

Fig. C-2 Pressure profile of a stationary shock wave in a gas. In the laboratory coordinates, the motion of the shock front is from right to left.

pressure drop at the discontinuity

$$P_s - P_u = \rho_u v_u^2 - \rho_s v_s^2 = -\rho_s^2 v_s^2 \left(\frac{1}{\rho_s} - \frac{1}{\rho_u} \right) \tag{C-2}$$

From the first law of thermodynamics,

$$\Delta E = Q - W - \Delta \text{KE} \tag{C-3}$$

where ΔKE is the change in the total kinetic energy of the system. For a unit mass of gas,

$$W = P_s V_s - P_u V_u = \frac{P_s}{\rho_s} - \frac{P_u}{\rho_u} \tag{C-4}$$

and since the process is assumed to be adiabatic, $Q = 0$, so that for the unit mass of gas Eq. (C-3) becomes

$$E_s - E_u = \frac{P_u}{\rho_u} - \frac{P_s}{\rho_s} + \frac{v_u^2}{2} - \frac{v_s^2}{2} = \tfrac{1}{2}(P_s + P_u)\left(\frac{1}{\rho_u} - \frac{1}{\rho_s}\right) \quad \text{(C-5)}$$

after v_u and v_s have been eliminated by use of Eqs. (C-1) and (C-2). Equation (C-5) is known as the Hugoniot equation for shocks. For the steady-state one-dimensional model which was assumed, it defines completely the state of the fluid on the two sides of the shock front in terms of thermodynamic quantities. Interpreted physically, it states that the change in internal energy across the shock front is due to the work done by the average pressure in performing the compression. The *Hugoniot function*, $\mathscr{H}$, can be obtained by rearranging Eq. (C-5) to give

$$\mathscr{H} = E_s - E_u - \tfrac{1}{2}(P_s + P_u)\left(\frac{1}{\rho_u} - \frac{1}{\rho_s}\right) \quad \text{(C-6)}$$

and it is this quantity which is equal to zero for the type of shock under consideration. Figure C-3 shows three types of PV curves for the expansion of an ideal monatomic gas ($\gamma = 1.67$) for a specified set of initial conditions, P_u, V_u, T_u. The shock expansion corresponds to $\mathscr{H} = 0$, the reversible adiabatic expansion to $\Delta S = 0$, and the isothermal expansion to $\Delta T = 0$. In general, a given volume change produces a larger change in pressure in a shock expansion than in a reversible adiabatic or isothermal expansion,

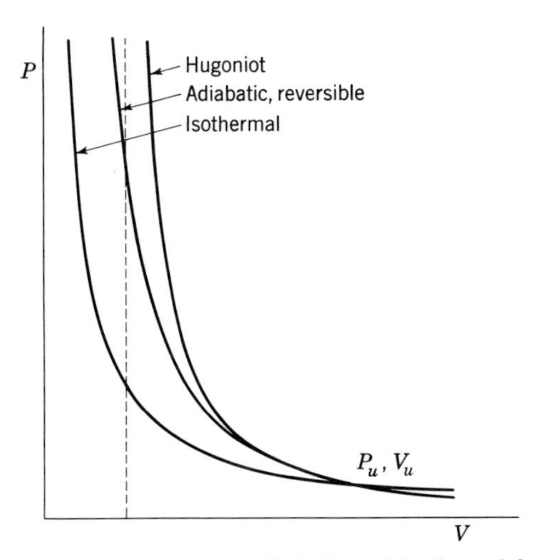

Fig. C-3 PV plots of Hugoniot, reversible adiabatic, and isothermal functions for an ideal monatomic gas ($\gamma = 1.67$). The dashed line is the asymptote of the Hugoniot curve.

although for weak shocks the expansion approaches that of the reversible adiabatic process.

To obtain additional information from the Hugoniot relation, it is necessary to specify in greater detail how the internal energy E varies with temperature and pressure. The simplest example to use is that of an ideal gas with constant heat capacity, in which case E is a function only of T, and

$$\Delta E = E_s - E_u = \int_{T_u}^{T_s} c_V\, dT = c_V(T_s - T_u) = \frac{1}{\gamma - 1}\left(\frac{P_s}{\rho_s} - \frac{P_u}{\rho_u}\right) \quad \text{(C-7)}$$

where c_V is the specific heat at constant volume. The ideal-gas relations

$$P = \rho RT/M \quad \text{(C-8)}$$

and

$$c_P - c_V = R/M \quad \text{(C-9)}$$

have been used to obtain the second relation for ΔE in Eq. (C-7).

Equations (C-7) and (C-8) may be combined with Eq. (C-5) to obtain the following expressions, which relate various parameters of the shocked gas to that of the unshocked gas:

$$\frac{\rho_s}{\rho_u} = \frac{P_s/P_u + \beta}{1 + \beta P_s/P_u} \quad \text{(C-10)}$$

$$\frac{T_s}{T_u} = \frac{1 + \beta P_s/P_u}{P_s/P_u + \beta}\frac{P_s}{P_u} = \frac{\rho_u}{\rho_s}\frac{P_s}{P_u} \quad \text{(C-11)}$$

where $\beta = (\gamma - 1)/(\gamma + 1)$. In strong shocks, which are characterized by $P_s/P_u \gg 1$, the density ratio ρ_s/ρ_u approaches a constant value of $1/\beta$, whereas T_s/T_u increases as P_s/P_u, approaching a value $\beta P_s/P_u$. The behavior of ρ_s/ρ_u results from the fact that as the density tends to increase with pressure, the tendency is offset by the increase in temperature across the shock front. The broken line in Fig. C-3 is a graphical representation of the tendency of the specific volume (reciprocal density) to reach a constant value as the pressure continues to rise.

From Eqs. (C-1) and (C-2) we find for the linear velocities of the unshocked and shocked gas relative to the shock front

$$v_u = \left(\frac{\rho_s}{\rho_u}\right)^{1/2}\left(\frac{P_s - P_u}{\rho_s - \rho_u}\right)^{1/2} = \left(\frac{\rho_s}{\rho_u}\right)^{1/2}\left(\frac{P_s/P_u - 1}{\rho_s/\rho_u - 1}\right)^{1/2}\left(\frac{P_u}{\rho_u}\right)^{1/2} \quad \text{(C-12)}$$

$$v_s = \frac{\rho_u}{\rho_s} v_u \quad \text{(C-13)}$$

For analytical convenience the relations developed up to this point have been derived in terms of linear velocities v_u and v_s with respect to the shock front. In practice, it is difficult to measure directly such properties of the shocked gas as v_s, ρ_s, P_s, and T_s. It is feasible, however, to measure the velocity U of the shock front and when necessary to convert from velocities with respect to the shock front to corresponding velocities in the laboratory-coordinate system. There are a number of satisfactory methods for measuring U, e.g., schlieren photography and piezoelectric detection.

For the type of system of present interest, where the gas ahead of the shock is stationary, the velocity of the shock front is equal in magnitude but opposite in sign to the velocity of the unshocked gas, or

$$|v_u| = |U| \tag{C-14}$$

This relationship and those previously derived can be used to obtain Eq. (6-16), which can be used to calculate the properties of shocked gases.

Reference

1. R. Becker, *Z. Phys.* (*Leipzig*) **8**, 321 (1922).

APPENDIX D

PHYSICAL CONSTANTS AND CONVERSION FACTORS

Electronic charge e	4.8029×10^{-10} esu
	(1.6022×10^{-19} coulomb)
Avogadro number N_0	6.0222×10^{23} molecules/mole
	(6.0222×10^{26} molecules/kmole)
Planck's constant h	6.6262×10^{-27} erg-sec
	(6.6262×10^{-34} joule-sec)
Boltzmann constant k	1.3806×10^{-16} erg molecule^{-1} °K^{-1}
	(1.3806×10^{-23} joule molecule^{-1} °K^{-1})
Gas constant R	8.3144×10^{7} ergs mole^{-1} °K^{-1}
	8.314×10^{3} joule kmol^{-1} °K^{-1}
	1.9872 cal mole^{-1} °K^{-1}
	82.057 cc atm mole^{-1} °K^{-1}

1 atm = 760 Torr = 1.01325×10^{6} dyne/cm^2
1 defined cal = 4.184000 joules
1 joule = 10^7 ergs = 1 volt-coulomb
1 electron volt = 1.60206×10^{-12} erg

APPENDIX **E**

SYMBOLS AND NOTATION

A number of entries have been omitted from this tabulation. These fall into the following groups: certain symbols with special subscripts for restricting, according to explanations in the text, the corresponding general symbols listed here; certain symbols defined in the text and used infrequently for special purposes, such as running variables introduced to reduce complicated integrals to simpler forms; certain mathematical symbols or notations which are considered elementary in character and therefore familiar to the reader.

A(B, C, D, G) generalized reactants or products; concentration of A, B, . . . , as in $d\mathrm{A}/dt$
Å angstrom unit, 10^{-8} cm
A cross-sectional area (in flow system); preexponential factor
A (B, C) coulombic contributions to potential energy of interaction; moments of inertia of transition-state complex
B generalized base
B $e^2\beta/2\varepsilon kT$ or $[e^3/(\varepsilon kT)^{3/2}](2\pi N_0/1000)^{1/2}$, a constant in the Debye–Hückel limiting law and in the limiting expressions for the primary and secondary salt effects in reactions between ions
$B(T)[C(T), \ldots]$ second (third, . . .) virial coefficients
C concentration (usually in units of mole per liter, M^{-1})
C_P molar heat capacity at constant pressure
C_V molar heat capacity at constant volume
D diffusion coefficient; spectroscopic energy of dissociation of a diatomic molecule (constant in Morse equation for potential energy of a diatomic molecule)
$\Delta D_0{}^0$ difference in dissociation energies of products and reactants measured from zero-point energy levels
E generalized enzyme
E initial kinetic energy of relative motion of reactants (units, energy mole^{-1}); energy
E_a activation energy per mole
E^0 standard energy content per mole
E' final kinetic energy of relative motion of reaction products (units, energy mole^{-1})
$\Delta E_0{}^0$ difference in standard energy contents at 0°K between products and reactants
$\Delta E_0{}^\ddagger$ difference in energy content between transition-state complex and reactants at 0°K

F	rotational–vibrational energy distribution function; momentum distribution function
G	constant in theory of general acid–base catalysis; free energy per mole
G^0	standard free energy per mole
H	Hamiltonian; enthalpy
H_{local}	magnetic field strength at nucleus of a molecule
H^0	standard enthalpy per mole
$\mathscr{H}$	Hugoniot function
I	generalized inhibitor
I	moment of inertia; flux (units, particles time^{-1}); pressure amplitude; beam intensity (units, particles area^{-1} time^{-1})
I_r	ratio of the rotational partition function of the activated complex to that of the reactant molecule, in RRKM theory
J	flux density (units, particles area^{-1} time^{-1})
$\mathbf{J}$	rotational angular momentum vector of reactants
$\mathbf{J}'$	rotational angular momentum vector of products
K	equilibrium constant; constant in expression for potential energy between mass points having inverse power repulsion (or attraction)
K_I	equilibrium constant for reaction between generalized enzyme and generalized inhibitor
K_P	Michaelis constant of product in reversible enzyme reactions; equilibrium constant at constant pressure
K_S	Michaelis constant of substrate in enzyme reactions
$\mathbf{L}$	orbital angular momentum vector of reactants
$\mathbf{L}'$	orbital angular momentum vector of products
M	mole liter^{-1}
$M^\ddagger$	transition-state complex
N	number of trajectories in trajectory calculations; energy-level density (number of energy levels per unit energy); population of energy level
N_0	Avogadro's number
P	generalized product in enzyme reaction
P	property of concentration (monotonic measure); pressure; probability of reaction between two (or more) molecules
$P(\theta)$	flux of reaction product at laboratory angle θ
P_M	fraction of ligand bound to nucleus M
Q	heat absorbed by a defined system
$Q^\ddagger$	nonequilibrium ratio of concentration of transition-state complex to product of concentration (to appropriate powers) of reactants
R	rate of reaction (units, concentration time^{-1}); gas constant per mole; rotational energy
S	generalized substrate
S	collision cross section for scattering; number of oscillators in a molecule containing s atoms; term symbol for atomic electronic energy state; entropy
S_R	total cross section for chemical reaction
S^0	standard entropy per mole
T	absolute temperature; kinetic energy in Hamiltonian; translational energy
T_1	spin–lattice relaxation time, in nuclear magnetic resonance
T_2	spin–spin relaxation time, in nuclear magnetic resonance
U	potential energy of interaction, either two-body or many-body; velocity of shock front
V	volume; vibrational energy

V_P maximum velocity of enzyme reactions in reverse direction (units, concentration time^{-1})

V_S maximum velocity of enzyme reactions in forward direction (units, concentration time^{-1})

W equilibrium fraction of molecules with energy in excess of a specified value; work done on surroundings

X reaction intermediate involving enzyme and substrate; generalized third body in intermolecular reactions

Z collision frequency per unit volume; internal energy of reactants (units, energy mole^{-1})

Z' internal energy of reaction products (units, energy mole^{-1})

a constant in equation of state; constant in Morse equation for potential energy of interaction; distance of closest approach of two solute molecules; area of aperture of beam source; elastic-hard-sphere binary-collision factor, $2\pi d^2(kT/\pi m)^{1/2}$ or $2\pi d_{12}^2(2kT/\pi m^*)^{1/2}$

a (b, c, d, g) stoichiometric coefficients associated with generalized reactants or products; initial concentration of A (B, C, D, G)

a_{ij} functions of rate constants and equilibrium concentrations in expressions for rates of reaction near equilibrium (unit, time^{-1})

b constant in equation of state; impact parameter

c concentration

c_P specific heat at constant pressure

c_V specific heat at constant volume

d diameter of a single molecule; collision radius for system of two molecules

e base of natural logarithms; electronic charge

f partition function; factor, involving interaction potential, in expression for rate of reaction in solution; speed and velocity distribution function

g quantum weight (degeneracy)

h Planck's constant

j total number of active vibrational quanta (having a common frequency) per molecule; rotational quantum number

k gas constant per molecule; square of electron overlap integral

$\mathscr{k}$ rate constant

$\mathscr{k}(E)$ microscopic rate constant

m constant, as in $P = mc + q$; minimum number of vibrational quanta per molecule required for activation in unimolecular decomposition; mass of a single molecule; fractional charge transfer, in theory of electron-transfer reactions

m^* reduced mass of two (or more) molecules

n number of moles (or molecules); exponent specifying order of reaction; number of square terms in classical expression for internal energy

p steric factor, correction for effects of orientation in binary collisions; magnitude of momentum; symmetry correction in theory of general acid–base catalysis; electronic state of atom with azimuthal quantum number equal to 1

pK $-\log K$

$\mathbf{p}$ initial-relative-momentum vector

q constant associated with m in expression for P; conjugate coordinate; ionic charge ze; symmetry correction in theory of general acid–base catalysis

r flux coefficient; internuclear distance; radius of a single molecule; radial coordinate in spherical coordinates; distance from beam source to detector

r_m — position of minimum in "exp-six" expression for U
s — constant in expression for U between mass points having inverse power repulsion (or attraction); number of atoms in a polyatomic molecule; ionic strength; electronic state of atom with azimuthal quantum number equal to 0
t — time
u — linear flow velocity
v — vibrational quantum number
v — molecular speed; initial velocity (rate) of enzyme reaction
$\mathbf{v}$ — vector velocity of a single molecule
$\mathbf{v}_c$ — vector velocity of center of mass of molecule 1 and molecule 2
$\mathbf{v}_r$ — relative vector velocity of molecule 1 with respect to molecule 2
w_{AB} — work required to bring ionic reactants A and B to within the reaction distance r_{AB}
x — reaction variable (decrease in concentration of reactant); distance (as in $u = dx/dt$)
$x_{A(i)}$ — fraction of molecules of specified type (A), in given quantum state (i)
y — fraction reacted; normal mode concentration variable
z — ionic valence

$\Gamma(n)$ — gamma function, $\int_0^\infty x^{n-1}e^x\,dx$
Ω — solid angle

α — constant in theory of rate of proton transfer; constant in theory of general acid–base catalysis; constant in "exp-six" expression for U; ultrasonic absorption coefficient
α (β, γ) — exchange-energy contributions to potential energy of interaction
β — constant, $(8\pi N_0 e^2 s/1000\varepsilon kT)^{1/2}$, in Debye–Hückel theory of electrolytes; ratio of free energy of activation to overall change in standard free energy, $\Delta G^{0\ddagger}/\Delta G^0$; function of ratio of heat capacities, $(\gamma - 1)/(\gamma + 1)$; constant in theory of general acid–base catalysis
γ — activity coefficient; nuclear gyromagnetic ratio; ratio of heat capacity at constant pressure to heat capacity at constant volume
δ — distance along reaction coordinate at top of potential barrier
ϵ — kinetic energy of relative motion per molecule; internal energy per molecule; depth of minimum in "exp-six" expression for U
ϵ_a — activation energy per molecule
ε — macroscopic dielectric constant
η — azimuthal angle of deflection
θ — polar angle in spherical coordinates; angle between axis of dipole and point charge
θ_S — angle of deflection in laboratory coordinates
θ_r — polar angle in relative spherical coordinates
κ — symmetry number (1, $\frac{1}{2}$) characterizing collisions between unlike or identical molecules
$\varkappa$ — transmission coefficient in transition-state theory
λ — wavelength; factor in theory of electron transfer reactions
μ — dipole moment
ν — frequency (often, vibrational frequency); stoichiometric coefficient
ρ — mass density (units, mass volume^{-1})
σ — differential cross section per unit solid angle
τ — relaxation time
τ_ϵ — average lifetime of molecule with energy ϵ
ϕ — azimuthal angle in spherical coordinates
ϕ_r — azimuthal angle in relative spherical coordinates

χ	polar relative angle of deflection
ψ	electrical potential of a single ion
ω	radial frequency, $2\pi\nu$
∇	gradient, $\mathbf{i}\dfrac{\partial}{\partial x} + \mathbf{j}\dfrac{\partial}{\partial y} + \mathbf{k}\dfrac{\partial}{\partial z}$
∇^2	operator, $\dfrac{\partial^2}{\partial x^2} + \dfrac{\partial^2}{\partial y^2} + \dfrac{\partial^2}{\partial z^2}$
Δ	constant time interval, in special integrated equations
( )	concentration
[ ]	activity

INDEX

N

P

R

A 8
B 9
C 0
D 1
E 2
F 3
G 4
H 5
I 6
J 7